Photon and Particle Impact Spectroscopy and Dynamics of Atoms, Molecules, and Clusters

Photon and Particle Impact Spectroscopy and Dynamics of Atoms, Molecules, and Clusters

Editors

Himadri S. Chakraborty
Hari R. Varma

Basel • Beijing • Wuhan • Barcelona • Belgrade • Novi Sad • Cluj • Manchester

Editors
Himadri S. Chakraborty
Department of Natural
Sciences
Dean L. Hubbard Center for
Innovation
Northwest Missouri State
University
Maryville, Missouri
USA

Hari R. Varma
School of Physical Sciences
Indian Institute of Technology
- Mandi
Kamand, Mandi, HP
India

Editorial Office
MDPI
St. Alban-Anlage 66
4052 Basel, Switzerland

This is a reprint of articles from the Special Issue published online in the open access journal *Atoms* (ISSN 2218-2004) (available at: https://www.mdpi.com/journal/atoms/special_issues/H34I1DD407).

For citation purposes, cite each article independently as indicated on the article page online and as indicated below:

Lastname, A.A.; Lastname, B.B. Article Title. *Journal Name* **Year**, *Volume Number*, Page Range.

ISBN 978-3-7258-0237-1 (Hbk)
ISBN 978-3-7258-0238-8 (PDF)
doi.org/10.3390/books978-3-7258-0238-8

Contents

 *atoms*

Editorial

Photon and Particle Impact Spectroscopy and Dynamics of Atoms, Molecules and Clusters

Himadri S. Chakraborty [1,*] and Hari R. Varma [2]

1   Department of Natural Sciences, Dean L. Hubbard Center for Innovation, Northwest Missouri State University, Maryville, MO 64468, USA
2   School of Physical Sciences, Indian Institute of Technology Mandi, Kamand 175075, India; hari@iitmandi.ac.in
*   Correspondence: himadri@nwmissouri.edu

1. Introduction

Atomic, molecular, and optical (AMO) physics is a vastly important sub-discipline. It provides insights into the fundamental nature of matter, drives technological innovation, and contributes to various scientific and applied fields across disciplines. Since the early days of quantum mechanics, AMO physics has allowed for the exploration of the fundamental processes that govern the behavior of atoms and molecules. This includes understanding the structure of atoms, the nature of chemical bonds, and the dynamics of AMO interactions. Many advanced processes were discovered and controlled, such as developments in lasers, spectroscopy, and quantum optics. These have had profound impacts on technology, ranging from medical imaging to communication systems [1]. AMO physics plays a vital role in the emerging field of quantum information science. Research in this field aims to harness the principles of quantum mechanics to manipulate and control quantum states to develop powerful quantum computers and communication systems [2]. Techniques developed in AMO physics are used for highly precise measurements, such as atomic clocks and global positioning systems [3]. AMO physics provided the ground, knowledge, and original field of applications in creating ultrashort pulses of light that can now measure the rapid processes in which electrons move or change energy in materials—a technology that was the topic of the 2023 Nobel Prize in Physics [4]. Advancing into materials science, AMO physics contributes to the development of new materials with specific properties, impacting areas, such as electronics, nanotechnology, and energy storage. Techniques from AMO physics are vital for medical diagnostics [5] that include advances in laser technology finding applications in medical treatments and surgeries. AMO physics contributes to the understanding of the universe, including the behavior of matter under extreme conditions found in stars and other astrophysical environments [6]. Coming back to the basics, AMO physics is still a powerhouse in addressing fundamental questions about the nature of matter and the universe. For example, studying ultracold atoms [7], quantum gases, and interactions involving antimatters [8] allows us to explore exotic states of matter and test the limits of our understanding of quantum mechanics and fundamental forces. In all the above, computational AMO physics offers essential machinery. It complements experiments, transforms the hypothesis to mathematical expectations, predicts and interprets measurements, forecasts new materials before synthesis, and contributes to the understanding of complex physical phenomena. Thus, computational AMO physics tackles the challenges posed by the complex nature of quantum systems, providing valuable insights and facilitating advancements in technological domains.

Therefore, research and development in AMO physics are vastly active and organically developing to invent many sub-fields of interest. The current Special Issue (SI), entitled *Photon and Particle Impact Spectroscopy and Dynamics of Atoms, Molecules, and Clusters*, is a modest but valuable effort to publish some novel research by renowned AMO research groups and scientists. A total of thirteen articles encompasses a collection

Citation: Chakraborty, H.S.; Varma, H.R. Photon and Particle Impact Spectroscopy and Dynamics of Atoms, Molecules and Clusters. *Atoms* **2023**, *11*, 156. https://doi.org/10.3390/atoms11120156

Received: 29 November 2023
Accepted: 7 December 2023
Published: 12 December 2023

of research findings in the interaction of charged particles and lights with varieties of AMO systems.

A Rundown of Articles

<u>Proton</u>: Two articles are published reporting measurements of the energetic proton impact fragmentation of molecules. The one by *Vinitha* et al. (contribution 1). measures multi-fragmentation of highly charged azulene and naphthalene ions produced in 50–150 keV proton collisions via the multi-hit time-of-flight experimental technique. Such spectroscopic results serve as crucial aspects in the diagnosis of hostile celestial regions, where polycyclic aromatic hydrocarbons are abundant and under constant bombardment by protons in the stellar wind. The other article, by *Duley and Kelkar* (contribution 7), considers the dissociation dynamics of carbon dioxide molecular ions impacted by 1 keV protons measured with a recoil-ion momentum spectrometer. Broadly speaking, such studies are important in plasma physics, hydrocarbon chemistry, as well as in modelling interstellar media.

<u>Electron</u>: Two articles report on theoretical electron-impact studies—both on fundamental interaction grounds. The one by *Msezane and Felfli* (contribution 10) studies the low-energy electron attachment process using a quantum mechanical effect of Ramsauer–Townsend minima based on the Regge pole analysis. This addresses the fundamental question of the origin and character of the electron affinity for large atoms: does the electron affinity of these systems characterize from the ground, metastable or excited states of their negative ions? The other study by Harris (contribution 6) examines a novel mechanism of selectively sculpted electron beams to collide with the helium atom in order to explore effects of the coherence length of the projectile wave packets. This study may lead to a new direction of measurements involving collisions with twisted beams of charged particles.

<u>Photon</u>: There are several articles that investigate the interactions of electromagnetic radiation with AMO systems. One article by *Hosea* et al. (contribution 2) theoretically explores the effect of the interference between the electric dipole and quadrupole order of coupling of the photon with a sodium atom for its valence photoionization. This is carried out near a spectrally sensitive feature as a function of energy called the Cooper minimum. The study paves a track to access even higher multipole effects. The other study, by *Shaik* et al. (contribution 3), focuses on the single-photon dipole photoionization of three magic-number sodium clusters in a framework of density functional theory to explore fascinating spectral features. The ultrafast timing of the photoionization process in attoseconds has been investigated on a fundamental theoretical track of Eisenbud–Wigner–Smith (EWS) in two papers. The one by *Grafstrom* and *Landman* (contribution 5) uses the relativistic random-phase approximation to calculate the time delay from outer subshells of various isoelectronic noble gas neutrals and anions. The other paper, by *Baral* et al. (contribution 8), applies the same theory to examine the modification in the EWS delay for the electron ejection from noble gas atoms within an optical dipole trap, a possible prototype for the qubit in quantum computing. A pair of theory articles have been published on research involving interactions with strong (multi-photon) radiation fields as well. In one of these, a study by *Simonovic* et al. (contribution 11) reports Rabi oscillation dynamics driven by intense, short, resonant laser pulses and invalidates a hypothesis about the origin of the multiple-peak pattern in the photoelectron energy distribution. The other study by *Schimmoller* et al. (contribution 9) performs simulations in the quantum trajectory Monte Carlo method and compares with experiments for a neutral diatomic molecule to demonstrate that the molecular ionization site in the strong-field spectra is insensitive to the pulse's carrier envelop phase.

<u>Plasma</u>: The research by *Biswas* et al. (contribution 4) utilizes a powerful relativistic coupled-cluster theory to study the plasma-field-induced structures and transitions of some high-charge cations of astrophysical relevance. It is expected that the results can be applied for celestial or laboratory plasma diagnostics.

Last but not at all least, we would like to mention two opinion articles published in this SI from two experienced and versatile researchers in AMO fields. The article by *Manson* (contribution 12) discusses the critical implications of an otherwise small force, the spin–orbit force. In fact, the importance of spin–orbit forces extends across branches of physics, from understanding the behavior of electrons in atoms to the structure of atomic nuclei and the properties of exotic matter. The inclusion of spin–orbit effects is crucial for accurate modeling and predicting the behavior of particles and systems in a wide range of physical scenarios. The other opinion article, by *Connerade* (contribution 13), delves into examples of contemporary physics questions to argue that atomic physics remains at the epicenter to test fundamental principles of physics, which are still inadequately understood. We re-quote Richard Feynman from this article. "When asked: should Armageddon occur, is there a simple, most important idea to preserve as a testament to human knowledge? The answer he suggested is the atomic hypothesis".

2. Future Research Prospects

We take this opportunity to comment on some future research roadmaps in AMO science that may emanate from the current landscape. The ongoing drive of strong pulsed-laser field research will increasingly empower the control, imaging, and manipulation of electron and ion-core dynamics on ultrafast timescales [9,10]. Such ultrafast spectroscopy and imaging, even for weak laser fields where the pristine electronic phenomena can be better captured, will have applications in studying chemical reactions, biological processes, and material properties [11,12]. This can further generate a focus direction of accessing the dephasing dynamics of plasmon resonances to enrich quantum plasmonic applications [13]. Future research on quantum information science may focus on the creation and manipulation of more robust and scalable qubits. This can explore new quantum algorithms, addressing challenges in quantum error correction. Likewise, AMO research on quantum optics may involve developing methods for quantum state engineering, communication, and sensing using techniques, like cavity quantum electrodynamics and laser cooling [14]. Continued exploration of ultracold quantum gases, such as Bose–Einstein condensates for quantum many-body physics and phase transitions, may investigate novel applications of ultracold atoms in precision measurements and quantum simulation [15,16]. Another particularly interesting direction is the integration of superconducting circuits with ion traps and AMO systems creating hybrid quantum systems [17]. This can lead to enhanced coherence times and improved quantum gates for applications in quantum technologies. Finally, investigating the interaction of AMO systems with emerging materials and interfaces [18] may explore novel phenomena and applications in areas like nanotechnology and condensed matter physics.

3. Conclusions

The original dream for the driving objective of this SI was that submissions should present novel effects, mechanisms, and phenomena in the energy response (spectroscopy) and the time evolution (dynamics) of excited target systems, highlighting new experimental techniques and powerful theoretical/computational methods and instigating novel questions to motivate future research and collaboration. On the other hand, today's AMO physics research dissemination is motivated by a dual commitment to featuring fundamental discoveries and to utilizing some of that knowledge for future technological applications. To that fantastically lofty goal, the current SI serves as a small but important leap forward.

Author Contributions: H.S.C. and H.R.V. Guest Editors. All authors have read and agreed to the published version of the manuscript.

Funding: This Editorial received no external funding.

Conflicts of Interest: The authors declare no conflict of interest.

List of Contributions

1. Vinitha, M.V.; Bhatt, P.; Safvan, C.P.; Vig, S.; Kadhane, U.R. Fragmentation of Multiply Charged C10H8 Isomers Produced in keV Range Proton Collision. *Atoms* **2023**, *11*, 138. https://doi.org/10.3390/atoms11110138.
2. Hosea, N.M.; Jose, J.; Varma, H.R.; Deshmukh, P.C.; Manson, S.T. Quadrupole Effects in the Photoionisation of Sodium 3s in the Vicinity of the Dipole Cooper Minimum. *Atoms* **2023**, *11*, 125. https://doi.org/10.3390/atoms11100125.
3. Shaik, R.; Varma, H.R.; Chakraborty, H.S. Density Functional Treatment of Photoionization of Sodium Clusters: Effects of Cluster Size and Exchange–Correlation Framework. *Atoms* **2023**, *11*, 114. https://doi.org/10.3390/atoms11080114.
4. Biswas, S.; Bhowmik, A.; Das, A.; Pal, R.R.; Majumder, S. Transitional Strength under Plasma: Precise Estimations of Astrophysically Relevant Electromagnetic Transitions of Ar^{7+}, Kr^{7+}, Xe^{7+}, and Rn^{7+} under Plasma Atmosphere. *Atoms* **2023**, *11*, 87. https://doi.org/10.3390/atoms11060087.
5. Grafstrom, B.; Landsman, A.S. Attosecond Time Delay Trends across the Isoelectronic Noble Gas Sequence. *Atoms* **2023**, *11*, 84. https://doi.org/10.3390/atoms11050084.
6. Harris, A.L. Projectile Coherence Effects in Twisted Electron Ionization of Helium. *Atoms* **2023**, *11*, 79. https://doi.org/10.3390/atoms11050079.
7. Duley, A.; Kelkar, A.H. Fragmentation Dynamics of COq+2 (q = 2, 3) in Collisions with 1 MeV Proton. *Atoms* **2023**, *11*, 75. https://doi.org/10.3390/atoms11050075.
8. Baral, S.; Easwaran, R.K.; Jose, J.; Ganesan, A.; Deshmukh, P.C. Temporal Response of Atoms Trapped in an Optical Dipole Trap: A Primer on Quantum Computing Speed. *Atoms* **2023**, *11*, 72. https://doi.org/10.3390/atoms11040072.
9. Schimmoller, A.; Pasquinilli, H.; Landsman, A.S. Does Carrier Envelope Phase Affect the Ionization Site in a Neutral Diatomic Molecule? *Atoms* **2023**, *11*, 67. https://doi.org/10.3390/atoms11040067.
10. Msezane, A.Z.; Felfli, Z. Rigorous Negative Ion Binding Energies in Low-Energy Electron Elastic Collisions with Heavy Multi-Electron Atoms and Fullerene Molecules: Validation of Electron Affinities. *Atoms* **2023**, *11*, 47. https://doi.org/10.3390/atoms11030047.
11. Simonović, N.S.; Popović, D.B.; Bunjac, A. Manifestations of Rabi Dynamics in the Photoelectron Energy Spectra at Resonant Two-Photon Ionization of Atom by Intense Short Laser Pulses. *Atoms* **2023**, *11*, 20. https://doi.org/10.3390/atoms11020020.
12. Manson, S.T. The Spin-Orbit Interaction: A Small Force with Large Implications. *Atoms* **2023**, *11*, 90. https://doi.org/10.3390/atoms11060090.
13. Connerade, J.-P. The Atom at the Heart of Physics. *Atoms* **2023**, *11*, 32. https://doi.org/10.3390/atoms11020032.

References

1. National Research Council. *Atoms, Molecules, and Light: AMO Science Enabling the Future*; The National Academies Press: Washington, DC, USA, 2002. [CrossRef]
2. Saffman, M. Quantum computing with atomic qubits and Rydberg interactions: Progress and challenges. *J. Phys. B At. Mol. Opt. Phys.* **2016**, *49*, 20200. [CrossRef]
3. Ludlow, A.D.; Boyd, M.M.; Ye, J.; Peik, E.; Schmidt, P. O. Optical atomic clocks. *Rev. Mod. Phys.* **2015**, *87*, 637. [CrossRef]
4. The Nobel Prize in Physics 2023. Available online: https://www.nobelprize.org/prizes/physics/2023/press-release/ (accessed on 28 November 2023).
5. Weber, S.; Wu, Y.; Wang, J. Recent progress in atomic and molecular physics for controlled fusion and astrophysics. *Matter Radiat. Extremes* **2021**, *6*, 023002. [CrossRef]
6. Higashi, Y.; Matsumoto, K.; Saitoh, H.; Shiro, A.; Ma, Y.; Laird, M. Iodine containing porous organosilica nanoparticles trigger tumor spheroids destruction upon monochromatic X-ray irradiation: DNA breaks and K-edge energy X-ray. *Sci. Rep.* **2021**, *11*, 14192. [CrossRef] [PubMed]
7. Brierley, R.; Li, Y.; Benini, L. Ultracold quantum technologies. *Nat. Phys.* **2021**, *17*, 1293. [CrossRef]
8. Adkinsa, G.S.; Cassidyb, D.B.; Pérez-Ríosc, J. Precision spectroscopy of positronium: Testing bound-state QED theory and the search for physics beyond the Standard Model. *Phys. Rep.* **2022**, *975*, 1–61. [CrossRef]
9. Tóth, A.; Csehi, A. Strong-field control by reverse engineering. *Phys. Rev. A* **2021**, *104*, 063102. [CrossRef]

10. Young, L.; Ueda, K.; Gühr, M.; Bucksbaum, P.H.; Simon, M.; Mukamel, S. Roadmap of ultrafast X-ray atomic and molecular physics. *J. Phys. B At. Mol. Opt. Phys.* **2018**, *51*, 032003. [CrossRef]
11. Maiuri, M.; Garavelli, M.; Cerullo, G. Ultrafast Spectroscopy: State of the Art and Open Challenges. *J. Am. Chem. Soc.* **2020**, *142*, 3–15. [CrossRef] [PubMed]
12. Hutchison, C.D.M.; Baxter, J.M.; Fitzpatrick, A.; Dorlhiac, G.; Fadini, A.; Perrett, S. Optical control of ultrafast structural dynamics in a fluorescent protein. *Nat. Chem.* **2023**, *15*, 1607–1615. [CrossRef] [PubMed]
13. Biswas, S.; Trabattoni, A.; Rupp, P.; Magrakvelidze, M.; Madjet, A.; De Giovannini, U.; Castrovilli, C.; Galli, M.; Liu, C.; Månsson, E.P.; et al. Attosecond correlated electron dynamics at C60 giant plasmon resonance. *arXiv* **2021**, arXiv:2111.14464.
14. Future Directions of Quantum Information Processing: A Workshop on the Emerging Science and Technology of Quantum Computation, Communication, and Measurement. Available online: https://basicresearch.defense.gov/Portals/61/Documents/future-directions/Future_Directions_Quantum.pdf?ver=2017-09-20-003031-450 (accessed on 28 November 2023).
15. Altuntaş, E.; Spielman, I.B. Weak-measurement-induced heating in Bose-Einstein condensates. *Phys. Rev. Res.* **2023**, *5*, 023185. [CrossRef] [PubMed]
16. Escudero, R.G.; Minář, J.; Pasquiou, B.; Bennetts, S.; Schreck, F. Continuous Bose–Einstein condensation, Chun-Chia Chen. *Nature* **2022**, *606*, 683–687. [CrossRef]
17. De Motte, D.; Grounds, A.R.; Rehák, M.; Rodriguez Blanco, A.; Lekitsch, B.; Giri, G.S.; Neilinger, P.; Oelsner, G.; Il'ichev, E.; Grajcar, M.; et al. Experimental system design for the integration of trapped-ion and superconducting qubit systems. *Quantum. Inf. Process* **2016**, *15*, 5385–5414. [CrossRef] [PubMed]
18. Becher, C.; Gao, W.; Kar, S.; Marciniak, C.D.; Monz, T.; Bartholomew, J.G.; Goldner, P.; Loh, H.; Marcellina, E.; Johnson Goh, K.E.; et al. 2023 roadmap for materials for quantum technologies. *Mater. Quantum. Technol.* **2023**, *3*, 012501. [CrossRef]

Article

Fragmentation of Multiply Charged $C_{10}H_8$ Isomers Produced in keV Range Proton Collision

Meloottayil V. Vinitha [1] Pragya Bhatt [2], Cholakka P. Safvan [2], Sarita Vig [3] and Umesh R. Kadhane [1,*]

[1] Department of Physics, Indian Institute of Space Science and Technology, Thiruvanathapuram 695 547, India; vinitha.meloottayil@mlib.ism.cnr

[2] Inter-University Accelerator Centre , Aruna Asaf Ali Marg, New Delhi 110 067, India; pbpragya@gmail.com (P.B.)

[3] Department of Earth and Space Science, Indian Institute of Space Science and Technology, Trivandrum 695 547, India

* Correspondence: umeshk@iist.ac.in

Abstract: The dissociation of multiply charged $C_{10}H_8$ isomers produced in fast proton collisions (velocities between 1.41 and 2.4 a.u.) is discussed in terms of their fundamental molecular dynamics, in particular the processes that produce different carbon clusters in such a collision. This aspect is assessed with the help of a multi-hit analysis of daughter ions detected in coincidence with the elimination of H^+ and CH_n^+ (n = 0 to 3). The elimination of H^+/C^+ is found to be significantly different from CH_3^+ loss. The loss of CH_3^+ proceeds through a cascade of momentum-correlated dissociations with the formation of heavy ions such as $C_9H_5^+$, $C_9H_5^{2+}$ and $C_7H_3^+$. The structure of such large fragment ions is predicted with the help of their calculated ground state electronic energies and the multi-hit time-of-flight (ToF) correlation between the second and third hit fragments if detected. Furthermore, we report experimentally the super-dehydrogenation of naphthalene and azulene targets, with evidence of complete dehydrogenation in a single collision.

Keywords: PAH; proton collision; multi-fragmentation

Citation: Vinitha, M.V.; Bhatt, P.; Safvan, C.P.; Vig, S.; Kadhane, U.R. Fragmentation of Multiply Charged $C_{10}H_8$ Isomers Produced in keV Range Proton Collision. *Atoms* **2023**, *11*, 138. https://doi.org/10.3390/atoms11110138

Academic Editors: Himadri S. Chakraborty and Hari R. Varma

Received: 1 August 2023
Revised: 2 October 2023
Accepted: 9 October 2023
Published: 25 October 2023

1. Introduction

The dissociation dynamics of a variety of polycyclic aromatic hydrocarbons (PAHs) have been investigated under irradiation with ultraviolet (UV), vacuum UV (VUV) or energetic charged particles in several works in the past [1–12]. Such measurements provide fundamental insight into the quantum chemistry of the organic molecules with PAHs as a convenient model system. A considerable body of this research domain has focused on the dissociation of monocations. The measurements of monocations are often accomplished with high levels of computational effort, and investigating the dissociation dynamics of highly charged ions is a challenge. The excess Coulomb energy stored in the di- or trication can radically change the nature of the dissociation dynamics. In addition, the complexities of the data assimilation, analysis and interpretations make the investigations on multiply charged ions a relatively uncommon endeavour.

The dissociative multiple ionisation of PAHs can occur in hostile astronomical regions by a number of excitation processes, such as multi-stage VUV absorption, interaction with X-rays or the low-energy component of cosmic rays, stellar wind protons and/or ions and other energetic particles in the interstellar medium [13–16]. The laboratory research in this domain is mainly conducted using extreme UV or X-ray photon, while charged particle collisions remain less explored [16–18]. In the last two decades, the study of the upper atmosphere of the jovian planets and their moons have led to the substantial understanding of how energetic charged particle radiation from the Sun can play a crucial role in their atmospheric composition and evolution [19–22]. In situ measurements by the Cassini–Huygens mission have demonstrated the importance of the 10 to 100 keV energy range of protons in the dynamics of Titan's ionosphere [21,22]. Such interactions are known to be more efficient in producing highly charged ions and diverse energetic fragment ions (C^+, CH_3^+ and $C_2H_2^+$ etc.), which may eventually induce a very rich and complex chemistry in Titan's ionosphere, atmosphere and haze.

Here, we compare the multi-fragmentation of highly charged azulene ($C_{10}H_8$) and naphthalene ($C_{10}H_8$) ions produced in fast proton collisions. Very little difference was observed between the multi-fragmentation mass spectrum of these two isomeric targets. We mainly focus on the emission of light fragments H^+, C^+, CH_3^+. The former two channels occur via violent multi-fragmentation, while the latter proceeds via an intermediate isomer that is often overlooked due to its relatively low intensity. Also, complete or partial dehydrogenation events are observed in this work under a single collision condition.

2. Computational Details

Multiply charged ions in the keV range collision are produced with a substantial amount of internal energy which assists in crossing various transition state barriers involved in isomerisation or dissociation processes. When a parent ion eliminates smaller neutral or ionic fragments like CH_3^+ or $C_2H_2^+$ from a precursor isomer, the smaller fragment will carry a substantial amount of vibrational energy per degree of freedom and the larger fragment will be produced close to the final equilibrium configuration of precursor isomer in the ground electronic state. Ground-state energies are therefore used here as a measure to predict the structure of large fragment ions. Various isomers of $C_9H_5^+$ and $C_9H_5^{2+}$, the large fragment ions formed after CH_3^+ loss, were optimised by density functional theory (DFT) calculations using the Dunning basis set cc-pVDZ to predict the total ground state electronic energy. This basis was chosen due to its larger set of basis functions compared to the one used for the complete $C_{10}H_8$ ions, since the respective fragment species are not stable or naturally occurring equilibrium structures. The energies of the doubly and triply charged $C_{10}H_8$ isomers that can emit CH_3^+ are also calculated. The relatively simple and stable ions of $C_{10}H_8$ were calculated using much simpler nonlocal hybrid B3LYP with 6-31G(d) basis, incorporated in the GAUSSIAN 09 package [23]. The numerical details of the structure and energies of all the relevant species are listed in the Supplementary Material.

3. Experimental Details and Analysis

A proton beam of energy ranging from 50 keV to 150 keV, in steps of 25 keV, was extracted from the electron cyclotron resonance ion source at the Low energy Ion Beam Facility (LEIBF) at the Inter-University Accelerator Centre (IUAC), New Delhi. These ions were made to interact with the target molecules in a ToF mass spectrometry system. The emitted electron (or electrons) as well as the possibly neutralised projectile in a given collision event were separately detected using channeltron detectors, one at the opposite direction of the recoil ToF tube and other at about 1.5 m distance from the interaction region. The recoil ions were accelerated and recorded using a 40 mm active diameter position-sensitive micro-channel plate detector (PSD) with a delay line anode. The data was recorded in common stop multi-hit mode. The secondary electron signal and the neutralised projectile signals were logically ORed and then delayed to be used as a common stop signal, with a relative delay adjusted to give priority to the projectile detection. The target molecules, naphthalene (nph) and azulene (az), were injected into the vacuum system using a long hypodermic needle connected to an external reservoir with a valve. The target samples were kept at room temperature and the effusion of the target vapour due to their vapour pressure at room temperature was found to be adequate to perform the experiment with an event rate of about 1 to 2 kHz. Entirely different target masses (not reported here) were used in between the two experiments and totally different supply lines were used for the two targets to avoid any cross contamination.

The recorded data were classified based on the electron emission and electron capture process and background noise was reduced using various sum conditions on the PSD. We report here the data sorted based on the electron emission mode. Although up to eight-hit events could be recorded, data up to the third hit were sufficient to evaluate the necessary fragmentation channels. The positive counts beyond the third hit were negligibly small. Typically, 76 % and 19 % of events recorded for the 50 keV proton beam collisions were purely single and double hits, respectively, and less than 4 % of events were triple hits. The exact percentage depends strongly on the collection and detection efficiencies for the ion signal. The extraction field was high enough to collect all the ions onto the detector, However, the detection efficiency varies from ~45 % for a single carbon ion to ~20 % for an intact molecular ion. The detection efficiency of the micro-channel plate detector was calculated using an empirical model reported in the literature [24].

4. Results and Discussion

4.1. Normalisation Process

The single ionisation cross section of az and nph are known to be identical for the studied collision energies [11]. The single hit data of two isomers are normalised to the sum of singly charged parent ions (m/q 128 and 129), neutral H loss (m/q 127) and C_2H_2 loss (m/q 102) ions from parent monocation of a given isomer. The normalisation procedure was chosen for two reasons. Firstly, because of the neural H loss contribution, although negligible, wait s difficult to separate from the parent peak. Secondly, it was seen that both isomers have same single ionisation cross sections, but az^+, being a high-energy isomer, experiences higher fraction of C_2H_2 emission than nph, the details of which have been published elsewhere [11]. All the one-dimensional (1D) spectra shown henceforth are normalised by this procedure and multiplied by an arbitrary factor of 1000 (arbitrary common factor). The two-dimensional (2D) spectra are shown as counts without normalisation.

4.2. Single Hit Analysis

The single hit mass spectrum of az and nph after normalisation are found to be similar, as shown in Figure 1. The mass spectra includes only the data wherein a single ionic species is detected in a given event; thus, the data shown is a mix of pure single ion events as well as events wherein only one of the fragment ions produced is detected. The peak at m/q 45 is acetone that was used for cleaning the target lines. The sharp peaks at m/q 20 and 40 are due to the addition of Ar to the background gas, and all the mass spectra analysed here were calibrated using these reference peaks. One striking difference between the two mass spectra is the 60 % excess yield at m/q 102 peak (loss of neutral C_2H_2 from monocation) in the az mass spectrum. This was explained on the basis of the ground-state energy difference between singly charged az and nph in our previous work [11]. Similarly, hydrocarbon fragments with six or fewer carbon atoms are produced in a slightly higher yield for the target az. This may again be due to the less stable configuration of az in higher charge states as suggested in a recent work (Figure 1 in reference [25]). However, the fragments in the region m/q 98 ($C_8H_m^+$ where 0 < m < 10) and 88 ($C_7H_m^+$, where 0 < m < 10) have similar yields for two isomers in the mass spectrum. Overall, there is little difference between the fragmentation mass spectra of az and nph in high-energy collisions, which may be due to the lower energy difference between multiply charged isomers compared to singly charged or neutral isomers. In a recent paper, Lee et al. [26] demonstrated that even at much lower internal energies than in the present study, an equilibrium can be established between nph^+ and az^+ before dissociation. In fact, the energy transfer to the molecule is expected to be very high (typically few tens of eV) in collisional excitation with fast proton, so that the intact molecular ion can explore all available isomers prior to dissociation, independently from the initially chosen isomer.

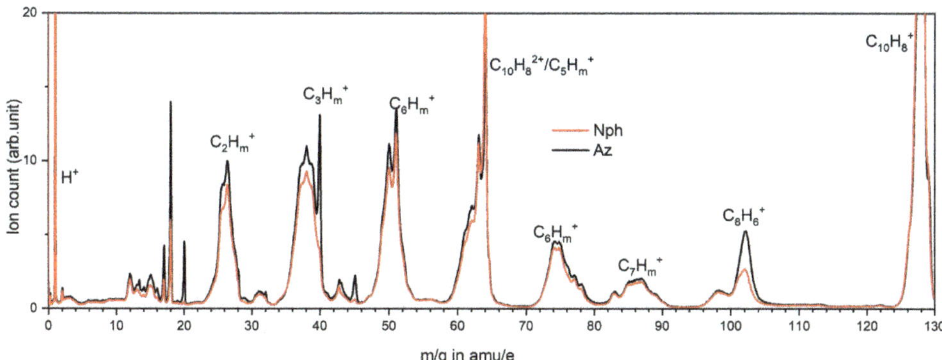

Figure 1. Normalised single hit spectrum of nph and az.

4.3. Coincidence Analysis of Double Hit Data

A 2D ToF coincidence map is plotted with first-hit ToF on the horizontal axis and second-hit ToF on the vertical axis as shown in Figure 2. The correlation patterns found are similar for az and nph. The coincidence map is rich in several ToF correlation patterns, most of which are broad and lack momentum correlation, representing multi-fragmentation or charge separation of highly charged ions with more than two fragment ions produced in a single event. The islands in coincidence with H^+ are the strongest channels in the 2D map. Most of the islands in the 2D spectrum are poorly resolved due to high momentum release, limiting our investigation to a few sets of channels. A few islands in the map are binary fragmentation channels, which are momentum-correlated, and most of them have an extended tail, indicating a fragmentation inside the extraction field of the ToF.

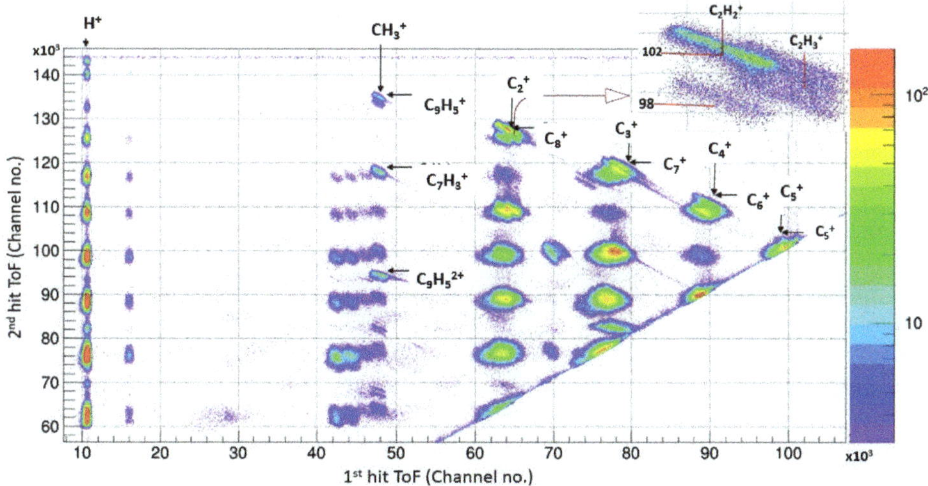

Figure 2. 1st hit vs. 2nd hit ToF coincidence map of az (data for all projectile energy are added). The position of the islands with H^+ or CH_3^+ in the first hit is indicated. Carbon-conserving fragmentation channels are marked and one of them is shown in the inset.

The carbon-conserving binary fragmentation islands that appear in the double-hit coincidence map (see Figure 2) are summarised below in Equation (1). Here, n = 8, 7 and 6 are major islands. An island with n = 5 also exists, but it is difficult to separate channels in this island due to the effect of pulse pair resolution. The island with n = 9 is not as intense as other channels. Some of the channels with n = 8 and n = 7 have an extended tail representing a dissociation within the extraction timescale of ToF. Meanwhile, no strong tail is observed for channels with n = 6, indicating fast dissociation before extraction.

$$C_{10}H_8^{x+} \implies C_n H_m^{y+} + C_{10-n} H_p^{(x-y)+} + (8 - m - p)H \tag{1}$$

Fragments with mainly few carbon atoms accompany with dehydrogenation. For instance, as one goes from n = 9 to n = 6, the channels shift from hydrogenated to dehydrogenated channels. Furthermore, there is a clear evidence of loss of H in multiples of two as observed in the previous studies [27,28]. One such example is the emission of $C_2H_n^+$, where the main channel follows $4H$ or $2H_2$ loss, not even $2H/H_2$ loss. The loss of $C_2H_3^+$, on the other hand, follows no such restriction. Channels on other islands cannot be summarised as Equation (1) because of the large incorporation of neutral and cationic hydrocarbon carbon fragments. Some interesting features are observed in such channels as discussed below in some examples.

4.4. H^+ Coincidence

As shown in Figure 3, the second-hit spectrum of nph with H^+ in the first hit is compared to its single-hit spectrum. All hydrocarbon fragments correlated to the first-hit ToF of H^+ are observed with higher intensity compared to the similar fragments in the single-hit spectrum, which can be related to the increased detection efficiency of H^+. Two quick observations can be made here: (i) none of the peaks in the second hit spectrum are sharp. This implies that all partner fragments are produced with some kinetic energy. Also, H^+ islands have no momentum correlation in the 2D. Both of these observations indicate that H^+ originates mainly from highly charged parent ions, with charge state greater than 2+, which is consistent with a previous theoretical results for multiply charged naphthalene [29]. (ii) the intensities of fragments containing an odd number of carbon atoms, especially CH_m^+ and $C_3H_m^+$, are substantially larger than the peaks in the single hit. This may be due to the oscillating binding energy of carbon clusters produced. The propensity of such cluster ions is usually decided by the energy of formation and the ionisation potential [30].

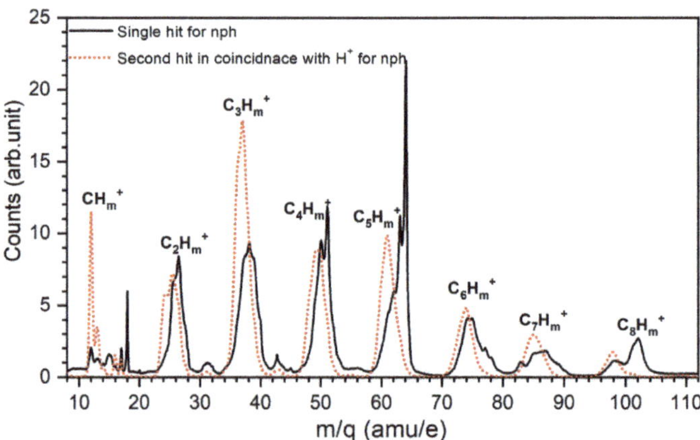

Figure 3. Single-hit mass spectrum of nph and second-hit mass spectrum with H^+.

The second-hit mass spectra of az and nph obtained in coincidence with H^+ are shown in Figure 4. For the peaks around n = 6, 7 and 8, there is no difference between az and nph. The intensity of all other fragments are slightly higher for az, which might be due to the inherently less-stable configuration of highly charged az compared to nph.

Another striking observation is the presence of super-dehydrogenated az$^+$ and nph$^+$ (m/q 120–127) as shown in Figure 5. For the first time, we report here the evidence of losing all hydrogen atoms from az/nph (*m/z* 120 in Figure 5a) in a single collision. The two isomers of $C_{10}H_8$ studied here prefer to lose 2, 6, 7 or 8 hydrogens, at least one of which is lost as H$^+$. These super-dehydrogenations and the fact that 3, 4 and 5 H loss channels are not favoured warrants a detailed investigation, which may be useful to understand the trend of atomic/molecular hydrogen in the astrophysical region where PAHs are irradiated by stellar wind protons.

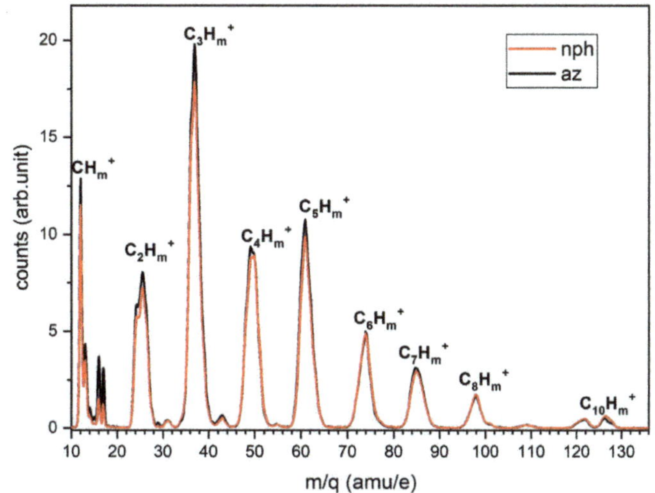

Figure 4. Second hit mass spectrum of az and nph with H$^+$ in the first hit.

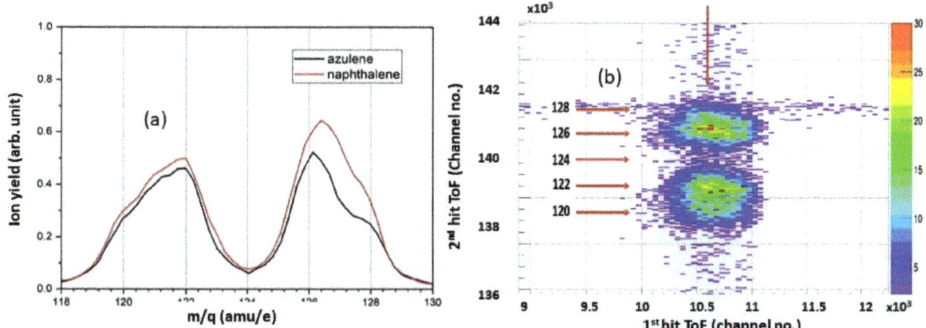

Figure 5. (**a**) Second hit mass spectrum of az and nph in the m/q 120–128 region for H$^+$ in the first hit, showing single and multiple H losses in singly charged parent ions, total dehydrogenation at m/q 120 can also be noted. (**b**) H$^+$ coincidence of az in m/q 120–128 region of the 2D ToF mass spectrum.

4.5. C^+, CH^+, CH_2^+ Coincidence

The second-hit mass spectrum with C$^+$ in the first hit of az and nph is shown in Figure 6. The second-hit mass spectrum of C$^+$ fragment is identical to the second-hit mass spectrum of H$^+$, except the fragments with 8 carbon atoms are missing. The second-hit mass spectrum of C$^+$ is similar to the second-hit mass spectrum of CH$^+$ and CH$_2^+$, which are mainly populated with lighter fragments. This indicates that these fragments are produced in the multi-fragmentation of a highly charged ion. The similarity between the second-hit mass spectrum of H^+, C^+, CH^+ and CH_2^+ may also suggest that these ions are produced in a single event, but H^+ is not detected; instead, C^+ is observed in the first hit.

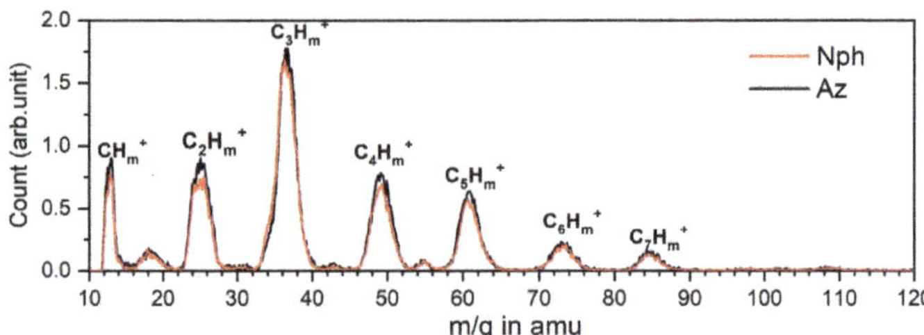

Figure 6. Second-hit mass spectrum of az and nph in coincidence with C^+.

4.6. CH_3^+ Coincidence

The ring-opening and ring-expansion mechanisms in smaller PAHs are pivotal to the understanding of formation of larger PAHs in ISM. To this end, the HACA (Hydrogen-Abstraction/acetylene-Addition) process [31] has been in the discussion for a quite some time and has been considered as a possible mechanism for the growth of larger PAHs. But, in recent years, laboratory experimental results have conclusively demonstrated that the HACA mechanism is not a favourable option, since it fails to account for the possible rate of formation of PAHs in the ISM [32]. A more plausible hypothesis for ring expansion and PAH growth has been proposed by Zhao et al. [32]. This mechanism involves the addition of a methyl group to a pentagon ring in the PAH so as to expand the ring to a hexagon structure and thus successively grow PAH size. In a laboratory setup, it is much easier to study the dissociation process to understand the reverse barriers associated with dissociative/associative reactions, than to experimentally look for the associative reaction itself. In this context, we looked at the second hit mass spectrum of CH_3^+ channel and compared it with C^+ and CH^+. We believe that an investigation of the parent conformers and associated reverse barriers will add significant value to the understanding of the methyl addition reactions in PAHs.

The second-hit mass spectrum of two isomers with CH_3^+ in the first hit are shown in Figure 7. Once again we see the similarity between az and nph. There is a clear difference between the second-hit mass spectrum of C^+ and CH_3^+ channels. $C_7H_3^+$, $C_9H_5^+$ and $C_9H_5^{2+}$ are the major fragments formed after CH_3^+ loss. These channels appear with a tail in the 2D ToF mass spectrum (see Figure 2). We were able to determine the number of H atoms in each of these fragments using ToF correlations in the 2D. The mass spectrum of az below m/q of 46 was heavily contaminated by the acetone, which is used for cleaning the target lines before introducing this isomer target. The second-hit mass spectrum of az is subtracted from nph and shown in Figure 7b. The differential mass spectrum is dominated by fragments from acetone (below m/q of 46). We further analysed the CH_m^+ (m/q 12 to 17) region in coincidence with H^+ in the 2D ToF map (Figure 8). It is observed that masses m/q 12, 13 and 14 (C^+, CH^+, CH_2^+) are clearly visible, but only a few counts at m/q 15 (CH_3^+) are present due to chance coincidence with H^+. This once more supports our argument that the production of H^+, C^+, CH^+ and CH_2^+ are related, whereas CH_3^+ is produced exclusively by a different mechanism.

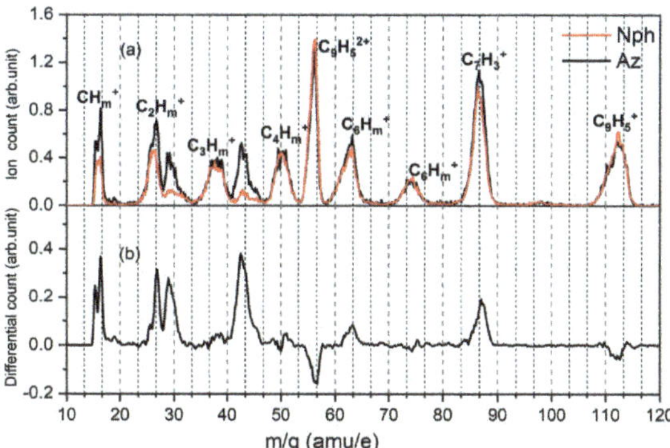

Figure 7. (**a**) Second- hit mass spectra of az and nph in coincidence with CH_3^+ ions, (**b**) the differential second-hit mass spectrum between the two isomers for CH_3^+ loss channel .

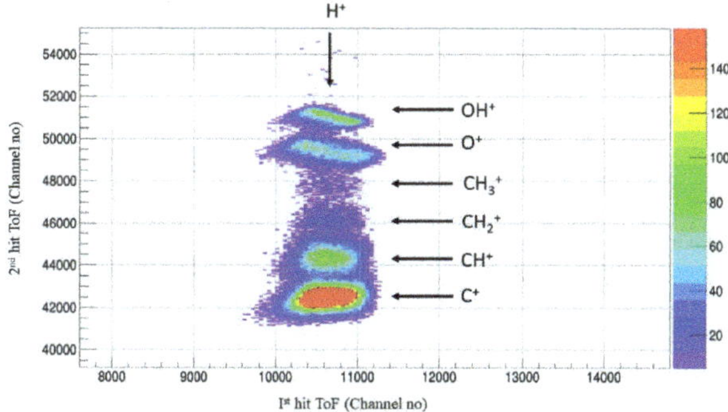

Figure 8. H^+ coincidence of az in m/q 12-17 region of the 2D ToF mass spectrum. m/q 15 region is filled with only chance coincidence.

The relative ground state energy of $C_{10}H_8^{2+}/C_{10}H_8^{3+}$ isomers, which can emit CH_3^+ are shown in table1. Structure A (see Table 1) was proposed as the precursor ion for $CH_3^++C_9H_5^+$ channel by Kingston et al. [33]. This was further investigated by Leach et al. [18] based on the kinetic energy release (KER) of CD_3^+ loss in dication of naphthalene-d_8 ($C_{10}D_8^{2+}$) [18]. The experimental KER reported by them was approximately 1 eV, whereas the structure A would correspond to about 2 eV KER and therefore Leach et al. [18] suggested that a linear geometry may be preferred over structure A. The most probable value of KER for this channel obtained in this work is 2.9 eV for nph and az targets. This value matches with the value measured recently by Reitsma et al. [34]. It suggests that structure A may be a common isomer of az and nph that can emit CH_3^+, as originally suggested by Kingston et al. [33]. But, when we performed DFT calculations for other possible structures, we observed that the dication structure A was about 3 eV higher in the energy with respect to nph^{2+}. On the other hand, structure B shown in Table 1 is found to be more favourable with 2.0 eV higher energy than nph^{2+}. Hence, we propose that the elimination of CH_3^+ might occur via a common dication conformer with structure B. Obviously, the exact transition of nph or az to the parent isomer of the CH_3^+ loss channel would occur through various complex intermediate as well as transition states. Such calculations have been carried out for some important dissociation channels of monocations in the past [35,36], but rarely attempted for di- or trications of nph and az . Calculations of such complexity are presently beyond the scope of this work. But it does not impede the present work in explaining the CH_3^+ elimination in az^{2+} and nph^{2+} via a common isomer, because the formation of a dication in the high energy proton collision proceeds via double plasmon excitation process, as suggested in our previous work [11]. This process is found to deposit an internal energy of about 13 eV in the resulting dication [11]. Thus, this internal energy will be sufficient to cross any possible transition state barrier the species may encounter on the way to producing CH_3^+ eliminating parent structure.

Table 1. Ground state energies of $C_{10}H_8^{2+}$ and $C_{10}H_8^{3+}$ isomers relative to the most stable isomer (ΔEs, eV), as calculated using DFT method with 631G(d) basis.

	Structure	ΔE of Dication	ΔE of Trication
Nph		0	0
Az		0.41	0.29
A		3.07	2.20
B		1.96	1.14

4.7. Multihit Analysis of CH_3^+ Channel

A ToF correlation between second and third hit fragments of C^+ and CH_3^+ channels are shown in Figure 9. Several broad islands are observed in the 2D coincidence map of the C^+ channel. Only three binary dissociation channels of $C_9H_5^{2+}$ are observed in the CH_3^+ emission and are summarised below.

$$C_9H_5^{2+} \Longrightarrow C_nH_m^+ + C_{9-n}H_{m-5}^+ + (5-m-n)H \tag{2}$$

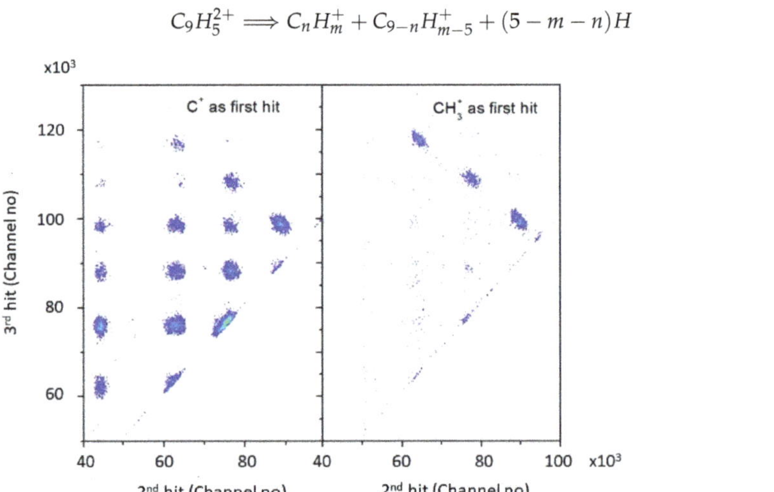

Figure 9. 2nd vs. 3rd hit ToF coincidence map of C^+ (**left**) and CH_3^+ (**right**) of az.

The third-hit mass spectra of the CH_3^+ channel for az and nph are shown in Figure 10. Both mass spectra are very similar, consisting of n = 2–9 fragments of moderate size. The formation of such neutral or ionic hydrocarbon species is of importance in the astronomical context [37]. For instance, the structural conformations of partially hydrogenated C_7 and C_9 neutrals as well as ions are proposed as the possible carriers of some weak diffuse interstellar bands in the ISM [38,39]. Steglich et al. [38] computationally identified stable structures of C_9H_5 radicals, which are found to feature visible absorption bands that coincide with a few very weak diffuse interstellar bands. We consider here the same structures to identify stable isomers of mono- and dications of C_9H_5. As mentioned earlier, the di- and trications are produced at high internal energies in fast proton collisions. A significant fraction of this energy can be utilised to overcome various transition state barriers to be able to produce stable fragment geometries. The proposed isomers of $C_9H_5^+$ and $C_9H_5^{2+}$ and their ground state energies relative to the most stable isomer are given in Table 2. According to this calculation, the most stable isomers of $C_9H_5^{2+}$ are A and E. The third hit mass spectrum of CH_3^+ channel indicates that an ensemble of $C_9H_5^{2+}$ isomers are formed, which can produce fragment partners, $C_2H_m^+-C_7H_m^+$, $C_3H_m^+-C_6H_m^+$ and $C_4H_m^+-C_5H_m^+$, in equal intensities. This observation combined with the energy of optimised structures suggests that structure A may predominate the ensemble of $C_9H_5^{2+}$ isomers. We also propose that structure D (the lowest energy structure of $C_9H_5^+$) might be an important fraction of $C_9H_5^+$ isomers formed after eliminating CH_3^+ from the final conformer of nph^{2+}/az^{2+}. The last two conclusions require more dedicated experimental and theoretical investigations.

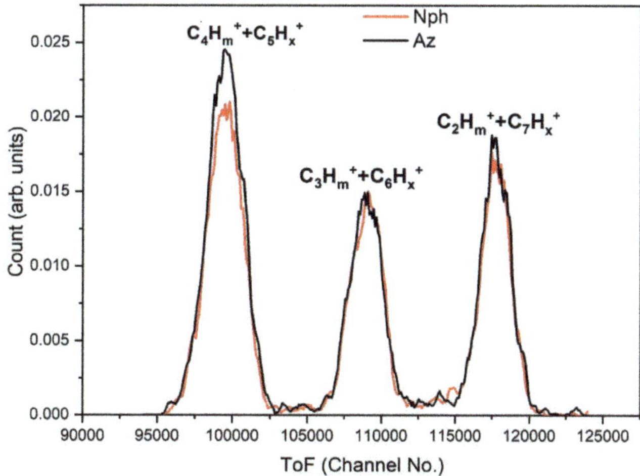

Figure 10. 3rd -hit ToF mass spectrum of az and nph for CH_3^+ channel.

Table 2. Ground-state energies of $C_9H_5^+$ and $C_9H_5^{2+}$ isomers relative to the most stable isomer (ΔEs, eV), as calculated using DFT method with cc-pVDZ basis.

	Structure	ΔE of Monocation	ΔE of Dication
A		0.9	0
B		1.11	0.45

Table 2. *Cont.*

	Structure	ΔE of Monocation	ΔE of Dication
C		1.33	0.52
D		0	0.49
E		0.27	0.02
F		0.22	1.17
G		1.54	0.2

5. Conclusions

Swift charged-particle-induced single and double ionisation of nph as well as az are known to be plasmon-dominated. These multiple ionisation processes deposit a high amount of excess energy in the resultant cations, which then can be considered to have a high probability for high-energy isomer formation, often followed by dissociation. One such mechanism of CH_3^+ elimination is found to progress via a common isomer of nph and az. Moreover, a cascade of dissociation is observed in multihit analysis up to a third hit of this channel. New parent dication as well as trication isomers of nph and az are proposed here based on the ground-state energies calculated by DFT theory, which can eliminate CH_3^+. $C_9H_5^+$ formed after the loss of CH_3^+ eliminates C_2H_2 to form $C_7H_3^+$. The dication fragment $C_9H_5^{2+}$ undergoes binary fragmentations such that the precursor ion is an isomer containing a long carbon chain. Various DFT optimised structural conformers of $C_9H_5^{2+}$ were compared and the lowest-energy structure is found to have a structure of a pentagon with a long linear chain.

In addition to the decay cascade of the CH_3^+ loss channel, a few other observations are worth noting. First, we report the super-dehydrogenation of nph and az in a single collision condition with a detectable intensity of total dehydrogenation. Second, the production of H^+, C^+, CH^+ and CH_2^+ are related, whereas CH_3^+ is produced exclusively by a different mechanism. These observations warrant more theoretical investigations.

Supplementary Materials: The following supporting information can be downloaded at: https://www.mdpi.com/article/10.3390/atoms11110138/s1, Table S1a: Structure A dication; Table S1b: Structure B dication; Table S1c: Structure A trication; Table S1d: Structure B trication; Table S2a: structure A monocation; Table S2b: structure B monocation; Table S2c: structure C monocation; Table S2d: structure D monocation; Table S2e: structure E monocation; Table S2f: structure F monocation; Table S2g: structure G monocation; Table S2h: structure A dication; Table S2i: structure B dication; Table S2j: structure C dication; Table S2k: structure D dication; Table S2l: structure E dication; Table S2m: structure F dication; Table S2n: structure G dication.

Author Contributions: Conceptualisation, U.R.K.; methodology, U.R.K., P.B., C.P.S. and S.V.; validation, M.V.V., U.R.K., C.P.S. and S.V.; formal analysis, M.V.V.; investigation, M.V.V. and U.R.K.; resources, U.R.K. and C.P.S.; data curation, M.V.V. and U.R.K.; writing—original draft preparation, M.V.V. and U.R.K.; writing—review and editing, M.V.V., U.R.K., C.P.S. and S.V.; visualisation, M.V.V. and U.R.K.; supervision, U.R.K.; project administration, U.R.K. All authors have read and agreed to the published version of the manuscript.

Funding: This research received no external funding.

Institutional Review Board Statement: Not applicable.

Informed Consent Statement: Not applicable.

Data Availability Statement: The data presented in this study are available in this article. The actual raw data presented in this study are available on request from the corresponding author.

Conflicts of Interest: The authors declare no conflict of interest.

References

1. Van Brunt, R.J.; Wacks, M.E. Electron-Impact Studies of Aromatic Hydrocarbons. III. Azulene and Naphthalene. *J. Chem. Phys.* **1964**, *41*, 3195–3199. [CrossRef]
2. Ruehl, E.; Price, S.D.; Leach, S. Single and double photoionization processes in naphthalene between 8 and 35 eV. *J. Phys. Chem.* **1989**, *93*, 6312–6321. [CrossRef]
3. Jochims, H.; Rasekh, H.; Rühl, E.; Baumgärtel, H.; Leach, S. The photofragmentation of naphthalene and azulene monocations in the energy range 7–22 eV. *Chem. Phys.* **1992**, *168*, 159–184. [CrossRef]
4. Jochims, H.; Baumgärtel, H.; Leach, S. Structure-dependent photostability of polycyclic aromatic hydrocarbon cations: Laboratory studies and astrophysical implications. *Astrophys. J.* **1999**, *512*, 500. [CrossRef]
5. Cui, W.; Hadas, B.; Cao, B.; Lifshitz, C. Time-resolved photodissociation (TRPD) of the naphthalene and azulene cations in an ion trap/reflectron. *J. Phys. Chem. A* **2000**, *104*, 6339–6344. [CrossRef]
6. West, B.; Joblin, C.; Blanchet, V.; Bodi, A.; Sztáray, B.; Mayer, P.M. On the dissociation of the naphthalene radical cation: New iPEPICO and tandem mass spectrometry results. *J. Phys. Chem. A* **2012**, *116*, 10999–11007. [CrossRef]
7. Reitsma, G.; Zettergren, H.; Martin, S.; Brédy, R.; Chen, L.; Bernard, J.; Hoekstra, R.; Schlathölter, T. Activation energies for fragmentation channels of anthracene dications—Experiment and theory. *J. Phys. B At. Mol. Opt. Phys.* **2012**, *45*, 215201. [CrossRef]
8. Postma, J.; Bari, S.; Hoekstra, R.; Tielens, A.; Schlathölter, T. Ionization and Fragmentation of Anthracene upon Interaction with keV Protons and α Particles. *Astrophys. J.* **2009**, *708*, 435. [CrossRef]
9. West, B.; Castillo, S.R.; Sit, A.; Mohamad, S.; Lowe, B.; Joblin, C.; Bodi, A.; Mayer, P.M. Unimolecular reaction energies for polycyclic aromatic hydrocarbon ions. *Phys. Chem. Chem. Phys.* **2018**, *20*, 7195–7205. [CrossRef]
10. Simon, A.; Champeaux, J.P.; Rapacioli, M.; Moretto Capelle, P.; Gadéa, F.X.; Sence, M. Dissociation of polycyclic aromatic hydrocarbons at high energy: MD/DFTB simulations versus collision experiments: Fragmentation paths, energy distribution and internal conversion: Test on the pyrene cation. *Theor. Chem. Accounts* **2018**, *137*, 1–11. [CrossRef]
11. Vinitha, M.V.; Najeeb, P.K.; Kala, A.; Bhatt, P.; Safvan, C.P.; Vig, S.; Kadhane, U.R. Plasmon excitation and subsequent isomerization dynamics in naphthalene and azulene under fast proton interaction. *J. Chem. Phys.* **2018**, *149*, 194303. [CrossRef]
12. Bagdia, C.; Biswas, S.; Mandal, A.; Bhattacharjee, S.; Tribedi, L.C. Ionization and fragmentation of fluorene upon 250 keV proton impact. *Eur. Phys. J. D* **2021**, *75*, 1–7.
13. Leach, S. The formation and destruction of doubly-charged polycyclic aromatic hydrocarbon cations in the interstellar medium. *J. Electron Spectrosc. Relat. Phenom.* **1986**, *41*, 427–438. [CrossRef]
14. Witt, A.N.; Gordon, K.D.; Vijh, U.P.; Sell, P.H.; Smith, T.L.; Xie, R.H. The excitation of extended red emission: New constraints on its carrier from hubble space telescope observations of NGC 7023. *Astrophys. J.* **2006**, *636*, 303. [CrossRef]
15. Malloci, G.; Joblin, C.; Mulas, G. Theoretical evaluation of PAH dication properties. *Astron. Astrophys.* **2007**, *462*, 627–635. [CrossRef]
16. Monfredini, T.; Quitián-Lara, H.M.; Fantuzzi, F.; Wolff, W.; Mendoza, E.; Lago, A.F.; Sales, D.A.; Pastoriza, M.G.; Boechat-Roberty, H.M. Destruction and multiple ionization of PAHs by X-rays in circumnuclear regions of AGNs. *Mon. Not. R. Astron. Soc.* **2019**, *488*, 451–469.
17. Leach, S.; Eland, J.; Price, S. Formation and dissociation of dications of naphthalene, azulene and related heterocyclic compounds. *J. Phys. Chem.* **1989**, *93*, 7575–7583. [CrossRef]
18. Leach, S.; Eland, J.; Price, S. Formation and dissociation of dications of naphthalene-d8. *J. Phys. Chem.* **1989**, *93*, 7583–7593. [CrossRef]
19. Sittler, E.C.; Hartle, R.; Bertucci, C.; Coates, A.; Cravens, T.; Dandouras, I.; Shemansky, D. Energy deposition processes in Titan's upper atmosphere and its induced magnetosphere. In *Titan from Cassini-Huygens*; Springer: New York, NY, USA, 2010; pp. 393–453.

20. Smith, H.; Mitchell, D.; Johnson, R.; Paranicas, C. Investigation of energetic proton penetration in Titan's atmosphere using the Cassini INCA instrument. *Planet. Space Sci.* **2009**, *57*, 1538–1546. [CrossRef]
21. Sillanpää, I.; Johnson, R.E. The role of ion-neutral collisions in Titan's magnetospheric interaction. *Planet. Space Sci.* **2015**, *108*, 73–86. [CrossRef]
22. Cravens, T.; Robertson, I.; Ledvina, S.; Mitchell, D.; Krimigis, S.; Waite, J., Jr. Energetic ion precipitation at Titan. *Geophys. Res. Lett.* **2008**, *35*, L03103.
23. Frisch, M.J.; Trucks, G.W.; Schlegel, H.B.; Scuseria, G.E.; Robb, M.A.; Cheeseman, J.R.; Scalmani, G.; Barone, V.; Mennucci, B.; Petersson, G.A.; et al. Gaussian 09: (Revision D.01). *Inc. Wallingford CT* **2009**, *121*, 150–166.
24. Krems, M.; Zirbel, J.; Thomason, M.; DuBois, R.D. Channel electron multiplier and channelplate efficiencies for detecting positive ions. *Rev. Sci. Instrum.* **2005**, *76*, 093305.
25. Santos, J.C.; Fantuzzi, F.; Quitián-Lara, H.M.; Martins-Franco, Y.; Menéndez-Delmestre, K.; Boechat-Roberty, H.M.; Oliveira, R.R. Multiply charged naphthalene and its C10H8 isomers: Bonding, spectroscopy, and implications in AGN environments. *Mon. Not. R. Astron. Soc.* **2022**, *512*, 4669–4682.
26. Lee, J.W.; Stockett, M.H.; Ashworth, E.K.; Navarro Navarrete, J.E.; Gougoula, E.; Garg, D.; Ji, M.; Zhu, B.; Indrajith, S.; Zettergren, H.; et al. Cooling dynamics of energized naphthalene and azulene radical cations. *J. Chem. Phys.* **2023**, *158*, 174305. [PubMed]
27. Ławicki, A.; Holm, A.I.; Rousseau, P.; Capron, M.; Maisonny, R.; Maclot, S.; Seitz, F.; Johansson, H.A.; Rosén, S.; Schmidt, H.T.; et al. Multiple ionization and fragmentation of isolated pyrene and coronene molecules in collision with ions. *Phys. Rev. A* **2011**, *83*, 022704.
28. Champeaux, J.P.; Moretto-Capelle, P.; Cafarelli, P.; Deville, C.; Sence, M.; Casta, R. Is the dissociation of coronene in stellar winds a source of molecular hydrogen? application to the HD 44179 nebula. *Mon. Not. R. Astron. Soc.* **2014**, *441*, 1479–1487. [CrossRef]
29. Holm, A.I.; Johansson, H.A.; Cederquist, H.; Zettergren, H. Dissociation and multiple ionization energies for five polycyclic aromatic hydrocarbon molecules. *J. Chem. Phys.* **2011**, *134*, 044301.
30. Martínez, J.I.; Alonso, J.A. Hydrogen quenches the size effects in carbon clusters. *Phys. Chem. Chem. Phys.* **2019**, *21*, 10402–10410.
31. Wang, H.; Frenklach, M. Calculations of rate coefficients for the chemically activated reactions of acetylene with vinylic and aromatic radicals. *J. Phys. Chem.* **1994**, *98*, 11465–11489. [CrossRef]
32. Zhao, L.; Kaiser, R.I.; Lu, W.; Xu, B.; Ahmed, M.; Morozov, A.N.; Mebel, A.M.; Howlader, A.H.; Wnuk, S.F. Molecular mass growth through ring expansion in polycyclic aromatic hydrocarbons via radical–radical reactions. *Nat. Commun.* **2019**, *10*, 3689.
33. Kingston, R.; Guilhaus, M.; Brenton, A.; Beynon, J. Multiple ionization, charge separation and charge stripping reactions involving polycyclic aromatic compounds. *Org. Mass Spectrom.* **1985**, *20*, 406–412. [CrossRef]
34. Reitsma, G.; Zettergren, H.; Boschman, L.; Bodewits, E.; Hoekstra, R.; Schlathölter, T. Ion–polycyclic aromatic hydrocarbon collisions: Kinetic energy releases for specific fragmentation channels. *J. Phys. At. Mol. Opt. Phys.* **2013**, *46*, 245201.
35. Dyakov, Y.A.; Ni, C.K.; Lin, S.; Lee, Y.; Mebel, A. Ab initio and RRKM study of photodissociation of azulene cation. *Phys. Chem. Chem. Phys.* **2006**, *8*, 1404–1415. [CrossRef]
36. Solano, E.A.; Mayer, P.M. A complete map of the ion chemistry of the naphthalene radical cation? DFT and RRKM modeling of a complex potential energy surface. *J. Chem. Phys.* **2015**, *143*, 104305. [PubMed]
37. Huang, J.; Oka, T. Constraining the size of the carrier of the $\lambda 5797$. 1 diffuse interstellar band. *Mol. Phys.* **2015**, *113*, 2159–2168.
38. Steglich, M.; Maity, S.; Maier, J.P. Visible absorptions of potential diffuse ISM hydrocarbons: C9H9 and C9H5 radicals. *Astrophys. J.* **2016**, *830*, 145.
39. Maity, S.; Steglich, M.; Maier, J.P. Gas Phase Detection of Benzocyclopropenyl. *J. Phys. Chem.* **2015**, *119*, 10849–10853.

Article

Quadrupole Effects in the Photoionisation of Sodium 3s in the Vicinity of the Dipole Cooper Minimum

Nishita M. Hosea [1], Jobin Jose [2], Hari R. Varma [1,*], Pranawa C. Deshmukh [3,4] and Steven T. Manson [5]

[1] School of Physical Sciences, Indian Institute of Technology Mandi, Kamand 175075, India; d18043@students.iitmandi.ac.in

[2] Department of Physics, Indian Institute of Technology Patna, Bihta 801103, India; jobinjosen@gmail.com

[3] Department of Physics and CAMOST, Indian Institute of Technology Tirupati, Tirupati 517619, India; pcd@iittp.ac.in

[4] Department of Physics, Dayananda Sagar University, Bengaluru 560078, India

[5] Department of Physics and Astronomy, Georgia State University, Atlanta, GA 30303, USA; smanson@gsu.edu

* Correspondence: hari@iitmandi.ac.in

Abstract: A procedure to obtain relativistic expressions for photoionisation angular distribution parameters using the helicity formulation is discussed for open-shell atoms. Electric dipole and quadrupole transition matrix elements were considered in the present work, to study the photoionisation dynamics of the 3s electron of the sodium atom in the vicinity of the dipole Cooper minimum. We studied dipole–quadrupole interference effects on the photoelectron angular distribution in the region of the dipole Cooper minimum. Interference with quadrupole transitions was found to alter the photoelectron angular distribution, even at rather low photon energies. The initial ground and final ionised state discrete wavefunctions of the atom were obtained in the present work using GRASP, and we employed RATIP with discrete wavefunctions, to construct continuum wavefunctions and to calculate transition amplitudes, total cross-sections and angular distribution asymmetry parameters.

Keywords: non-dipole interactions; photoelectron angular distributions; open-shell atomic systems; Cooper minimum; GRASP; RATIP

Citation: Hosea, N.M.; Jose, J.; Varma, H.R.; Deshmukh, P.C.; Manson, S.T. Quadrupole Effects in the Photoionisation of Sodium 3s in the Vicinity of the Dipole Cooper Minimum. *Atoms* **2023**, *11*, 125. https://doi.org/10.3390/atoms11100125

Academic Editor: Emmanouil P. Benis

Received: 8 August 2023
Revised: 19 September 2023
Accepted: 21 September 2023
Published: 28 September 2023

1. Introduction

In the majority of studies of light–matter interaction, the dipole approximation is used. It is generally applicable when electromagnetic radiation has a wavelength much larger than the size of the atomic or molecular system. In the dipole approximation ($e^{ikr} \sim 1$), where k is the wavenumber of the incident photon, one neglects the spatial variation of the electromagnetic field over the target system. Non-dipole effects are important at short wavelengths, and have prompted several atomic and molecular studies [1–5] in condensed matter physics [6] and astrophysics [7]. The emergence of intense laser light sources, such as the free-electron laser (FEL) [8–10], have further revealed the importance of non-dipole interactions in explaining photoelectron spectra, especially in relation to non-linear absorption and time-resolved studies.

The importance of non-dipole effects has been highlighted by several authors, in both experimental and theoretical works [3,11–25]. These studies have revealed that dipole–quadrupole (E1–E2) interference affects the angular distribution of photoelectrons, due to first-order corrections to the dipole approximation, even at rather low energies. Numerous studies are available for closed-shell systems, but those on open-shell systems, especially using relativistic methodologies, are few [26,27]. To the best of our knowledge, relativistic calculations, including interchannel coupling, are not available for open-shell atoms.

Higher multipole corrections to total subshell cross-sections become important for photon energies more than a few keV above the ionisation threshold. However, a number of situations exist in photoionisation processes that demand going beyond the dipole approximation, even at energies as low as a few eV [28–40]. Instances where the quadrupole transition amplitudes are comparatively larger than the electric dipole transition amplitudes occur in regions of the dipole Cooper minimum, dipole/quadrupole autoionisation resonances, etc. The present work was motivated by an earlier work by Pradhan et al. [37], which showed the importance of non-dipole effects in the case of Mg 3s photoionisation at rather low photon energies, due to the presence of the Cooper minimum in the 3s dipole ionisation channel. We explored a similar situation for the case of a typical open-shell atom, viz., sodium, by studying the photoionisation of its valence shell in the photon energy range 5.14 eV to 7 eV. The 3s dipole photoionisation goes through the Cooper minimum in this region. The required non-dipole angular distribution parameters were obtained, following an earlier work by Huang [41,42], which used helicity eigenstates to study the dynamics. This formulation is applicable to both open- and closed-shell systems [41,42]. On the other hand, the methodology described in Derevianko et al. [43] is applicable only to closed-shell systems. Below, we briefly present an overview of the procedure to include non-dipole effects in the photoelectron angular distribution asymmetry parameters for s-subshell photoionisation. The required transition amplitudes were determined in the present work by using a combination of two computational algorithms, namely, the General-Purpose Relativistic Atomic Structure Program (GRASP) [44–46] and the Relativistic calculations of Atomic Transition, Ionisation and recombination Properties (RATIP) [47]. The combination of GRASP and RATIP has already been successfully applied in a number of cases studying atomic structure and dynamics [48–51]. In the present work, a single configuration initial state of photoionisation was considered, but a multi-configuration initial state could also be considered.

In Section 2, details of the helicity formalism [41,42], along with the important steps involved in the derivation of the required photoelectron angular distribution parameters by a linearly polarised light, are discussed. The results of our calculations are discussed in Section 3. The important findings of this work are summarised in Section 4.

2. Theory

This section is divided into three sub-sections. In Section 2.1, the salient features of the helicity formulation of photoelectron angular distribution from references [41,42] are summarised. References [41,42] provide the form of β, and we explicitly discuss the various steps involved in arriving at the equations in Section 2.2. The general expression for the differential cross-section is available in the work of Huang [41,42], but not the expressions for non-dipole angular distribution asymmetry parameters. Explicit expressions for angular distribution asymmetry parameters inclusive of the quadrupole terms are developed and presented in Section 2.3 for the first time, to the best of our knowledge. Also provided is a brief discussion of second-order non-dipole photoelectron angular distribution parameters.

2.1. Photoionisation Dynamics Based on Helicity Formalism

Conventionally, the photoionisation transition matrix element is constructed using angular-momentum eigenstates. However, in the helicity formalism, angular-momentum eigenstates are transformed to helicity eigenstates. This approach was first adopted by Lee [52] for the non-relativistic formulation of photoionisation processes in the electric dipole approximation. This was extended to the relativistic regime by Huang [41,42]. In this work, the reduced matrix element $D_\alpha(Ej)$, for photoionisation in the Coulomb gauge for an electric 2^j- pole transition (Ej), is given by

$$D^{(Ej)}(\kappa_\alpha) = i^{-\ell_\alpha} e^{i\delta_{\kappa_\alpha}} \left\langle \alpha^- J \left\| \sum_{i=1}^{N} \vec{\alpha} \cdot \vec{A}^{(Ej)}(\vec{r}_i) \right\| J_0 \right\rangle. \tag{1}$$

Here, δ_{κ_α} is the Coulomb phase shift of the photoelectron in the particular channel $\kappa_\alpha = (\ell_\alpha j_\alpha)$, and $\vec{A}^{(Ej)}$ is the normalised electric multipole vector potential, while j and J_0 are, respectively, the total angular momenta of the photon and the initial states of the atom. J represents the total angular momentum of the photoelectron plus the ionised atom system. A similar expression of the reduced matrix element can be obtained for the magnetic 2^j-pole transitions, and is defined by $D^{(Mj)}(\kappa_\alpha)$.

The expression for the angle-dependent differential cross-section, including all multipole transitions in helicity formalism [41,42], is

$$\frac{d\sigma(\theta,\phi)}{d\Omega} = \frac{\sigma}{4\pi}F(\theta,\phi), \tag{2}$$

where σ is the total photoionisation cross-section and $F(\theta,\phi)$ is an angular distribution function given by

$$F(\theta,\phi) = 1 + \sum_{\ell \geq 1} B_{0\ell}d^\ell_{00} + (S_x\cos(2\phi) + S_y\sin(2\phi))\sum_{\ell \geq 2} B_{1\ell}d^\ell_{20}, \tag{3}$$

where S_x, S_y and S_z are the Stokes parameters of the incident light, d^ℓ_{mn} denotes the 'd' functions of the rotation matrices and θ and ϕ are the polar and azimuthal angles of the emitted photoelectron (of total angular momentum j_α), with respect to the incident photon direction $\hat{k}$, as shown in Figure 1a. In the above expression,

$$B_{0\ell} = \sum_{j'J'j'_\alpha jJj_\alpha} \frac{(-1)^{J_0 - J_\alpha + 1/2}}{\bar{\sigma}}[jJj_\alpha][j'J'j'_\alpha][\ell]^2 \begin{Bmatrix} J & J' & \ell \\ j'_\alpha & j_\alpha & J_\alpha \end{Bmatrix} \begin{Bmatrix} J & J' & \ell \\ j' & j & J_0 \end{Bmatrix} \begin{pmatrix} j'_\alpha & j_\alpha & \ell \\ 1/2 & -1/2 & 0 \end{pmatrix} \begin{pmatrix} j' & j & \ell \\ -1 & 1 & 0 \end{pmatrix}$$
$$\times \left\{ \pi_{\ell+}\left[\pi_{\ell+}(-1)^{\ell/2}(E'E + M'M) + \pi_{\ell-}(-1)^{(\ell+1)/2}(E'M - M'E)\right]\cos(\delta_{\alpha'} - \delta_\alpha) \right.$$
$$\left. + \pi_{\ell-}\left[\pi_{\ell-}(-1)^{(\ell+1)/2}(E'E + M'M) - \pi_{\ell+}(-1)^{\ell/2}(E'M - M'E)\right]\sin(\delta_{\alpha'} - \delta_\alpha) \right\} \tag{4}$$

and,

$$B_{1\ell} = \sum_{j'J'j'_\alpha jJj_\alpha} \frac{(-1)^{J_0 - J_\alpha + 1/2}}{\bar{\sigma}}[jJj_\alpha][j'J'j'_\alpha][\ell]^2 \begin{Bmatrix} J & J' & \ell \\ j'_\alpha & j_\alpha & J_\alpha \end{Bmatrix} \begin{Bmatrix} J & J' & \ell \\ j' & j & J_0 \end{Bmatrix} \begin{pmatrix} j'_\alpha & j_\alpha & \ell \\ 1/2 & -1/2 & 0 \end{pmatrix} \begin{pmatrix} j' & j & \ell \\ -1 & -1 & 2 \end{pmatrix}$$
$$\times \left\{ \pi_{\ell+}\left[\pi_{\ell+}(-1)^{\ell/2}(E'E - M'M) - \pi_{\ell-}(-1)^{(\ell+1)/2}(E'M + M'E)\right]\cos(\delta_{\alpha'} - \delta_\alpha) \right.$$
$$\left. + \pi_{\ell-}\left[\pi_{\ell-}(-1)^{(\ell+1)/2}(E'E - M'M) + \pi_{\ell+}(-1)^{\ell/2}(E'M + M'E)\right]\sin(\delta_{\alpha'} - \delta_\alpha) \right\}. \tag{5}$$

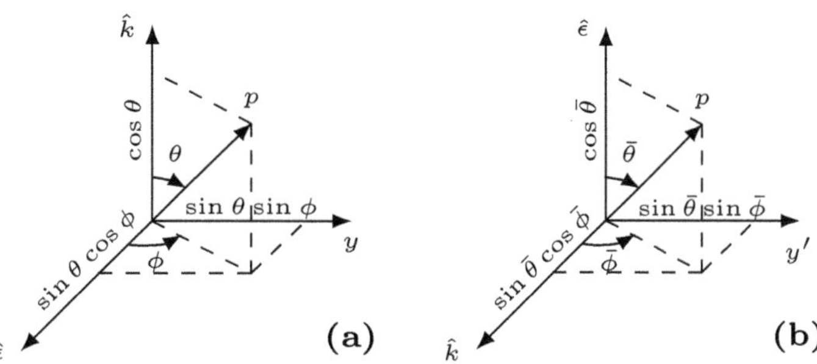

Figure 1. Transforming from the co-ordinate system (θ,ϕ) to the co-ordinate system $(\bar{\theta},\bar{\phi})$. The co-ordinate $(\bar{\theta},\bar{\phi})$ is obtained by rotating the co-ordinate (θ,ϕ) by 180° about the z-axis and 90° about the y-axis.

The summation $\sum_{j'J'j'_\alpha jJj_\alpha}$ takes into account the interference between various transition matrix elements, and δ_α is the phase of the reduced matrix element, $D^{(Ej)}(\kappa_\alpha)$. In the above expressions, J_α is the total angular momentum of the ionised state of the atom.

B_{0l} and B_{1l} correspond to various multipole terms arising from electric–electric, electric–magnetic and magnetic–magnetic interactions. Here, $\bar{\sigma} = \sum_{j\bar{j}\kappa_\alpha} \left[|D^{(Ej)}(\kappa_\alpha)|^2 + |D^{(Mj)}(\kappa_\alpha)|^2 \right]$, $\sigma = 8\pi^4 c/\omega[J_0]^2\bar{\sigma}$, and $E \equiv |D^{(Ej)}(\kappa_\alpha)|$. Similarly, $M \equiv |D^{(Mj)}(\kappa_\alpha)|$. Note that $[j] = \sqrt{2j+1}$. The effects of magnetic interactions are very weak compared to electric interactions, and hence are neglected in the present work. Furthermore, $\pi_{\ell\pm}$ and $\pi_{\hbar\pm}$ in Equations (4) and (5) are defined as follows:

$$\pi_{\ell+}(\pi_{\ell-}) = \begin{cases} 1(0) & \ell \text{ is even} \\ 0(1) & \ell \text{ is odd} \end{cases} \tag{6}$$

$$\pi_{\hbar+}(\pi_{\hbar-}) = \begin{cases} 1(0) & \hbar \text{ is even} \\ 0(1) & \hbar \text{ is odd,} \end{cases} \tag{7}$$

where ℓ is the summation index in Equation (3), $\hbar = j' - j$ and '$+/-$' correspond to even/odd.

A photon with linear momentum vector $\vec{k}$ is not in an eigenstate of the angular momentum j. However, being a massless particle, it has a definite value of helicity, which is the component of the angular momentum in the direction of the photon momentum. Now, for the electromagnetic waves (being transverse) the total angular momentum can take the values $j = 1, 2, 3, \ldots$ [53]. The infinite series in Equation (3) can be truncated, depending on the level of approximation considered, by making use of the Wigner $3j$ selection rules. The truncation procedure at the level of dipole, quadrupole and octupole approximations are discussed below.

2.2. Dipole Approximation

In the electric dipole approximation, $j = j' = 1$. The ℓ values in the Wigner $3j$ symbols, $\begin{pmatrix} j' & j & \ell \\ -1 & 1 & 0 \end{pmatrix}$ and $\begin{pmatrix} j' & j & \ell \\ -1 & -1 & 2 \end{pmatrix}$, of Equations (4) and (5) range from $|j' - j|$ to $j' + j$, giving $\ell = 0, 1$ and 2. The summations begin from $\ell = 1$ for the second term and $\ell = 2$ for the third term of Equation (3). Hence, under the dipole approximation, ℓ only takes values 1 and 2. In this particular case, the variable $\hbar$ in Equations (4) and (5) is an even number ($\pi_{\hbar+}$), because $\hbar$ is given by $j' - j$, which is zero. Since ℓ is odd and $\hbar$ is even, the only term that needs to be considered in Equation (4) is the one that involves the electric and magnetic interactions. However, in the dipole approximation, magnetic interactions do not appear and, hence, $B_{01} = 0$. Therefore, only $\ell = 2$ contributes in the dipole approximation giving rise to B_{02} and B_{12}. The Wigner $3j$ symbols of Equations (4) and (5), $\begin{pmatrix} j' & j & \ell \\ -1 & 1 & 0 \end{pmatrix}$ and $\begin{pmatrix} j' & j & \ell \\ -1 & -1 & 2 \end{pmatrix}$ give $1/\sqrt{30}$ and $1/\sqrt{5}$, respectively, for $j = j' = 1$ and $\ell = 2$. Using B_{01} and B_{02} along with Equation (3), Equation (2) reduces to

$$\frac{d\sigma}{d\Omega} = \frac{\sigma}{4\pi} \left[1 + B_{02}d_{00}^2 + \left(S_x \cos 2\phi + S_y \sin 2\phi \right) B_{12}d_{20}^2 \right]. \tag{8}$$

The right-hand side of Equations (4) and (5) for $\ell = 2$ can be written in terms of a single parameter, β_1 (a dipole asymmetry parameter), as follows: $B_{02} = -\beta_1/2$ and $B_{12} = -\sqrt{3/2}\,\beta_1$, where β_1 is

$$\beta_1 = -\sqrt{30}\frac{(-1)^{J_0 - J_\alpha + 1/2}}{\bar{\sigma}} \sum_{j'J'j'_\alpha} \sum_{jJj_\alpha} [J j_\alpha J' j'_\alpha] \begin{Bmatrix} J & J' & 2 \\ j'_\alpha & j_\alpha & J_\alpha \end{Bmatrix} \begin{Bmatrix} J & J' & 2 \\ 1 & 1 & J_0 \end{Bmatrix} \begin{pmatrix} j'_\alpha & j_\alpha & 2 \\ 1/2 & -1/2 & 0 \end{pmatrix} E'E \cos(\delta_{\alpha'} - \delta_\alpha). \tag{9}$$

The term $S_1 \cos 2\phi + S_2 \sin 2\phi$ in (8) is expressed as $-p \cos 2\alpha \cos(2(\phi - \gamma))$. Here, the parameters p, α and γ can be understood as follows. Consider a coordinate system XYZ, such that the Z axis is in the direction of the photon flux, as shown in Figure 2. The X axis is chosen conveniently to determine the photon polarisation. The parameter γ specifies the azimuthal orientation of the polarisation. When the photon polarisation vector coincides with the X axis, $\gamma = 0$. The angle between the electric field vectors at their successive crests is α. In case of linearly polarised light, $\alpha = 0$. The probability p ($0 \leq p \leq 1$) of complete polarisation is referred to as the degree of polarisation of the photon. The degree of polarisation $p = 1$ for pure linearly polarised incident photons. In Equation (8), $d^2_{00} = P_2(\cos\theta)$ and $d^2_{20} = \frac{\sqrt{3}}{2\sqrt{2}}\sin^2\theta$. Using these above relations, Equation (8) can be further reduced to

$$\frac{d\sigma}{d\Omega} = \frac{\sigma}{4\pi}\left\{1 - \frac{1}{2}\beta_1\left[P_2(\cos\theta) - \frac{3}{2}p\cos 2\alpha\cos(2(\phi - \gamma))\sin^2\theta\right]\right\}. \tag{10}$$

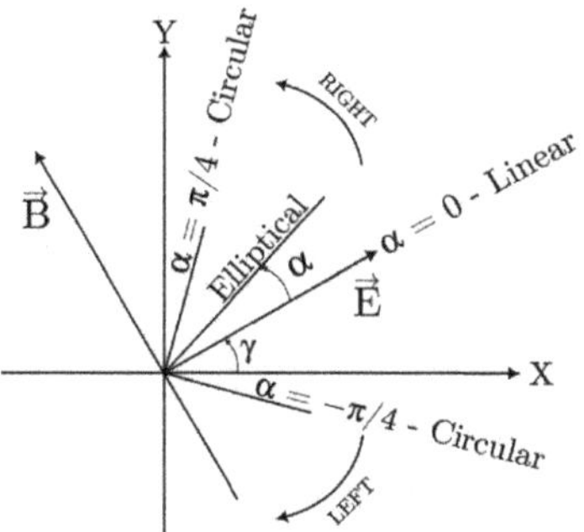

Figure 2. Types of polarisation.

It is often convenient to express the above equation in an alternate co-ordinate system, as shown in Figure 1b, where $\bar{\theta}$ corresponds to the angle between the photoelectron momentum and the polarisation direction, and where $\bar{\phi}$ corresponds to the angle between the propagation vector and the projection of momentum vector in the X-Y plane. Under this new co-ordinate system, Equation (10) reduces to

$$\frac{d\sigma}{d\Omega} = \frac{\sigma}{4\pi}\left\{1 + \beta_1 P_2(\cos\bar{\theta})\right\} \tag{11}$$

for linearly polarised light, where $\alpha = 0$ and $p = 1$.

2.3. Beyond Dipole Approximation

Including quadrupole interactions, j and j' can take values 1 and 2. Consequently, ℓ takes values $1 \leq \ell \leq 3$. The variable $k = j' - j$ can be -1 and 1, which implies that k is odd for the dipole and quadrupole interference terms. The even ℓ and the odd k can contribute only via electric–magnetic interactions. Since they are neglected in the present work, $B_{02} = 0$. Therefore, the only terms that contribute to the quadrupole approximation are B_{01}, B_{03} and B_{13}, which are rewritten in terms of Γ_1 and Γ_3 as follows:

$$\Gamma_1 = -3\sqrt{\frac{3}{2}}\frac{(-1)^{J_0 - J_\alpha + 1/2}}{\bar{\sigma}}\sum_{j'J'j'_\alpha}\sum_{jJj_\alpha}[J j_\alpha J' j'_\alpha]\begin{Bmatrix} J & J' & 1 \\ j'_\alpha & j_\alpha & J_\alpha \end{Bmatrix}\begin{Bmatrix} J & J' & 1 \\ j' & j & J_0 \end{Bmatrix}\begin{pmatrix} j'_\alpha & j_\alpha & 1 \\ 1/2 & -1/2 & 0 \end{pmatrix}(-1)^{(k+1)/2}E'E\sin(\delta_{\alpha'} - \delta_\alpha), \tag{12}$$

$$\Gamma_3 = -\sqrt{21}\frac{(-1)^{J_0 - J_\alpha + 1/2}}{\bar{\sigma}}\sum_{j'J'j'_\alpha}\sum_{jJj_\alpha}[J j_\alpha J' j'_\alpha]\begin{Bmatrix} J & J' & 3 \\ j'_\alpha & j_\alpha & J_\alpha \end{Bmatrix}\begin{Bmatrix} J & J' & 3 \\ j' & j & J_0 \end{Bmatrix}\begin{pmatrix} j'_\alpha & j_\alpha & 3 \\ 1/2 & -1/2 & 0 \end{pmatrix}(-1)^{(k+1)/2}E'E\sin(\delta_{\alpha'} - \delta_\alpha), \tag{13}$$

where $\Gamma_1 = B_{01}$ and $\Gamma_3 = B_{03} = \sqrt{3/10}B_{13}$. Using these, the differential cross-section for the photoionisation, given in Equation (3), with the inclusion for first-order quadrupole terms, can be expressed as

$$\frac{d\sigma}{d\Omega} = \frac{\sigma}{4\pi}\left\{1 + \Gamma_1 P_1(\cos\theta) - \frac{1}{2}\beta_2 P_2(\cos\theta) + \Gamma_1 P_3(\cos\theta) + \left[\frac{3}{4}\beta_2 - \frac{5}{2}\Gamma_3\cos\theta\right]p\cos 2\alpha\cos(2(\phi - \gamma))\sin^2\theta\right\}. \tag{14}$$

Note that the dipole asymmetry parameter β_1 is now written as β_2 under the quadrupole approximation. The expression for β_2 differs from β_1 only in the $\bar{\sigma}$ term in the denominator of Equation (9). With the inclusion of non-dipole interactions, $\bar{\sigma} = \sum_{J\kappa_\alpha} \left[|D^{(E1)}(\kappa_\alpha)|^2 + |D^{(E2)}(\kappa_\alpha)|^2 \right]$. In general, the dipole amplitude dominates the quadrupole amplitude. It can be, therefore, easily seen that this additional term in $\bar{\sigma}$ only plays a major role when the dipole amplitude goes through a minima. We again transform the representation of $d\sigma/d\Omega$ from $(\theta, \phi) \rightarrow (\bar{\theta}, \bar{\phi})$:

$$\frac{d\sigma}{d\Omega} = \frac{\sigma}{4\pi} \left\{ 1 + \beta_2 P_2(\cos\bar{\theta}) + \left[\Gamma_1 + \Gamma_3 - 5\Gamma_3 \cos^2\bar{\theta} \right] \sin\bar{\theta} \cos\bar{\phi} \right\}. \tag{15}$$

The usual experimental scheme to measure these non-dipole parameters is to set $\bar{\theta} = 54.7°$ and $\bar{\phi} = 0°$. The differential cross-section at these angles is

$$\frac{d\sigma}{d\Omega} = \frac{\sigma}{4\pi} \left\{ 1 + \sqrt{\frac{2}{3}} \left(\Gamma_1 + \Gamma_3 - \frac{5}{3}\Gamma_3 \right) \right\}. \tag{16}$$

It is convenient to write $\delta = \Gamma_1 + \Gamma_3$ and $\gamma = -5\Gamma_3$, so that

$$\frac{d\sigma}{d\Omega} = \frac{\sigma}{4\pi} \left\{ 1 + \sqrt{\frac{2}{27}} (3\delta + \gamma) \right\}, \tag{17}$$

where the combined quantity $3\delta + \gamma$ can be extracted from a measurement. Equation (15) is re-written in terms of δ and γ:

$$\frac{d\sigma}{d\Omega} = \frac{\sigma}{4\pi} \left\{ 1 + \beta_2 P_2(\cos\bar{\theta}) + \left[\delta + \gamma \cos^2\bar{\theta} \right] \sin\bar{\theta} \cos\bar{\phi} \right\}. \tag{18}$$

This is in agreement with Derevianko et al.'s [43] expression for $d\sigma/d\Omega$ while considering only the electric dipole and lowest order quadrupole interactions, which works for the closed-shell system. The above expression can also be written as $d\sigma/d\Omega = \sigma\{1 + A(\bar{\theta}, \bar{\phi})\}/4\pi$, where $A(\bar{\theta}, \bar{\phi})$ provides the angular distribution associated with the photoelectron ejection.

The above procedure can be further extended to higher-order terms. If we include the second-order correction, the differential cross-section can be expressed as follows:

$$\begin{aligned}
\frac{d\sigma}{d\Omega} = \frac{\sigma}{4\pi} \Big\{ &1 + \left(B_{01}^{E1,E2} \right) d_{00}^1 + \left(B_{02}^{E1,E1} + B_{02}^{E1,E3} + B_{02}^{E2,E2} \right) \\
&\times d_{00}^2 + B_{03}^{E1,E2} d_{00}^3 + \left(B_{04}^{E1,E3} + B_{04}^{E2,E2} \right) d_{00}^4 + (S_x \cos 2\phi - S_y \sin 2\phi) \times \\
&\left[\left(B_{12}^{E1,E1} + B_{12}^{E1,E3} + B_{12}^{E2,E2} \right) d_{20}^2 + B_{13}^{E1,E2} d_{20}^3 + \left(B_{14}^{E1,E3} + B_{14}^{E2,E2} \right) d_{20}^4 \right] \Big\}.
\end{aligned}$$

In Section 3, we present the results obtained using the above-mentioned procedures; in particular, the quadrupole effects in the photoionisation of Na 3s. The calculations considered are those of single configurations. The photon–atom interaction resulting in the emission of an electron with a residual ion, for dipole transitions, is expressed as $\hbar\omega(j = 1) + Na(1s^2 2s^2 2p^6 3s^1)^2 S_{J_0=1/2} \rightarrow Na^+(1s^2 2s^2 2p^6)^1 S_{J_\alpha=0} + \epsilon p_{j_\alpha=1/2}, \epsilon p_{j_\alpha=3/2}$. Similarly, for the quadrupole transitions, $\hbar\omega(j = 2) + Na(1s^2 2s^2 2p^6 3s^1)^2 S_{J_0=1/2} \rightarrow Na^+(1s^2 2s^2 2p^6)^1 S_{J_\alpha=0} + \epsilon d_{j_\alpha=3/2}, \epsilon d_{j_\alpha=5/2}$. The transition amplitudes required for these calculations were calculated using a combination of two software packages, GRASP and RATIP.

3. Results and Discussion

3.1. Cross-Section

Figure 3a shows the Na 3s cross-section, calculated at the dipole approximation level in the length and velocity gauges, which is in reasonable agreement with the experimental data in the energy range 5.14 to 7 eV, even at the single configuration level of calculation. This region is particularly interesting, because of the presence of the dipole Cooper minimum here. Above 7 eV, there is a disagreement with the experimental data, and the agreement between length and velocity deteriorates. This may be due to the absence of initial state correlation in the present calculation. However, the present work was aimed at the Cooper minimum region below 7 eV, where there is a reasonable agreement between theory and experiment. Figure 3b shows the dipole and quadrupole cross-sections in the length gauge. This shows that the quadrupole cross-section was larger than the dipole cross-section at the dipole minimum over a small range of photon energies ∼0.03 eV. It is to be noted that although the cross-section is going through the Cooper minimum, the cross-section was not zero, even at the single configuration level. This was because of the relativistic interactions resulting from the $s \to \epsilon p_{1/2}$ and $s \to \epsilon p_{3/2}$ transitions, which underwent their respective minima at slightly different energies. For simplicity, these final states are denoted ϵp_+ and ϵp_-, respectively [54].

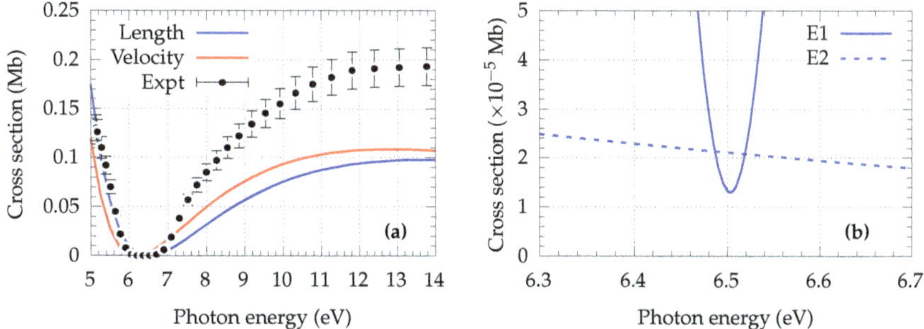

Figure 3. (**a**) Total cross-section of sodium 3s in length and the velocity gauges compared to the experimental data [55], (**b**) dipole and quadrupole cross-sections in length gauge in the region of the Cooper minimum.

Direct experimental measurements provide information only about the sum of the dipole (E1) and quadrupole (E2) cross-sections. Information about the relative magnitudes of E1 and E2 individually is not available from experiment. However, this information can be extracted from angular distribution studies. Previous studies have shown that, at low energies, the effect of non-dipole interactions is more significant on the angular distribution parameters than the cross-section [28–40]. To understand the effect of quadrupole transitions on photoelectron angular distributions, we examine the asymmetry parameters.

3.2. Dipole Parameter, β

The equation for the asymmetry parameter for a half-filled ns subshell at the dipole approximation level can be deduced from Equation (9), by making appropriate substitutions of angular momentum values. We denote the transition matrix elements by $D_\alpha \equiv D^{(E1)}(\kappa_\alpha)$ for dipole, and $Q_\alpha \equiv D^{(E2)}(\kappa_\alpha)$ for quadrupole terms. This turns out to be

$$\beta_1 = \frac{|D_{\epsilon p+}|^2 - 2\sqrt{2}|D_{\epsilon p-}||D_{\epsilon p+}|\cos(\delta_{p-} - \delta_{p+})}{|D_{\epsilon p-}|^2 + |D_{\epsilon p+}|^2}. \tag{19}$$

Subscript 1 is used in β, to indicate that this determines the angular distribution parameter in the dipole approximation. In the absence of relativistic effects, $\beta_1 = 2$. This can be seen by re-writing Equation (19) in terms of radial matrix elements. The reduced matrix elements and the radial matrix elements are related as follows: $D_{\epsilon p-} = -\sqrt{2/3}R_{\epsilon p-} = \sqrt{2/3}|R_{\epsilon p-}|e^{i(\delta'_{\epsilon p-}+\pi)}$ and $D_{\epsilon p+} = +\sqrt{4/3}R_{\epsilon p+} = \sqrt{4/3}|R_{\epsilon p+}|e^{i\delta'_{\epsilon p+}}$. Here, $\delta'_{\epsilon p\pm}$ represents the phase of the radial matrix elements, $R_{\epsilon p\pm}$. The term π is included, along with $\delta'_{\epsilon p-}$, to account for the negative sign accompanying the radial matrix element $R_{\epsilon p-}$. Note that $\delta_{\epsilon p-} \equiv \delta'_{\epsilon p-} + \pi$ and $\delta_{\epsilon p+} \equiv \delta'_{\epsilon p+}$. In the non-relativistic limit, $|R_{\epsilon p-}| = |R_{\epsilon p+}|$ and

$\delta'_{\epsilon p-} = \delta'_{\epsilon p+}$, resulting in $\cos(\delta_{\epsilon p-} - \delta_{\epsilon p+}) = \cos(\pi)$, which reduces Equation (19) to its non-relativistic value, 2.

Figure 4 shows rapid variation of the asymmetry parameter β_1 in the region of the Cooper minimum. Under the dipole approximation, the value of β_1 is ≈ 2 over most of the energy range, except in the region of the Cooper minimum, where it undergoes a dip and takes a value close to -1, as expected [56]. However, when the quadrupole interactions are taken into account, the above formula is modified, as discussed in Section 2. The expression for the asymmetry parameter is modified, with an additional term in the denominator, which is denoted as β_2. It is easy to show that these two parameters are related via the following equation:

$$\beta_2 = \beta_1 \times \sum_{\kappa_\alpha} \frac{|D_{\kappa_\alpha}|^2}{|D_{\kappa_\alpha}|^2 + |Q_{\kappa_\alpha}|^2}. \tag{20}$$

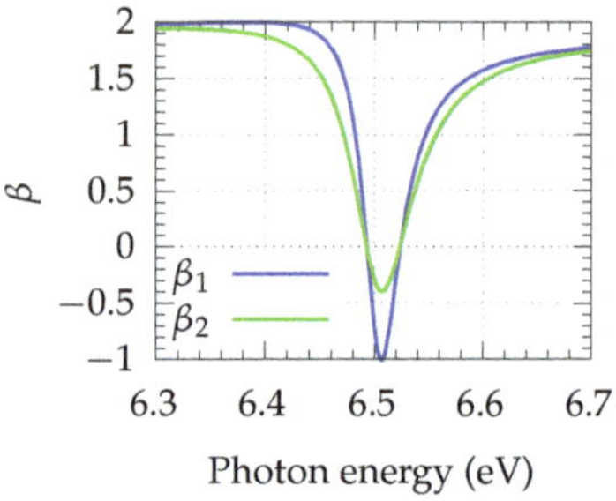

Figure 4. Asymmetry parameter β_1 and the effect of quadrupole transitions on β is represented as β_2.

From the above equation, it can be deduced that the factor multiplied by β_1 can take values unity or less. Thus, the deviation in the region of the Cooper minimum becomes shallower when non-dipole interactions are present, as shown in Figure 4. In the absence of non-dipole interactions, β_1 goes all the way to -1 in the Cooper minimum region. This amounts to a stronger yield in the direction perpendicular to the polarisation of the photon. The inclusion of non-dipole terms causes the dip in the Cooper minimum region to be close to zero ($\beta_2 \approx 0$), which corresponds to the photoelectron angular distribution being roughly isotropic in comparison to β_1. The present work, therefore, shows the importance of non-dipole interactions in determining the photoelectron yield in different directions, even at low energies.

3.3. Quadrupole Parameters Γ_1 and Γ_3

By making suitable substitutions in Equations (12) and (13), we arrive at the expressions for Γ_1 and Γ_3 for the photoionisation of a half-filled ns subshell:

$$\Gamma_1 = \frac{-2}{\sigma}\left(\sqrt{\frac{3}{2}}|Q_{3/2}||D_{1/2}|\sin(\delta_{3/2} - \delta_{1/2}) - \frac{\sqrt{3}}{10}|Q_{3/2}||D_{3/2}|\sin(\delta_{3/2} - \delta_{3/2}) + \frac{9}{5\sqrt{2}}|Q_{5/2}||D_{3/2}|\sin(\delta_{5/2} - \delta_{3/2})\right), \tag{21}$$

$$\Gamma_3 = \frac{2}{\sigma}\left(-|Q_{5/2}||D_{1/2}|\sin(\delta_{5/2} - \delta_{1/2}) - \frac{3\sqrt{3}}{5}|Q_{3/2}||D_{3/2}|\sin(\delta_{3/2} - \delta_{3/2}) + \frac{4}{5\sqrt{2}}|Q_{5/2}||D_{3/2}|\sin(\delta_{5/2} - \delta_{3/2})\right). \tag{22}$$

These parameters, Γ_1 and Γ_3, are plotted in Figure 5. They are nearly zero in the region away from the Cooper minimum, and they show rapid variation near the Cooper minimum. Recall from Section 2 that $\Gamma_1 + \Gamma_3 = \delta$ (also shown in Figure 5) and $\gamma = -5\Gamma_3$. The features of Γ_1 and Γ_3 can be better understood in their non-relativistic limits. The relations between the quadrupole reduced matrix elements and radial matrix elements are $Q_{\epsilon d-} = \sqrt{4/5}R_{\epsilon d-} = \sqrt{4/5}|R_{\epsilon d-}|e^{i\delta'_{d-}}$ and $Q_{\epsilon d+} = -\sqrt{6/5}R_{\epsilon d+} = \sqrt{6/5}|R_{\epsilon d+}|e^{i(\delta'_{d+} + \pi)}$.

The corresponding relations for the dipole matrix elements are discussed in Section 3.2. Using these relations, Equations (21) and (22) can be reduced to their non-relativistic limits:

$$\Gamma_1 = \frac{6\sqrt{6}}{6|D|^2 + 5|Q^2|}|Q||D|\sin(\delta^Q - \delta^D), \tag{23}$$

$$\Gamma_3 = -\frac{6\sqrt{6}}{6|D|^2 + 5|Q^2|}|Q||D|\sin(\delta^Q - \delta^D). \tag{24}$$

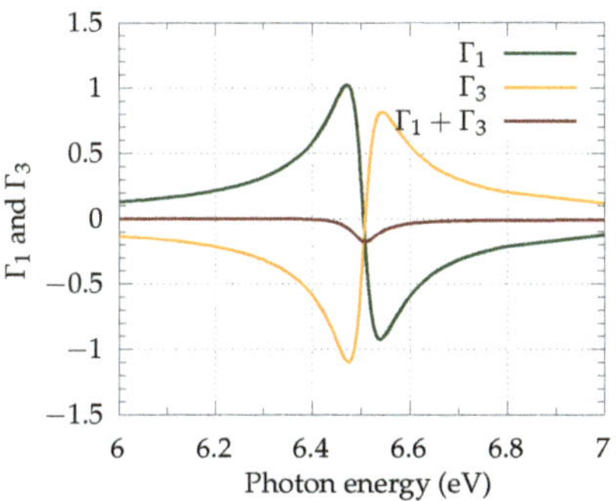

Figure 5. Quadrupole parameters Γ_1 and Γ_3 in the vicinity of the Cooper minimum.

It is easily seen that the non-relativistic δ vanishes, since $\Gamma_1 = -\Gamma_3$ for the *ns* subshell. In the regions away from the CM, $Q \ll D$. Hence, it becomes evident from the above expressions that Γ_1 and Γ_3 are directly proportional to Q/D, resulting in their values being nearly zero. In the region where the dipole amplitude, D, undergoes the Cooper minimum, the ratio becomes larger. At the dipole CM, there is a phase jump of π, leading to the sign flip of Γ_1 and Γ_3. All these features are preserved, even in the relativistic Γ_1 and Γ_3, except for the fact that $\delta = \Gamma_1 + \Gamma_3$ has non-zero values in the region of the CM. This can be attributed to the fact that relativistic dipole channels are undergoing the CM at slightly different energies, which deviates the δ from its non-relativistic value.

As discussed in Section 2.3, one of the experimentally relevant parameters is $\gamma = -5\Gamma_3$. The significant values of γ and its strong dependence on photon energy show that the photoelectron angular distribution (PAD) is very sensitive to photon energy in the region of the Cooper minimum. To illustrate this, the shape of the photoelectron angular distribution, $1 + A(\bar{\theta}, \bar{\phi})$, is plotted for a few selected energies, and is shown in Figure 6 in 2D (the xz plane) and 3D. In order to bring out the role of the quadrupole effects, the PAD obtained with and without the inclusion of the quadrupole interactions (denoted as $1 + A(\bar{\theta}, \bar{\phi})_{E1E2}$ and $1 + A(\bar{\theta}, \bar{\phi})_{E1}$, respectively) are shown. In the absence of relativistic and/or non-dipole effects, the angular distribution is essentially $\cos^2 \bar{\theta}$. Here, the preferential direction (or direction of maximum yield) of photoelectron ejection is along the polarisation direction $\hat{e}$. This is seen at a photon energy of 5.25 eV. Both $1 + A(\bar{\theta}, \bar{\phi})_{E1}$ and $1 + A(\bar{\theta}, \bar{\phi})_{E1E2}$ show a dipolar distribution, since this is well below the Cooper minimum, where non-dipole interactions do not play any significant role in the dynamics. The values of $\beta_1 \approx \beta_2 = 1.93$ (close to the non-relativistic value 2), $\Gamma_1 = 0.05$ and $\Gamma_3 = -0.05$ are smaller, because Q is significantly less than D, as seen in Figure 3.

Figure 6b,d show the PAD at photon energies 6.48 eV and 6.54 eV, where both Γ_1 and Γ_3 take extremum values, due to the presence of the Cooper minimum. At 6.48 eV, $\Gamma_1 = 0.97$, $\Gamma_3 = -1.07$, $\beta_1 = 1.14$ and $\beta_2 = 0.65$, whereas at 6.54 eV the values are $\Gamma_1 = -0.92$, $\Gamma_3 = 0.82$, $\beta_1 = 0.80$ and $\beta_2 = 0.58$. As a result, $1 + A(\bar{\theta}, \bar{\phi})_{E1E2}$ at these energies significantly differs from $1 + A(\bar{\theta}, \bar{\phi})_{E1}$. The direction of the maximum photoelectron yield moves away from $\hat{e}$ at these energies for $1 + A(\bar{\theta}, \bar{\phi})_{E1E2}$, i.e., with the inclusion of quadrupole effects.

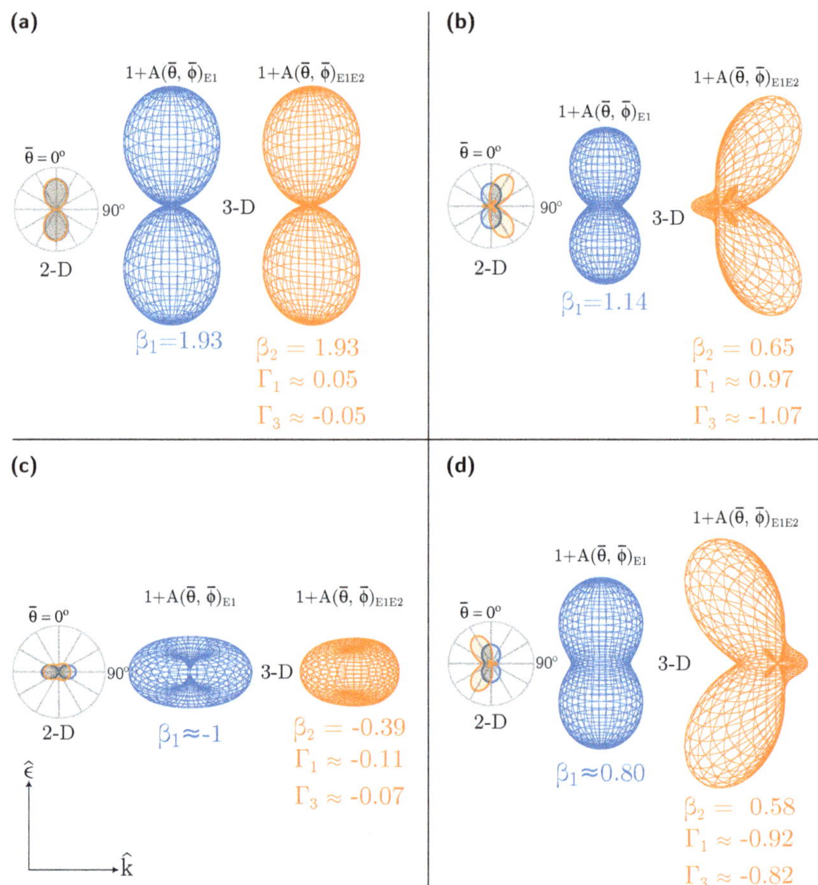

Figure 6. Shape of the PAD in 2D and 3D, showing $(1 + A(\bar{\theta}, \bar{\phi}))_{E1}$, including only the dipole–dipole interactions (blue) and $(1 + A(\bar{\theta}, \bar{\phi}))_{E1E2}$, including both the dipole–dipole and dipole–quadrupole interactions (orange) at photon energies (**a**) PE = 5.25 eV (**b**), PE = 6.48 eV (**c**) PE = 6.51 eV and (**d**) PE = 6.54 eV.

Also shown, in Figure 6c, is the PAD at a photon energy of 6.51 eV, where $\Gamma_1 + \Gamma_3$ is at its minimum. The shape of PAD $1 + A(\bar{\theta}, \bar{\phi})_{E1E2}$ significantly deviates from PAD $1 + A(\bar{\theta}, \bar{\phi})_{E1}$. For $1 + A(\bar{\theta}, \bar{\phi})_{E1}$, the yield is zero along the $\hat{e}$; however, it is non-zero when the quadrupole effects are considered, although the preferential direction remains the same in both cases. It is important to note that the above determination of the PAD employs only the first-order non-dipole parameters in the present work, which is valid, as long as $D \gg Q$. However, as seen from Figure 3b, the quadrupole cross-section (and, thus, the matrix element Q) is larger than the dipole matrix element D between 6.48 eV and 6.52 eV. In this region, the second-order E2–E2 interference terms will not only be important but dominant. Thus, Figure 6b–d should only be considered representative of the effects of first-order non-dipole corrections, but do not represent physical reality, since the E2–E2 terms are not included in the description of the PAD. For example, the small petal-like structures in Figure 6b,d are artefacts of the first-order approximations. They will no longer appear if the second corrections are included. The methodology developed here can be extended, to incorporate the E2–E2 interference effect, and work in this direction is in progress.

4. Conclusions

Following the earlier work on photoionisation dynamics based on helicity formalism [41,42], explicit relativistic formulae for angular distribution parameters, including the dipole and quadrupole interference effects, were derived for the *ns* subshells for cases of open-shell atomic systems. Using the formulae obtained, the photoionisation dynamics of Na 3s were studied in the region of the dipole Cooper minimum, which demonstrated the importance of quadrupole transitions in determining the angular distribution at low photon energies ($\approx$7 eV). Although the calculations were done at the level of single-particle approximation, they could be extended, to include multi-electron effects, by replacing the matrix elements obtained by single configuration calculation with that of multi-configuration calculation, using GRASP and RATIP. The methodology developed here could also be extended to higher-order multi-pole interactions and to other subshells. We hope the current work will stimulate photoionisation dynamics studies of open-shell systems, studies that would highlight both relativistic and multi-electron effects.

Author Contributions: Conceptualization, N.M.H. and H.R.V.; Methodology, N.M.H. and H.R.V.; Software, N.M.H. and J.J.; Formal analysis, N.M.H., J.J., H.R.V., P.C.D. and S.T.M.; Investigation, N.M.H., J.J., H.R.V., P.C.D. and S.T.M.; Writing—original draft, N.M.H.; Writing—review & editing, J.J., H.R.V., P.C.D. and S.T.M.; Supervision, H.R.V. All authors have read and agreed to the published version of the manuscript.

Funding: Two of the authors, Nishita M. Hosea and Hari R. Varma, extend their gratitude to the Science and Engineering Research Board, Department of Science and Technology, Government of India, for funding this work, Grant Number CRG/2022/002309. The work of Steven T. Manson was supported by the US Department of Energy, Office of Basic Sciences, Division of Chemical Science, Geosciences and Biosciences, under Grant No. DE-FG02-03ER15428.

Data Availability Statement: The data generated and/or analysed during the current study are available from the corresponding author on a reasonable request.

Conflicts of Interest: The authors declare no conflict of interest.

References

1. Hemmers, O.; Guillemin, R.; Lindle, D.W. Nondipole effects in soft X-ray photoemission. *Radiat. Phys. Chem.* **2004**, *70*, 123–147. [CrossRef]
2. Guillemin, R.; Hemmers, O.; Lindle, D.W.; Manson, S.T. Experimental investigation of nondipole effects in photoemission at the advanced light source. *Radiat. Phys. Chem.* **2006**, *75*, 2258–2274. [CrossRef]
3. Hemmers, O.; Guillemin, R.; Rolles, D.; Wolska, A.; Lindle, D.W.; Kanter, E.P.; Krässig, B.; Southworth, S.H.; Wehlitz, R.; Zimmermann, B.; et al. Low-energy nondipole effects in molecular nitrogen valence-shell photoionization. *Phys. Rev. Lett.* **2006**, *97*, 103006. [CrossRef] [PubMed]
4. Cherepkov, N.A.; Semenov, S.K. Non-dipole effects in spin polarization of photoelectrons from Xe 4p and 5p shells. *J. Phys. B At. Mol. Opt. Phys.* **2001**, *34*, L211. [CrossRef]
5. Khalil, T.; Schmidtke, B.; Drescher, M.; Müller, N.; Heinzmann, U. Experimental verification of quadrupole-dipole interference in spin-resolved photoionization. *Phys. Rev. Lett.* **2002**, *89*, 053001. [CrossRef]
6. Jensen, S.V.B.; Madsen, L.B. Propagation time and nondipole contributions to intraband high-order harmonic generation. *Phys. Rev. A* **2022**, *105*, L021101. [CrossRef]
7. Tyndall, N.B.; Ramsbottom, C.A.; Ballance, C.P.; Hibbert, A. Photoionization of Co^+ and electron-impact excitation of Co^{2+} using the Dirac R-matrix method. *Mon. Not. R. Astron. Soc.* **2016**, *462*, 3350–3360.
8. Emma, P.; Akre, R.; Arthur, J.; Bionta, R.; Bostedt, C.; Bozek, J.; Brachmann, A.; Bucksbaum, P.; Coffee, R.; Decker, F.J.; et al. First lasing and operation of an ångstrom-wavelength free-electron laser. *Nat. Photonics* **2010**, *4*, 641–647. [CrossRef]
9. McNeil, B.W.; Thompson, N.R. X-ray free-electron lasers. *Nat. Photonics* **2010**, *4*, 814–821. [CrossRef]
10. Ishikawa, T.; Aoyagi, H.; Asaka, T.; Asano, Y.; Azumi, N.; Bizen, T.; Ego, H.; Fukami, K.; Fukui, T.; Furukawa, Y.; et al. A compact X-ray free-electron laser emitting in the sub-ångström region. *Nat. Photonics* **2012**, *6*, 540–544. [CrossRef]
11. Lindle, D.W.; Hemmers, O. Breakdown of the dipole approximation in soft-X-ray photoemission. *J. Electron Spectrosc. Relat. Phenom.* **1999**, *100*, 297–311. [CrossRef]
12. Bechler, A.; Pratt, R. Higher retardation and multipole corrections to the dipole angular distribution of 1s photoelectrons at low energies. *Phys. Rev. A* **1989**, *39*, 1774. [CrossRef]
13. Bechler, A.; Pratt, R. Higher multipole and retardation corrections to the dipole angular distributions of L-shell photoelectrons ejected by polarized photons. *Phys. Rev. A* **1990**, *42*, 6400. [CrossRef] [PubMed]

14. Cooper, J.W. Multipole corrections to the angular distribution of photoelectrons at low energies. *Phys. Rev. A* **1990**, *42*, 6942. [CrossRef] [PubMed]
15. Cooper, J.W. Erratum: Multipole corrections to the angular distribution of photoelectrons at low energies [Phys. Rev. A 42, 6942 (1990)]. *Phys. Rev. A* **1992**, *45*, 3362. [CrossRef] [PubMed]
16. Cooper, J. Photoelectron-angular-distribution parameters for rare-gas subshells. *Phys. Rev. A* **1993**, *47*, 1841. [CrossRef]
17. Krässig, B.; Jung, M.; Gemmell, D.; Kanter, E.; LeBrun, T.; Southworth, S.; Young, L. Nondipolar asymmetries of photoelectron angular distributions. *Phys. Rev. Lett.* **1995**, *75*, 4736. [CrossRef]
18. Hemmers, O.; Fisher, G.; Glans, P.; Hansen, D.; Wang, H.; Whitfield, S.; Wehlitz, R.; Levin, J.; Sellin, I.; Perera, R.C.; et al. Beyond the dipole approximation: Angular-distribution effects in valence photoemission. *J. Phys. B At. Mol. Opt. Phys.* **1997**, *30*, L727. [CrossRef]
19. Dolmatov, V.K.; Manson, S.T. Enhanced nondipole effects in low energy photoionization. *Phys. Rev. Lett.* **1999**, *83*, 939. [CrossRef]
20. Amusia, M.Y.; Baltenkov, A.; Felfli, Z.; Msezane, A. Large nondipole correlation effects near atomic photoionization thresholds. *Phys. Rev. A* **1999**, *59*, R2544. [CrossRef]
21. Derevianko, A.; Hemmers, O.; Oblad, S.; Glans, P.; Wang, H.; Whitfield, S.B.; Wehlitz, R.; Sellin, I.A.; Johnson, W.; Lindle, D.W. Electric-octupole and pure-electric-quadrupole effects in soft-X-ray photoemission. *Phys. Rev. Lett.* **2000**, *84*, 2116. [CrossRef] [PubMed]
22. Amusia, M.Y.; Baltenkov, A.; Chernysheva, L.; Felfli, Z.; Msezane, A. Nondipole parameters in angular distributions of electrons in photoionization of noble-gas atoms. *Phys. Rev. A* **2001**, *63*, 052506. [CrossRef]
23. Johnson, W.R.; Cheng, K. Strong nondipole effects in low-energy photoionization of the 5 s and 5 p subshells of xenon. *Phys. Rev. A* **2001**, *63*, 022504. [CrossRef]
24. Cherepkov, N.A.; Semenov, S.K. On quadrupole resonances in atomic photoionization. *J. Phys. B At. Mol. Opt. Phys.* **2001**, *34*, L495. [CrossRef]
25. Hemmers, O.; Guillemin, R.; Kanter, E.; Krässig, B.; Lindle, D.W.; Southworth, S.; Wehlitz, R.; Baker, J.; Hudson, A.; Lotrakul, M.; et al. Dramatic Nondipole Effects in Low-Energy Photoionization: Experimental and Theoretical Study of Xe 5 s. *Phys. Rev. Lett.* **2003**, *91*, 053002. [CrossRef]
26. Trzhaskovskaya, M.; Nikulin, V.; Nefedov, V.; Yarzhemsky, V. Non-dipole second order parameters of the photoelectron angular distribution for elements Z = 1–100 in the photoelectron energy range 1–10 keV. *At. Data Nucl. Data Tables* **2006**, *92*, 245–304. [CrossRef]
27. Trzhaskovskaya, M.; Nefedov, V.; Yarzhemsky, V. Photoelectron angular distribution parameters for elements Z = 1 to Z = 54 in the photoelectron energy range 100–5000 eV. *At. Data Nucl. Data Tables* **2001**, *77*, 97–159. [CrossRef]
28. Leuchs, G.; Smith, S.; Dixit, S.; Lambropoulos, P. Observation of interference between quadrupole and dipole transitions in low-energy (2-eV) photoionization from a sodium Rydberg state. *Phys. Rev. Lett.* **1986**, *56*, 708. [CrossRef]
29. Martin, N.; Thompson, D.; Bauman, R.; Caldwell, C.; Krause, M.; Frigo, S.; Wilson, M. Electric-dipole–quadrupole interference of overlapping autoionizing levels in photoelectron energy spectra. *Phys. Rev. Lett.* **1998**, *81*, 1199. [CrossRef]
30. Grum-Grzhimailo, A. Non-dipole effects in magnetic dichroism in atomic photoionization. *J. Phys. B At. Mol. Opt. Phys.* **2001**, *34*, L359. [CrossRef]
31. Krässig, B.; Kanter, E.; Southworth, S.; Guillemin, R.; Hemmers, O.; Lindle, D.W.; Wehlitz, R.; Martin, N. Photoexcitation of a dipole-forbidden resonance in helium. *Phys. Rev. Lett.* **2002**, *88*, 203002. [CrossRef] [PubMed]
32. Kanter, E.; Krässig, B.; Southworth, S.; Guillemin, R.; Hemmers, O.; Lindle, D.W.; Wehlitz, R.; Amusia, M.Y.; Chernysheva, L.; Martin, N. E 1-E 2 interference in the vuv photoionization of He. *Phys. Rev. A* **2003**, *68*, 012714. [CrossRef]
33. Lépine, F.; Zamith, S.; de Snaijer, A.; Bordas, C.; Vrakking, M. Observation of large quadrupolar effects in a slow photoelectron imaging experiment. *Phys. Rev. Lett.* **2004**, *93*, 233003. [CrossRef] [PubMed]
34. Dolmatov, V.; Bailey, D.; Manson, S. Gigantic enhancement of atomic nondipole effects: The 3 s→ 3 d resonance in Ca. *Phys. Rev. A* **2005**, *72*, 022718. [CrossRef]
35. Deshmukh, P.; Banerjee, T.; Varma, H.R.; Hemmers, O.; Guillemin, R.; Rolles, D.; Wolska, A.; Yu, S.; Lindle, D.W.; Johnson, W.; et al. Theoretical and experimental demonstrations of the existence of quadrupole Cooper minima. *J. Phys. B At. Mol. Opt. Phys.* **2008**, *41*, 021002. [CrossRef]
36. Argenti, L.; Moccia, R. Nondipole effects in helium photoionization. *J. Phys. B At. Mol. Opt. Phys.* **2010**, *43*, 235006. [CrossRef]
37. Pradhan, G.; Jose, J.; Deshmukh, P.; LaJohn, L.; Pratt, R.; Manson, S. Cooper minima: A window on nondipole photoionization at low energy. *J. Phys. B At. Mol. Opt. Phys.* **2011**, *44*, 201001. [CrossRef]
38. Gryzlova, E.; Grum-Grzhimailo, A.; Strakhova, S.; Meyer, M. Non-dipole effects in the angular distribution of photoelectrons in sequential two-photon double ionization: Argon and neon. *J. Phys. B At. Mol. Opt. Phys.* **2013**, *46*, 164014. [CrossRef]
39. Gryzlova, E.; Grum-Grzhimailo, A.; Kuzmina, E.; Strakhova, S. Sequential two-photon double ionization of noble gases by circularly polarized XUV radiation. *J. Phys. B At. Mol. Opt. Phys.* **2014**, *47*, 195601. [CrossRef]
40. Ilchen, M.; Hartmann, G.; Gryzlova, E.; Achner, A.; Allaria, E.; Beckmann, A.; Braune, M.; Buck, J.; Callegari, C.; Coffee, R.; et al. Symmetry breakdown of electron emission in extreme ultraviolet photoionization of argon. *Nat. Commun.* **2018**, *9*, 4659. [CrossRef]
41. Huang, K.N. Theory of angular distribution and spin polarization of photoelectrons. *Phys. Rev. A* **1980**, *22*, 223. [CrossRef]

42. Huang, K.N. Addendum to "Theory of angular distribution and spin polarization of photoelectrons". *Phys. Rev. A* **1982**, *26*, 3676. [CrossRef]
43. Derevianko, A.; Johnson, W.; Cheng, K. Non-dipole effects in photoelectron angular distributions for rare gas atoms. *At. Data Nucl. Data Tables* **1999**, *73*, 153–211. [CrossRef]
44. Dyall, K.; Grant, I.; Johnson, C.; Parpia, F.; Plummer, E. GRASP: A general-purpose relativistic atomic structure program. *Comput. Phys. Commun.* **1989**, *55*, 425–456. [CrossRef]
45. Parpia, F.A.; Fischer, C.F.; Grant, I.P. GRASP92: A package for large-scale relativistic atomic structure calculations. *Comput. Phys. Commun.* **1996**, *94*, 249–271. [CrossRef]
46. Jönsson, P.; Gaigalas, G.; Bieroń, J.; Fischer, C.F.; Grant, I. New version: Grasp2K relativistic atomic structure package. *Comput. Phys. Commun.* **2013**, *184*, 2197–2203. [CrossRef]
47. Fritzsche, S. The Ratip program for relativistic calculations of atomic transition, ionization and recombination properties. *Comput. Phys. Commun.* **2012**, *183*, 1525–1559. [CrossRef]
48. Hütten, K.; Mittermair, M.; Stock, S.O.; Beerwerth, R.; Shirvanyan, V.; Riemensberger, J.; Duensing, A.; Heider, R.; Wagner, M.S.; Guggenmos, A.; et al. Ultrafast quantum control of ionization dynamics in krypton. *Nat. Commun.* **2018**, *9*, 719. [CrossRef] [PubMed]
49. Perry-Sassmannshausen, A.; Buhr, T.; Borovik, A., Jr.; Martins, M.; Reinwardt, S.; Ricz, S.; Stock, S.; Trinter, F.; Müller, A.; Fritzsche, S.; et al. Multiple photodetachment of carbon anions via single and double core-hole creation. *Phys. Rev. Lett.* **2020**, *124*, 083203. [CrossRef]
50. Schippers, S.; Beerwerth, R.; Bari, S.; Buhr, T.; Holste, K.; Kilcoyne, A.D.; Perry-Sassmannshausen, A.; Phaneuf, R.A.; Reinwardt, S.; Savin, D.W.; et al. Near L-edge single and multiple photoionization of doubly charged iron ions. *Astrophys. J.* **2021**, *908*, 52. [CrossRef]
51. Hosea, N.M.; Jose, J.; Varma, H.R. Near-threshold Cooper minimum in the photoionisation of the 2p subshell of sodium atom and its impact on the angular distribution parameter. *J. Phys. B At. Mol. Opt. Phys.* **2022**, *55*, 135001. [CrossRef]
52. Lee, C. Spin polarization and angular distribution of photoelectrons in the Jacob-Wick helicity formalism. Application to autoionization resonances. *Phys. Rev. A* **1974**, *10*, 1598. [CrossRef]
53. Landau, L.; Lifshitz, E. *A Shorter Course of Theoretical Physics. Vol. 2: Quantum Mechanics*; Pergamon: Oxford, UK, 1974.
54. Seaton, M.J. A comparison of theory and experiment for photo-ionization cross-sections II. Sodium and the alkali metals. *Proc. R. Soc. Lond. Ser. A. Math. Phys. Sci.* **1951**, *208*, 418–430.
55. Hudson, R.D.; Carter, V.L. Atomic absorption cross sections of lithium and sodium between 600 and 1000 Å. *J. Opt. Soc. Am.* **1967**, *57*, 651–654. [CrossRef]
56. Manson, S.T.; Starace, A.F. Photoelectron angular distributions: Energy dependence for s subshells. *Rev. Mod. Phys.* **1982**, *54*, 389. [CrossRef]

Article

Density Functional Treatment of Photoionization of Sodium Clusters: Effects of Cluster Size and Exchange–Correlation Framework

Rasheed Shaik [1], Hari R. Varma [1,*] and Himadri S. Chakraborty [2,*]

[1] School of Physical Sciences, Indian Institute of Technology Mandi, Kamand 175075, India; di1602@students.iitmandi.ac.in

[2] Department of Natural Sciences, Dean L. Hubbard Center for Innovation and Entrepreneurship, Northwest Missouri State University, Maryville, MO 64468, USA

* Correspondence: hari@iitmandi.ac.in (H.R.V.); himadri@nwmissouri.edu (H.S.C.)

Abstract: The ground state and photoionization properties of Na_x (x = 20, 40, and 92) clusters are investigated using a method based on density functional theory (DFT) in a spherical jellium frame. Two different exchange–correlation treatments with the Gunnarsson–Lundqvist parametrization are used: (i) the electron self-interaction correction (SIC) scheme and (ii) the van Leeuwen–Baerends (LB94) scheme based on the gradient of the electron density. The shapes of the mean-field potentials and bound state properties, obtained in the two schemes, qualitatively agree, but differ in the details. The effect of the schemes on the photoionization dynamics, calculated in linear response time-dependent DFT is compared, in which the broader features are found to be universal. The general similarity of the results in SIC and LB94 demonstrates the reliability of DFT treatments. The study further elucidates the evolution of the ground state and ionization description as a function of the cluster size.

Keywords: collective effects in photoionization; sodium clusters; plasmon resonances; correlation minimum; effects of exchange–correlation functionals in photoionization

Citation: Shaik, R.; Varma, H.R.; Chakraborty, H.S. Density Functional Treatment of Photoionization of Sodium Clusters: Effects of Cluster Size and Exchange–Correlation Framework. *Atoms* **2023**, *11*, 114. https://doi.org/10.3390/atoms11080114

Academic Editor: Eugene T. Kennedy

Received: 10 July 2023
Revised: 10 August 2023
Accepted: 11 August 2023
Published: 18 August 2023

1. Introduction

Atomic and molecular cluster physics has emerged as a distinct area of research over the last decades. The area has evolved into an important field by bridging the gap between the atomic/molecular and the bulk domain. This opened new pathways to characterize and control parts of the nano-world. The research that emerged has intertwined physics, chemistry, astronomy, and biology, thus making cluster studies an interdisciplinary topic [1–3]. Theoretical models for the description of atomic clusters based on *ab initio* principles along with the experimental studies have unravelled intriguing features, especially resulting from the interaction of a cluster with light [4,5]. Such photo-induced processes include, for instance, the plasmon resonances [6–9], Auger-type Fano resonances [10,11], inter-Coulombic decay (ICD) resonances [12–15], and modulations in the photoelectron intensity due to the diffraction from cluster edges [16–18]. In addition to their role as "spectral laboratories" to probe many-electron effects, cluster studies provide impetus for a wide variety of applications, such as, in nano-optical devices [19], in solar energy harvesting [20], and in chemical and biological sensors [21–23]. Therefore, testing the efficacy of theoretical models, taking into account the many-body and quantum phenomena, is particularly valuable.

Experiments suggest that the details of the ionic core configuration of metal clusters, such as sodium clusters, play a less significant role in extracting structural and dynamical information [24,25]. In addition, the loosely bound valence electrons in the clusters can be approximated as delocalized and confined within a broad potential well. The jellium model, which makes use of these facts, replaces the metallic ion core by a uniform charge distribution, which provides an electrostatic attraction to the valence electron cloud. The electronic structure may then be determined by applying a mean-field approximation to the interacting electrons that includes static exchange and correlation (XC) effects in addition to the direct electron–electron Coulomb repulsion.

An accurate description of the XC potential is crucial in the above model. The Kohn–Sham (KS) density functional theory (DFT) method is known to have some limitations in handling the electron exchange, which is intrinsically a fully non-local attribute, by treating it in a local frame [26]. As a result, the exchange functional cannot perfectly cancel the self-interaction present in the direct (Hartree) term. This causes the XC potential for a finite system to approach zero exponentially as $r \rightarrow \infty$ rather than to residual $-1/r$, which is the correct long-distance behavior for a neutral system. Certain structural and dynamical properties are affected due to this inaccurate description of the asymptotic behavior. It leads to errors in the determination of properties, such as the ionization potential and cross-section. In order to overcome this limitation, for instance, one of the following approximate methods can be used: (i) the self-interaction correction (SIC) or (ii) the generalized gradient approximation (GGA).

In method (i), introduced by Perdew and Zunger [27], the self-interactions are subtracted from the potential in the KS equations and iterated until a self-consistent solution is obtained. It improves the asymptotic behavior of the DFT potential, although the resulting KS Hamiltonian becomes state-dependent. This approach was found to be useful for a wide range of compact atomic or molecular systems [28], and especially so for explaining the absorption spectra of alkali metal clusters [29].

Method (ii), which is more intrinsic to the formalism, is within the GGA class and was developed by van Leeuwen and Baerends [30]. In this approach, known as LB94, the issue with self-interaction is addressed by introducing a term that is dependent on the gradient of electron density by using the Becke GGA construction for the modeling of the XC potential. LB94 produces a state-independent potential, thus offering a relatively easy and inexpensive implementation in the computer code. The study of fullerene molecules using LB94 is found to show a somewhat better agreement with the experimental results [31]. However, in a recent ICD study of fullerene plasmon resonance in $Na_{20}@C_{240}$, the SIC ground state structure complied better with quantum chemical results [15].

The photoresponse of alkali metal clusters energetically below the ionization threshold is dominated by the giant surface plasmon resonance excitation. The photospectra of clusters are more robust than their bulk counterpart due to the existence of a higher-energy volume plasmon [6]. Our previous theoretical study on Na-clusters [32] predicted the spillover of a smaller remnant of the giant surface plasmon and most of the volume plasmon out to the ionization region and showed a reasonable agreement with measurements. It further predicted a feature called the correlation minimum in the ionization spectra of Na clusters. However, in [32], the LB94 scheme was employed while modeling the XC interaction and a linear response time-dependent DFT (LR-TDDFT) formalism was used to calculate the response properties. Since the photoionization spectra are expected to be somewhat sensitive to the XC model, in this study, we compare the effect of the two schemes, i.e., SIC and LB94, on the details of the photoresponse behavior using LR-TDDFT. Furthermore, in order to explicate the evolution trend of the results on the cluster size, three cluster systems are selected for the study.

2. Theoretical Methodology

2.1. DFT Exchange–Correlation Functionals

The details of the method are in line with the framework discussed in Ref. [33]. We adopted a jellium-based DFT approach to explore the ground state electronic structure of Na_x ($x = 20$, 40, and 92) clusters in a spherical model. The jellium potential, V_{jel}, replaces the ionic core of N ($N = 20$, 40, and 92) Na^+ ions with a potential created by homogeneously smearing their total charges over a sphere. The radial component of the spherically symmetric potential generated by this distribution is the following:

$$V_{jel}(r) = \begin{cases} -\frac{N}{2R_c}\left(3 - \left(\frac{r}{R_c}\right)^2\right), & r \leq R_c \\ -\frac{N}{r}, & r > R_c \end{cases}$$

The radius of each cluster is determined by $R_c = r_s N^{1/3}$, where $r_s = 3.93$ a.u. is the Wigner–Seitz radius of the Na atom. The KS equations for N number of delocalized valence electrons, i.e., the $3s^1$ electron from each Na atom, are solved to obtain the electronic structure. It is to be noted that a constant pseudo-potential is added to match the first ionization threshold with the experimentally known values [34].

The ground state self-consistent field DFT potential can be written in terms of the single-particle density $\rho(\mathbf{r})$ as,

$$V_{DFT}(\mathbf{r}) = V_{jel}(\mathbf{r}) + \int d\mathbf{r}' \frac{\rho(\mathbf{r}')}{|\mathbf{r} - \mathbf{r}'|} + V_{XC}[\rho(\mathbf{r})], \tag{1}$$

where the second and third terms on the RHS are known as the direct and the XC components, respectively. We consider the following formula to initially parameterize V_{XC} by using $\rho(\mathbf{r})$ [35]:

$$V_{XC}[\rho(\mathbf{r})] = -\left(\frac{3\rho(\mathbf{r})}{\pi}\right)^{1/3} - 0.0333 \log\left[1 + 11.4\left(\frac{4\pi\rho(\mathbf{r})}{3}\right)^{1/3}\right]. \tag{2}$$

Equation (2) qualifies the method as a local density approximation (LDA) approach. By the standard variational technique, it is possible to exactly derive the first term on the right hand side of Equation (2) from the Hartree–Fock (HF) exchange energy of a homogeneous electron system that has a uniform positively charged background. The second term is the so-called correlation potential, which is not accounted for in the HF formalism. As mentioned earlier, the localization of the potential leads to the non-cancellation of self-interactions. A corrective scheme is therefore adopted from the outset to artificially eliminate unphysical self-interactions for each ith occupied subshell. This leaves the LDA potential orbital-specific, but it approximately captures the asymptotic properties of the electron. We describe this model, referred to as DFT-SIC, in the equation below:

$$V_{DFT\text{-}SIC}^{i}(\mathbf{r}) = V_{jel}(\mathbf{r}) + \int d\mathbf{r}' \frac{\rho(\mathbf{r}') - \rho_i(\mathbf{r}')}{|\mathbf{r} - \mathbf{r}'|} + V_{XC}[\rho(\mathbf{r})] - V_{XC}[\rho_i(\mathbf{r})]. \tag{3}$$

DFT-SIC thus mimics two desirable XC functional features: it cancels out the self-interaction part of the Hartree energy and it vanishes for a one-electron system.

An alternative method to correct the XC functional makes use of Equation (2) but refines it further by the addition of a parameterized potential defined in terms of the reduced density and its gradient $\nabla\rho$, as follows:

$$V_{LB} = -\beta[\rho(\mathbf{r})]^{1/3}\frac{(\zeta X)^2}{1 + 3\beta\zeta X \sinh^{-1}(\zeta X)}. \tag{4}$$

In Equation (4), $\beta = 0.01$ and is empirical, while $X = [\nabla\rho]/\rho^{4/3}$. The parameter ζ is a factor that arises in transition from the spin-polarized to spin-unpolarized form [36]. Such a gradient correction approach to the XC functional is built more into the theory and, hence, is less artificial than SIC. We refer to this model of non-local correction as DFT-LB94. It improves the asymptotic behavior of the electron when compared to the exact KS potentials computed from correlated densities. This model potential holds the unitary transformation of all occupied and unoccupied orbitals that are eigenstates of the KS equation with one

DFT-LB94 potential. In contrast, the description of DFT-SIC unoccupied orbitals is a bit ambiguous due to the ad hoc nature of the potential in Equation (3).

However, accurate excitation and ionization energies for atoms, molecules, and small clusters [37] are produced using asymptotically correct functionals. Thus, it was anticipated that both DFT-SIC and DFT-LB94 would compete to significantly improve the quality of ground, excited, and continuum spectra in the current study.

2.2. LR-TDDFT Dynamical Response

A time-dependent DFT (TDDFT) approach is used to compute the dynamical response of the clusters to the electromagnetic radiation [38]. The system's behavior is studied in response to a time-dependent weak external perturbation. The external perturbation z, which represents dipole interaction with linearly polarized light, induces a frequency-dependent complex change in the electron density, $\delta\rho(\mathbf{r};\omega)$. Thus, the linear response of the system can be determined using the density–density response function χ by

$$\delta\rho(\mathbf{r};\omega) = \int \chi(\mathbf{r},\mathbf{r}';\omega)z'd\mathbf{r}', \tag{5}$$

where the full susceptibility χ includes the electrons' dynamical correlations. In the auxiliary KS system the same induced density can be equivalently calculated using

$$\delta\rho(\mathbf{r};\omega) = \int \chi_0(\mathbf{r},\mathbf{r}';\omega)\delta V_{eff}(\mathbf{r}';\omega)d\mathbf{r}', \tag{6}$$

where δV_{eff} includes the external field, plus the induced Hartree and induced XC potentials as follows

$$\delta V_{eff}(\mathbf{r}';\omega) = z' + V_{ind}(\mathbf{r}';\omega) \tag{7}$$

with

$$V_{ind}(\mathbf{r}';\omega) = \int \frac{\delta\rho(\mathbf{r};\omega)}{|\mathbf{r}-\mathbf{r}'|}d\mathbf{r} + \left[\frac{\partial V_{XC}}{\partial\rho}\right]_{\rho=\rho_0}\delta\rho(\mathbf{r}';\omega). \tag{8}$$

The response of non-interacting electrons, that is the independent particle (IP) susceptibility, is described by the KS response function χ_0, which can be expressed in terms of the ground state KS eigenvalues ϵ_i and eigenfunctions ϕ_i as

$$\chi_0(\mathbf{r},\mathbf{r}';\omega) = \sum_i \phi_i^*(\mathbf{r})\phi_i(\mathbf{r}')G(\mathbf{r},\mathbf{r}';\epsilon_i+\omega) + \sum_i \phi_i(\mathbf{r})\phi_i^*(\mathbf{r}')G^*(\mathbf{r},\mathbf{r}';\epsilon_i-\omega). \tag{9}$$

Here the index i runs over the occupied states only. For a spherically symmetric atomic cluster, the Green's function for a parameter E can be expanded in the spherical basis as $G(\mathbf{r},\mathbf{r}';E) = \sum_{lm} G_{lm}(r,r';E)Y_{lm}^*(\Omega)Y_{lm}(\Omega')$, where the radial component $G_{lm}(r,r';E)$ satisfies the radial equation

$$\left(\frac{1}{r^2}\frac{\partial}{\partial r}r^2\frac{\partial}{\partial r} - \frac{l(l+1)}{r^2} - V_{DFT} + E\right)G_{lm}(r,r';E) = \frac{\delta(r-r')}{r^2} \tag{10}$$

with $G_{lm}(r,r';E) = \frac{j_l(r_<;E)h_l(r_>;E)}{W[j_l,h_l]_{r=c}}$, where j_l and h_l are homogenous solutions of Equation (10) satisfying boundary conditions at $r=0$ and $r=\infty$, respectively, and W is the Wronskian, which is independent of an arbitrary constant c.

A Dyson-like equation for the interacting response function, χ, can be easily derived from Equations (5) and (6) as

$$\chi(\mathbf{r},\mathbf{r}';\omega) = \chi_0(\mathbf{r},\mathbf{r}';\omega) + \int d\mathbf{r}''d\mathbf{r}'''\chi_0(\mathbf{r},\mathbf{r}'';\omega)\left(\frac{1}{|\mathbf{r}''-\mathbf{r}'''|} + f_{XC}(\mathbf{r}'',\mathbf{r}''';\omega)\right)\chi(\mathbf{r}''',\mathbf{r}';\omega), \tag{11}$$

where f_{XC} is the so-called time-dependent XC kernel evaluated with approximate XC functional at ground state density ρ_0 and expressed as,

$$f_{XC}[\rho](\mathbf{r}, \mathbf{r}'; \omega) = \left. \frac{\delta V_{XC}[\rho](\mathbf{r}; \omega)}{\delta\rho(\mathbf{r}'; \omega)} \right|_{\rho=\rho_0}. \tag{12}$$

Equation (11) can be solved self-consistently for χ at any desired photon energy. The χ in Equation (11) can be computed using matrix notation as

$$\chi = \left(1 - \frac{\partial V}{\partial \rho}\chi_0\right)^{-1}\chi_0. \tag{13}$$

where V refers to the ground state potential. Equation (13) can then be solved for χ using the matrix inversion method [39]. $\delta\rho$ and, hence, δV_{eff} can be directly obtained via Equations (5) and (7), respectively.

In LR-TDDFT formalism, the photoionization (PI) cross-section associated with a bound-to-continuum dipole transition is then computed as the sum of independent subshell cross-sections, $\sigma_{nl \to kl'}$, and is given by:

$$\sigma_{PI}(\omega) = \sum_{nl} \sigma_{nl \to kl'} \sim \sum_{nl} 2(2l+1)|\langle\psi_{kl'}|\delta V_{eff}|\psi_{nl}\rangle|^2. \tag{14}$$

The radial component $R_{kl'}$ of the final continuum wavefunction $\psi_{kl'}$ has the appropriate asymptotic behavior:

$$\lim_{r\to\infty} R_{kl'} \sim \lim_{r\to\infty}\left[cos(\delta_{l'})f_{l'}(kr) + sin(\delta_{l'})g_{l'}(kr))\right] = sin(kr - \frac{l'\pi}{2} + \frac{z}{k}ln(2kr) + \zeta_{l'} + \delta_{l'}), \tag{15}$$

where f_l and g_l represent the regular and irregular spherical Coulomb functions, respectively, and δ_l and $\zeta_l = arg\Gamma(l+1-iz/k)$ are, respectively, the short-range and Coulomb phase shifts seen by the ejected electron.

In addition to the external perturbation z, Equation (14) also includes the complex induced field V_{ind} produced by the many-electron interactions. Evidently, the IP level DFT cross-section that disregards correlations is obtained by setting $\delta V_{eff} = z$. This approach makes it simple to compare DFT and TDDFT in order to investigate the role of the many-electron effects during the photoionization process. In this work, we employ the two XC kernels to calculate the PI cross-sections: one by a global averaging procedure, $f_{XC}^{SIC} = \frac{N-1}{N}V_{XC}^{LDA}$ with V_{XC} given in Equation (2). We refer to the PI cross-section calculated in this regime as LR-TDDFT-SIC. The other XC kernel with V_{XC} in Equation (2), augmented by Equation (4), yields f_{XC}^{LB94}, which in turn is used to evaluate PI cross-sections in LR-TDDFT-LB94.

3. Results and Discussion

3.1. Ground State Structure

In Figure 1, we show the ground state radial potentials of Na_{20}, Na_{40}, and Na_{92} calculated using DFT-SIC and DFT-LB94. The DFT-SIC curves are obtained by taking an occupancy-weighted average over all the subshells. As the cluster size grows, the potential depth remains roughly unchanged, predominantly since the average density $\rho(r)$ remains nearly the same. As a consequence, the electron energy levels should become denser with increasing size [40]. This is seen in both SIC and LB94. The shapes of the radial orbitals of the two outer levels (HOMO and HOMO-1), shown in the inset, are almost the same in LB94 and SIC. For a metal cluster, the potential is expected to remain flat in the interior region of delocalized quasi-free electrons while exhibiting strong screening at the edge around R_c (cluster radius). These characteristics are exhibited by DFT-LB94, while DFT-SIC shows some spatial variations in the interior region. In SIC, these variations are due to the electrons in each orbital experiencing a distinct potential. Additionally, in DFT-SIC, there appear unphysical cusp-like structures below R_c caused from the radial nodes of occupied subshells [41]. These structures are absent in DFT-LB94, as it is more of an ab initio approach compared to the artificial vacancy-elimination technique in SIC.

The result presented in Figure 1 clearly indicates that in the region outside the radius, the DFT-LB94 potential is somewhat deeper (attractive) and better approximates the $-1/r$ behavior toward the asymptotic region compared to DFT-SIC. In other words, the DFT-LB94 potential has a more accurate asymptotic representation resulting in a slower decay beyond R_c. Such a deeper potential shape allows DFT-LB94 to generate an extended number of virtual unoccupied KS orbitals compared to SIC. On the other hand, it may be noted that the occupied levels in SIC in Na_{20} are slightly more bound compared to LB94 . However, as the cluster size increases, this difference tends to diminish as an indication of a gradual irrelevance of a particular choice of *XC* with increasing size. Recall that appropriate pseudo-potentials are added to match the ionization thresholds with the experimental values, which is evident from the identical binding energies of HOMO levels obtained in SIC and LB94 in each cluster.

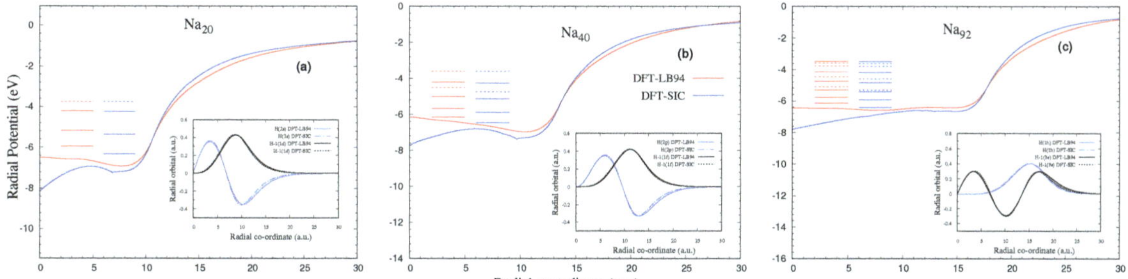

Figure 1. Ground state radial potential and wavefunctions (*inset*) for HOMO and HOMO-1 levels calculated for Na_{20} (**a**), Na_{40} (**b**), and Na_{92} (**c**) using DFT-SIC and DFT-LB94. The horizontal lines indicate occupied levels (red for DFT-LB94 and blue for DFT-SIC), the nodeless orbitals are represented with thick lines and orbitals with nodes are with dotted lines).

3.2. Total Photoionization Cross-Section

Figure 2 presents the LR-TDDFT photoionization cross-sections for three clusters, Na_{20}, Na_{40}, and Na_{92}, along with the IP results obtained in LR-DFT. TDDFT and DFT calculations are performed using the two *XC* schemes. The graphs are displayed on a logarithmic scale to emphasize the characteristics at the low-energy region near the ionization threshold. For energies exceeding 20 eV, there are agreements between the TDDFT and DFT results for all three clusters. Furthermore, collective effects disappear leading to discernible oscillations caused by the photoelectron with momentum k being diffracted from the cluster edges [41]. Since this characteristic is related to the geometry of the cluster, which is an IP attribute, they are also observed in the DFT results. Equivalently, the vanishing electron correlations at higher energies is evident from the convergence of oscillations in TDDFT and DFT. The specific choice of an *XC* functional that is insensitive to diffraction also has no significance at these energies. As the cluster size increases, the frequency of oscillations is also found to increase—a fact that can be explained using a Woods–Saxon model potential. Using this model, the dipole partial photoionization cross-section can be obtained as in [42]

$$\sigma_{nl \to kl'}(k) \propto \frac{e^{-ak}}{\omega^{5/2}}(1 + \cos(2kR_c - l'\pi)).$$

The decay that follows an exponential pattern is a consequence of the steepness (represented by the parameter "a") around the edge (refer to Figure 1). Note that potentials being "soft" at the cluster edge is intrinsic in DFT calculations. Based on the equation above, therefore, the state-selected cross-sections $\sigma_{nl \to kl'}$ will exhibit oscillations in k (or, equivalently, in photon energy) with a frequency of $2R_c$. Hence, this suggests an increase in oscillations when the cluster size (R_c) increases as seen in the total cross-sections in Figure 2.

When photon energies are in proximity to the ionization threshold and below 10 eV, notable differences emerge between the TDDFT and DFT cross-section profiles as a function of photon energy. These disparities encompass significant enhancements right above the threshold, a host of narrow autoionization resonances, and the occurrence of a correlation minimum around 6–7 eV, all in LR-TDDFT for the three clusters. The tiny and discrete jumps seen in DFT total cross-sections in Figure 2 are due to the opening of new subshell ionization channels.

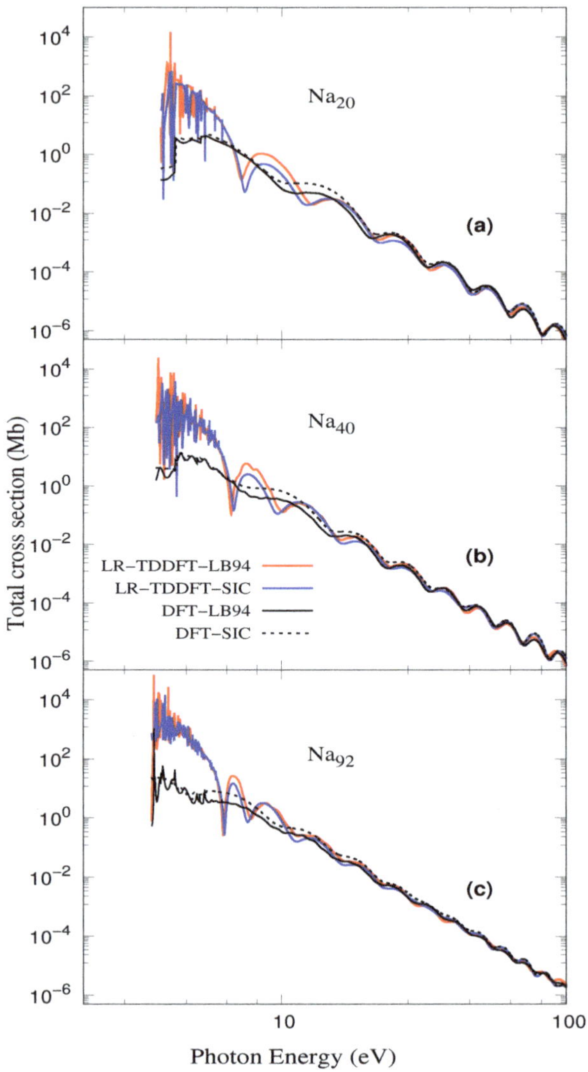

Figure 2. LR-TDDFT and LR-DFT photoionization cross-sections of Na_{20} (**a**), Na_{40} (**b**), and Na_{92} (**c**) using LB94 and SIC.

As noted, the LR-TDDFT cross-sections, regardless of the XC potential chosen, exhibit a strong enhancement in the vicinity of the threshold region for the three clusters compared to LR-DFT. This enhancement can be attributed to the spillover effect of plasmon resonances, which arise from correlated collective electronic motion. Our previous work already characterized this enhancement as the sum of an extension of the surface plasmon resonance from the absorption spectrum and a bulk portion of the volume plasmon resonance [32]. Overall, both the LB94 and SIC approaches exhibit qualitatively similar enhancements. This spillover from discrete to continuum spectra is accompanied by several narrow spikes known as the autoionization resonances, which arise due to the degeneracy between the ionization channels and inner-level single-electron excitation channels.

The phenomenon of plasmon resonance is the result of an in-phase coherence mechanism arising from the coupling of various degenerate ionization channels that are present. In Fano's framework based on first-order perturbation theory [43], the correlation-modified (LR-TDDFT) matrix element, $M_{nl \to \epsilon\lambda}(E)$, of a dipole ionization channel $nl \to \epsilon\lambda$ can be written as

$$M_{n\ell \to \epsilon\lambda}(E) = D_{n\ell \to \epsilon\lambda}(E) + \sum_{n'\ell' \neq n\ell} \int dE' \frac{\langle \psi_{n'\ell' \to \epsilon'\lambda'}(E')| \frac{1}{|\mathbf{r}_{n\ell} - \mathbf{r}_{n'\ell'}|} |\psi_{n\ell \to \epsilon\lambda}(E)\rangle}{E - E'} \times D_{n'\ell' \to \epsilon'\lambda'}(E'), \tag{16}$$

where $D_{n\ell \to \epsilon\lambda}$ refers to the unperturbed (DFT) matrix element. The wavefunctions of the interacting continuum channels are represented by $|\psi\rangle$, and the summation is carried over to all degenerate continuum channels with the exception of the $n\ell$ channel. The significance of electron correlations in enhancing the plasmon resonance spillover in the $n\ell \to \epsilon\lambda$ channel can be seen in the second term of Equation (16), which is referred as the interchannel coupling matrix element. $\mathbf{r}_{n\ell}$ and $\mathbf{r}_{n'\ell'}$ are the spatial co-ordinates of photoelectrons in the interacting continuum channels from initially occupied sub-shells $n\ell$ and $n'\ell'$, respectively. The summation over all subshells exhibits a coherent mixing primarily due to the bound wavefunctions occupying similar spatial regions. This results in a significant increase in the LR-TDDFT cross-section, as shown in Figure 2. However, as illustrated in Figure 2, the DFT predictions that disregard electron correlations do not exhibit such enhancement.

The interpretation of narrow resonances within the Fano framework requires coupling between the bound (excited) and continuum channels, resulting in a modification of the coupling matrix element in Equation (16) as

$$\langle \psi_{n'\ell' \to \eta'\lambda'}(E')| \frac{1}{|\mathbf{r}_{n\ell} - \mathbf{r}_{n'\ell'}|} |\psi_{n\ell \to \epsilon\lambda}(E)\rangle$$

where $n'\ell' \to \eta'\lambda'$ denote discrete excitation channels. The detailed profiles of these resonant structures are seen in Figure 2 to be sensitive to the scheme of XC functional employed. Resonance positions and shapes differ between SIC and LB94. This happens as a result of the levels of Na_{20} and Na_{40} in SIC being energetically slightly deeper compared to their LB94 counterparts, as seen in Figure 1. Additionally, the descriptions of the unoccupied excited states, which depend on the potential's long-range behavior, differ as well. Since the LB94 potential shows a slightly better long-range behavior, the resonances in this scheme are expected to be somewhat more accurate. The density of the unoccupied levels of both LB94 and SIC being increased as the cluster grows larger (noted in Figure 2) results in a higher number of resonances with a growing size.

3.3. Comparison with Experiments

It is useful, in particular, to assess the effect of the XC schemes on PI below 8 eV, where the collective effect dominates. However, as seen, this region has a complicated spectra because of the presence of autoionization resonances mentioned above. These narrow resonances are usually not present in experimental spectra due to the finite temperature effect experienced by the metal clusters under experimental conditions. In fact, the measured spectrum displays an incoherent mixture of spectra from various satellite configurations driven by the temperature, acquiring, effectively, a specific width [44,45]. This width can camouflage the narrow spikes by smearing them. To simulate this thermal effect, our theoretical data are convoluted with a Gaussian of width 0.4 eV. These are shown for the two clusters in Figure 3, and compared with available measurements. Note, however, that for Na_{40}, experimental data are not available. It can clearly be seen from Figure 3a,b that the smoothed LR-TDDFT cross-sections calculated both in LB94 and SIC show a reasonable agreement with the experimental observations that describe the spillover contribution of the plasmon resonances. Interestingly, SIC produces a slightly better overall agreement, even though its ground state long-range behavior is found to be somewhat less accurate. This suggests a lesser sensitivity of asymptotic properties to the collective behavior.

As can be seen from Figure 3a,b, the peaks of the smoothed LR-TDDFT curve, which must be predominantly induced by the collective volume resonance, are slowly red-shifting as the cluster size increases irrespective of the *XC* choice. However, the peak position is consistently more red-shifted in LB94 in comparison to SIC for each cluster.

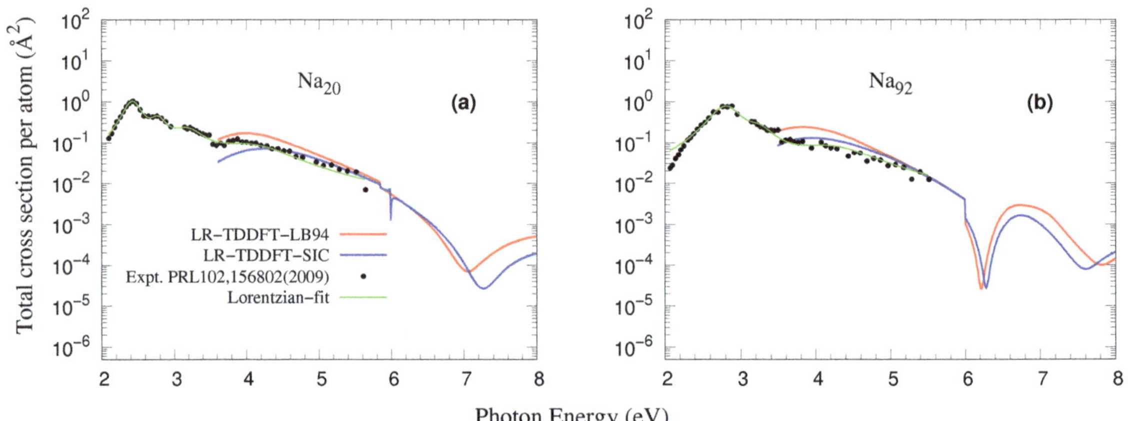

Figure 3. LR-TDDFT cross-sections for Na_{20} (**a**) and Na_{92} (**b**) compared with the available experimental data from [6]. The cross-sections from experiments are fitted using a Lorentzian.

It should be further noticed that the correlation minimum determined in SIC occurs at somewhat higher energies as compared to that in LB94 for all the three systems. As the cluster size increases, this energy separation between the minima from SIC versus LB94 diminishes, while their shapes become narrower. The discontinuities seen around 6 eV in the cross-section results are numerical artifacts, since the convolution is applied until around 6 eV, and therefore, are innocuous. Extending the region of convolution will modify the shape of the correlation minimum that we preferred to avoid.

The present work primarily focuses on the cross-section calculations. It may be interesting to see the effect of XC potentials on the angular distribution parameter since this quantity involves the phase of transition matrix elements. The present work does not make an attempt to study angular distribution because there are no related experimental data on neutral Na clusters. However the angular distribution and the photoelectron spectra measurements on anionic Na_n^- clusters were reported in the past [46,47], and subsequent theoretical studies using RPAE with the jellium model have shown a good agreement with the experiment [48,49].

3.4. Self-Consistent Induced Potential

Important electron correlations are embodied in the dipole matrix element in Equation (14) through the complex induced field, V_{ind}, computed using Equation (8). These correlations play a crucial role for the plasmonic enhancement in the cross-section. The behavior of V_{ind} delineates the detailed dynamics by visualizing a plasmon as a driven collective-electron oscillator with damping [15,50]. In this model, the real part of V_{ind}, $Re(V_{ind})$, represents the effective driving field, whereas the imaginary part, $Im(V_{ind})$, denotes the collective response of the system. Generally, at energies below the resonance peak, $Re(V_{ind}) < 0$ screens the external field so the electrons can build the collective motion and, simultaneously, the deepening of $Im(V_{ind})$ favors resonant binding. At the resonance peak, the $Re(V_{ind})$ becomes zero, making the field irrelevant, where $Im(V_{ind})$ offers the maximum binding so the collective response perfectly dominates. Above the peak energy, the plasmon decays with increasing $Im(V_{ind})$ since the field switches to the damping mode (anti-screening) as $Re(V_{ind}) > 0$. We expect this behavior in our current results but they are dominantly governed by the volume plasmon as the surface plasmon rules only below the ionization threshold.

Figure 4 shows the 3D plots of the real and imaginary components of the radial part of V_{ind} for all three clusters. Over the plasmon spillover region just above the threshold, $Im(V_{ind})$ shows a broad well-type shape from the collective dynamics, resulting in a transient attractive force that an emerging photoelectron will feel. In addition, the $Re(V_{ind})$ switches sign over this range for all three clusters as expected, although the negative $Re(V_{ind})$ range is shorter, as expected. The details of these results are seen to be sensitive to the form of the *XC* functional used. Within the graph scale, the magnitudes of the V_{ind} are slightly higher in LB94 than SIC. This may explain a slightly higher value of the LB94 cross-sections in Figure 3. Moreover, it is observed that the $Im(V_{ind})$ for SIC are a bit broader. This subsequently sustains the dynamics for a wider energy range to push the correlation minimum to higher energies in SIC, as also seen in Figure 3.

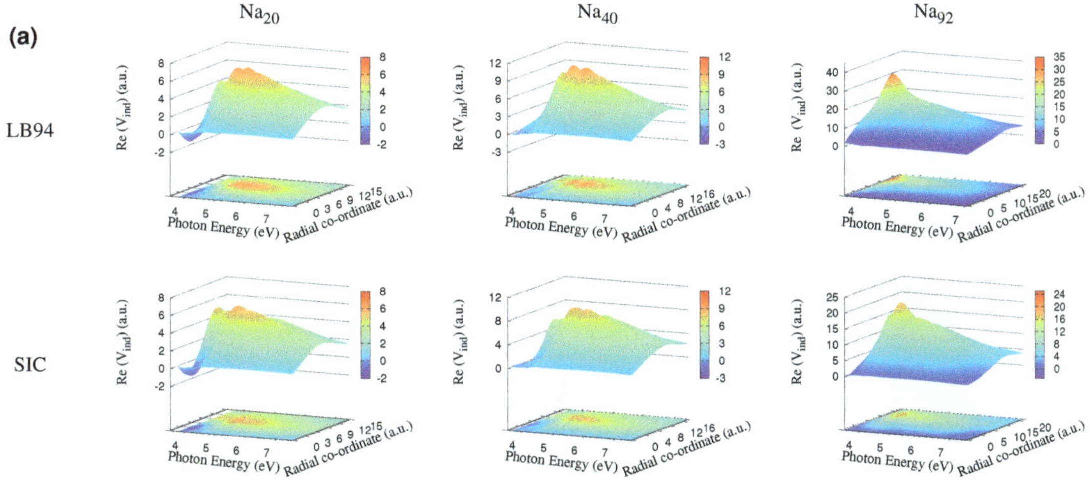

Figure 4. *Cont.*

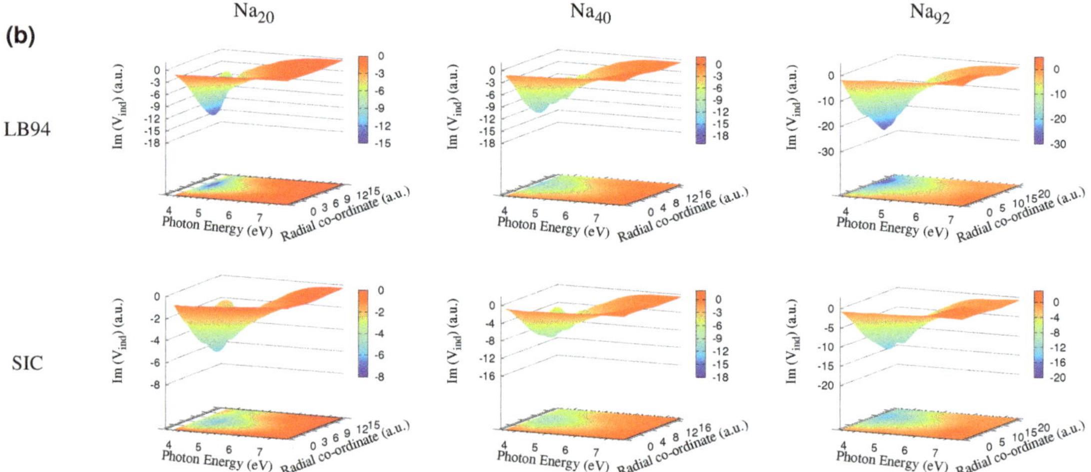

Figure 4. The real (**a**) and imaginary (**b**) components of the radial self-consistent field potential, V_{ind}, using LB94 and SIC within the LR-TDDFT frame are shown for Na_{20}, Na_{40}, and Na_{92}. For visual aid, some smoothing techniques have been applied.

4. Summary and Conclusions

The ground state structures and the photoionization properties of three sodium cluster systems, Na_x (x = 20, 40, and 92), were studied in a jellium-based LR-TDDFT methodology. Two previously successful schemes, SIC and LB94, of the XC functional implementation were employed. The energy levels obtained from the two schemes differ to some degree, while this difference is found to be at the maximum for Na_{20}. However, the difference reduces as the cluster size increases. The occupied level distributions remain roughly the same in the two schemes for each cluster. Both of the XC schemes show features such as the above-threshold enhancement from the plasmon resonant spillover, the narrow single-electron autoionizing resonances, the appearance of the correlation minimum, and the diffractive oscillations at higher energies. However, the XC functionals compete in producing the detailed characteristics of these features. The plasmon-enhanced spillover spectra in both XC formalisms show reasonable agreement with the available experiments. The results further indicate that the cluster size alters the ground state and ionization properties fairly monotonically, but the incongruence between SIC and LB94 tends to close as the size grows.

Even though the quantitative details are found to be sensitive to the particular XC scheme chosen, the significant qualitative similarities provide confidence on the accuracy of a DFT-level description of a metal cluster's static and dynamical properties. We hope the present work will motivate new experiments in the cluster science, particularly using the photoelectron spectroscopic techniques, which can provide further information in optimizing computational methods. This will help in extending calculations to address, for instance, the noble metal clusters and even to technologically relevant complexes with metal clusters as seed components. The current study may also motivate molecular-level calculations, going beyond the jellium model, to test the outcomes specifically and to increase the accuracy in general, although describing the continuum in such a frame will be a steep challenge to overcome.

Author Contributions: Conceptualization, H.S.C. and H.R.V.; methodology, R.S., H.R.V. and H.S.C.; software, R.S., H.S.C.; validation, R.S., H.R.V. and H.S.C.; formal analysis, R.S., H.R.V. and H.S.C.; investigation, R.S., H.R.V. and H.S.C.; resources, R.S., H.R.V. and H.S.C.; data curation, R.S.; writing—original draft preparation, R.S.; writing—review and editing, H.R.V. and H.S.C.; visualization, R.S., H.R.V. and H.S.C.; supervision, H.R.V. and H.S.C.; project administration, H.R.V.; funding acquisition, H.R.V. and H.S.C. All authors have read and agreed to the published version of the manuscript.

Funding: DST-SERB-CRG Project No. CRG/2022/002309, US National Science Foundation Grants No. PHY-1806206 (H.S.C.), No. PHY-2110318 (H.S.C.).

Data Availability Statement: All data are included. For specifics please contact authors.

Acknowledgments: The research is supported by the DST-SERB-CRG Project No. CRG/2022/002309, India (H.R.V.), and by the US National Science Foundation Grants No. PHY-1806206 (H.S.C.), No. PHY-2110318 (H.S.C.).

Conflicts of Interest: The authors declare no conflict of interest.

References

1. Reinhard, P.G.; Suraud, E.; Dinh, P.M. *An Introduction to Cluster Science*; John Wiley & Sons: Hoboken, NJ, USA, 2013.
2. Jena, P.; Sun, Q. Super atomic clusters: Design rules and potential for building blocks of materials. *Chem. Rev.* **2018**, *118*, 5755–5870.
3. Castleman, A., Jr.; Khanna, S. Clusters, superatoms, and building blocks of new materials. *J. Phys. Chem. C* **2009**, *113*, 2664–2675. [CrossRef]
4. Kawabata, A.; Kubo, R. Electronic properties of fine metallic particles. II. Plasma resonance absorption. *J. Phys. Soc. Jpn.* **1966**, *21*, 1765–1772. [CrossRef]
5. Kreibig, U.; Vollmer, M. *Optical Properties of Metal Clusters*; Springer Science & Business Media: Berlin/Heidelberg, Germany, 2013; Volume 25.
6. Xia, C.; Yin, C.; Kresin, V.V. Photoabsorption by volume plasmons in metal nanoclusters. *Phys. Rev. Lett.* **2009**, *102*, 156802. [CrossRef] [PubMed]
7. Madjet, M.E.A.; Chakraborty, H. Collective resonances in the photoresponse of metallic nanoclusters. *J. Phys. Conf. Ser.* **2009**, *194*, 022103. [CrossRef]
8. Ekardt, W. Size-dependent photoabsorption and photoemission of small metal particles. *Phys. Rev. B* **1985**, *31*, 6360. [CrossRef]
9. Sun, W.G.; Wang, J.J.; Lu, C.; Xia, X.X.; Kuang, X.Y.; Hermann, A. Evolution of the structural and electronic properties of medium-sized sodium clusters: A honeycomb-like Na20 cluster. *Inorg. Chem.* **2017**, *56*, 1241–1248. [CrossRef] [PubMed]
10. Miroshnichenko, A.E.; Flach, S.; Kivshar, Y.S. Fano resonances in nanoscale structures. *Rev. Mod. Phys.* **2010**, *82*, 2257. [CrossRef]
11. Luk'Yanchuk, B.; Zheludev, N.I.; Maier, S.A.; Halas, N.J.; Nordlander, P.; Giessen, H.; Chong, C.T. The Fano resonance in plasmonic nanostructures and metamaterials. *Nat. Mater.* **2010**, *9*, 707–715. [CrossRef] [PubMed]
12. Cederbaum, L.; Zobeley, J.; Tarantelli, F. Giant intermolecular decay and fragmentation of clusters. *Phys. Rev. Lett.* **1997**, *79*, 4778. [CrossRef]
13. Marburger, S.; Kugeler, O.; Hergenhahn, U.; Möller, T. Experimental evidence for interatomic Coulombic decay in Ne clusters. *Phys. Rev. Lett.* **2003**, *90*, 203401. [CrossRef]
14. De, R.; Magrakvelidze, M.; Madjet, M.E.; Manson, S.T.; Chakraborty, H.S. First prediction of inter-Coulombic decay of C_{60} inner vacancies through the continuum of confined atoms. *J. Phys. B At. Mol. Opt. Phys.* **2016**, *49*, 11LT01. [CrossRef]
15. Shaik, R.; Varma, H.R.; Madjet, M.E.A.; Zheng, F.; Frauenheim, T.; Chakraborty, H.S. Plasmonic Resonant Intercluster Coulombic Decay. *Phys. Rev. Lett.* **2023**, *130*, 233201. [CrossRef] [PubMed]
16. Jänkälä, K.; Tchaplyguine, M.; Mikkelä, M.H.; Björneholm, O.; Huttula, M. Photon energy dependent valence band response of metallic nanoparticles. *Phys. Rev. Lett.* **2011**, *107*, 183401. [CrossRef]
17. Frank, O.; Rost, J.M. From collectivity to the single-particle picture in the photoionization of clusters. *Phys. Rev. A* **1999**, *60*, 392. [CrossRef]
18. Frank, O.; Rost, J.M. Diffraction effects in the photoionization of clusters. *Chem. Phys. Lett.* **1997**, *271*, 367–371. [CrossRef]
19. Mirin, N.A.; Bao, K.; Nordlander, P. Fano resonances in plasmonic nanoparticle aggregates. *J. Phys. Chem. A* **2009**, *113*, 4028–4034. [CrossRef]
20. Cole, J.R.; Halas, N. Optimized plasmonic nanoparticle distributions for solar spectrum harvesting. *Appl. Phys. Lett.* **2006**, *89*, 153120. [CrossRef]
21. Dragan, A.I.; Geddes, C.D. Metal-enhanced fluorescence: The role of quantum yield, Q, in enhanced fluorescence. *Appl. Phys. Lett.* **2012**, *100*, 093115. [CrossRef]
22. Liao, H.; Nehl, C.L.; Hafner, J.H. *Biomedical Applications of Plasmon Resonant Metal Nanoparticles*; Nanomedicine: London, UK, 2006; Volume 1, pp. 201–208.
23. Huang, X.; Jain, P.K.; El-Sayed, I.H.; El-Sayed, M.A. Plasmonic photothermal therapy (PPTT) using gold nanoparticles. *Lasers Med. Sci.* **2008**, *23*, 217–228. [CrossRef]

24. De Heer, W.A. The physics of simple metal clusters: Experimental aspects and simple models. *Rev. Mod. Phys.* **1993**, *65*, 611. [CrossRef]
25. Brack, M. The physics of simple metal clusters: Self-consistent jellium model and semiclassical approaches. *Rev. Mod. Phys.* **1993**, *65*, 677. [CrossRef]
26. Kohn, W.; Sham, L.J. Self-consistent equations including exchange and correlation effects. *Phys. Rev.* **1965**, *140*, A1133. [CrossRef]
27. Perdew, J.P.; Wang, Y. Accurate and simple analytic representation of the electron-gas correlation energy. *Phys. Rev. B* **1992**, *45*, 13244. [CrossRef] [PubMed]
28. Klüpfel, P.; Dinh, P.M.; Reinhard, P.G.; Suraud, E. Koopmans' condition in self-interaction-corrected density-functional theory. *Phys. Rev. A* **2013**, *88*, 052501. [CrossRef]
29. Saito, S.; Bertsch, G.F.; Tománek, D. Collective electronic excitations in small metal clusters. *Phys. Rev. B* **1991**, *43*, 6804. [CrossRef]
30. Van Leeuwen, R.; Baerends, E. Exchange-correlation potential with correct asymptotic behavior. *Phys. Rev. A* **1994**, *49*, 2421. [CrossRef]
31. Choi, J.; Chang, E.; Anstine, D.M.; Madjet, M.E.A.; Chakraborty, H.S. Effects of exchange-correlation potentials on the density-functional description of C_{60} versus C_{240} photoionization. *Phys. Rev. A* **2017**, *95*, 023404. [CrossRef]
32. Shaik, R.; Varma, H.R.; Chakraborty, H.S. Collective effects in photoionization of sodium clusters: Plasmon resonance spill, induced attractive force and correlation minimum. *J. Phys. B At. Mol. Opt. Phys.* **2021**, *54*, 125101. [CrossRef]
33. Madjet, M.E.; Chakraborty, H.S.; Rost, J.M.; Manson, S.T. Photoionization of C_{60}: A model study. *J. Phys. B At. Mol. Opt. Phys.* **2008**, *41*, 105101. [CrossRef]
34. Chandezon, F.; Bjørnholm, S.; Borggreen, J.; Hansen, K. Electronic shell energies and deformations in large sodium clusters from evaporation spectra. *Phys. Rev. B* **1997**, *55*, 5485. [CrossRef]
35. Gunnarsson, O.; Lundqvist, B.I. Exchange and correlation in atoms, molecules, and solids by the spin-density-functional formalism. *Phys. Rev. B* **1976**, *13*, 4274. [CrossRef]
36. Oliver, G.; Perdew, J. Spin-density gradient expansion for the kinetic energy. *Phys. Rev. A* **1979**, *20*, 397. [CrossRef]
37. Marques, M.A.; Castro, A.; Rubio, A. Assessment of exchange-correlation functionals for the calculation of dynamical properties of small clusters in time-dependent density functional theory. *J. Chem. Phys.* **2001**, *115*, 3006–3014. [CrossRef]
38. Petersilka, M.; Gossmann, U.; Gross, E. Excitation energies from time-dependent density-functional theory. *Phys. Rev. Lett.* **1996**, *76*, 1212. [CrossRef]
39. Bertsch, G. An RPA program for jellium spheres. *Comput. Phys. Commun.* **1990**, *60*, 247–255. [CrossRef]
40. Parr, R.G.; Weitao, Y. *Density Functional Theory of Atoms and Molecules*; International series of monographs on Chemistry. 16; Oxford Science Publications: Oxford, UK, 1989; pp. 271–272.
41. Madjet, M.; Chakraborty, H.S.; Rost, J.M. Spurious oscillations from local self-interaction correction in high-energy photoionization calculations for metal clusters. *J. Phys. B At. Mol. Opt. Phys.* **2001**, *34*, L345. [CrossRef]
42. Bachau, H.; Frank, O.; Rost, J.M. Photoionization of alkali metal clusters. *Z. Phys. D At. Mol. Clust.* **1996**, *38*, 59–64.
43. Fano, U. Effects of configuration interaction on intensities and phase shifts. *Phys. Rev.* **1961**, *124*, 1866. [CrossRef]
44. Bertsch, G.; Tománek, D. Thermal line broadening in small metal clusters. *Phys. Rev. B* **1989**, *40*, 2749. [CrossRef]
45. Koskinen, M.; Manninen, M. Photoionization of metal clusters. *Phys. Rev. B* **1996**, *54*, 14796. [CrossRef] [PubMed]
46. Bartels, C.; Hock, C.; Huwer, J.; Kuhnen, R.; Schwobel, J.; Von Issendorff, B. Probing the angular momentum character of the valence orbitals of free sodium nanoclusters. *Science* **2009**, *323*, 1323–1327. [CrossRef]
47. Wrigge, G.; Hoffmann, M.A.; Issendorff, B.V. Photoelectron spectroscopy of sodium clusters: Direct observation of the electronic shell structure. *Phys. Rev. A* **2002**, *65*, 063201. [CrossRef]
48. Solov'yov, A.V.; Polozkov, R.G.; Ivanov, V.K. Angle-resolved photoelectron spectra of metal cluster anions within a many-body-theory approach. *Phys. Rev. A* **2010**, *81*, 021202. [CrossRef]
49. Polozkov, R.; Ivanov, V.; Verkhovtsev, A.; Korol, A.; Solov'yov, A. New applications of the jellium model for the study of atomic clusters. *J. Phys. Conf. Ser.* **2013**, *438*, 012009. [CrossRef]
50. Zangwill, A.; Soven, P. Local field effects in photoabsorption. *J. Vac. Sci. Technol.* **1980**, *17*, 159–163. [CrossRef]

Article

Transitional Strength Under Plasma: Precise Estimations of Astrophysically Relevant Electromagnetic Transitions of Ar^{7+}, Kr^{7+}, Xe^{7+}, and Rn^{7+} Under Plasma Atmosphere

Swapan Biswas [1], Anal Bhowmik [2,3], Arghya Das [4], Radha Raman Pal [1] and Sonjoy Majumder [4,*]

[1] Department of Physics, Vidyasagar University, Midnapore 721102, India; biswas.swapan25@gmail.com (S.B.); rrpal@mail.vidyasagar.ac.in (R.R.P.)
[2] Department of Physics, University of Haifa, Haifa 3498838, Israel
[3] Haifa Research Center for Theoretical Physics and Astrophysics, University of Haifa, Haifa 3498838, Israel
[4] Department of Physics, Indian Institute of Technology Kharagpur, Kharagpur 721302, India
* Correspondence: sonjoym@phy.iitkgp.ac.in

Highlights:

What are the main findings?
- Atomic spectroscopy for Ar^{7+}, Kr^{7+}, Xe^{7+}, and Rn^{7+} ions with high accuracy.
- Plasma screened ionization potential, atomic transition amplitudes and rates.
- Ionisation potential depression parameters.

What is the implication of the main finding?
- Properties of the astrophysical medium.
- Properties of laboratory plasma.

Citation: Biswas, S.; Bhowmik, A.; Das, A.; Pal, R.R.; Majumder, S. Transitional Strength Under Plasma: Precise Estimations of Astrophysically Relevant Electromagnetic Transitions of Ar^{7+}, Kr^{7+}, Xe^{7+}, and Rn^{7+} Under Plasma Atmosphere. *Atoms* **2023**, *11*, 87. https://doi.org/10.3390/atoms11060087

Academic Editors: Himadri S. Chakraborty and Hari R. Varma

Received: 26 March 2023
Revised: 9 May 2023
Accepted: 16 May 2023
Published: 25 May 2023

Abstract: The growing interest in atomic structures of moderately stripped alkali-like ions in the diagnostic study and modeling of astrophysical and laboratory plasma makes an accurate many-body study of atomic properties inevitable. This work presents transition line parameters in the absence or presence of plasma atmosphere for astrophysically important candidates Ar^{7+}, Kr^{7+}, Xe^{7+}, and Rn^{7+}. We employ relativistic coupled-cluster (RCC) theory, a well-known correlation exhaustive method. In the case of a plasma environment, we use the Debye Model. Our calculations agree with experiments available in the literature for ionization potentials, transition strengths of allowed and forbidden selections, and lifetimes of several low-lying states. The unit ratios of length and velocity forms of transition matrix elements are the critical estimation of the accuracy of the transition data presented here, especially for a few presented for the first time in the literature. We do compare our findings with the available recent theoretical results. Our reported data can be helpful to the astronomer in estimating the density of the plasma environment around the astronomical objects or in the discovery of observational spectra corrected by that environment. The present results should be advantageous in the modeling and diagnostics laboratory plasma, whereas the calculated ionization potential depression parameters reveal important characteristics of atomic structure.

Keywords: atomic data; transition probability; oscillator strength; lifetime; plasma density

1. Introduction

Barlow et al. [1] first observed noble gas molecules in the interstellar medium. The other detections of noble gas elements, either in diatomic [2–5] or ionic forms [6] in space at UV and IR spectra, motivate further observations of these species in the universe. It is well known that the atomic and spectroscopic processes are valuable diagnostics for plasma atmosphere in the laboratory or Astronomy. Noble gas atoms are known to be chemically inactive and require high energy to ionize. However, once ionized, their reaction rates are rather fast. Over the years, spectroscopic properties of ionized noble gas atoms have become popular, and observers have started to detect them in space [7]. On the other hand, alkali-like ions have emerged as the standard test beds for detailed investigation of current relativistic atomic calculations due to their adequately simple but highly correlated electronic structures [8–12]. Accurate theoretical and experimental determinations of the transition line parameters and excited-state lifetimes of highly stripped ions are collaborative with the astronomer to investigate dynamics, chemical compositions, opacity, density, and temperature distributions of the distant galaxy [13], planetary nebulae, and even the entire interstellar medium [14–24]. Furthermore, one requires the accurate atomic data of different isotopes of noble gas elements to understand the production of heavy elements in the stellar medium by radiative r- and s-processes [25,26]. The data of energy spectra of moderate to high-stripped ions are required for precise astrophysical and laboratory plasma modeling. All these physical facts and figures motivate us to investigate the transition line parameters and lifetimes of septuple ionized astrophysically pertinent inert gases, such as Ar^{7+}, Kr^{7+}, Xe^{7+}, and Rn^{7+}.

In the series, Ar^{7+} is well studied in the literature. Berry et al. [27] observed 74 lines of Ar^{7+} using the beam-foil technique. In 1982, Striganov and Odintsova [28] published the observed lines of Ar^{+} through Ar^{8+}. The authors of [29–31] applied the multi-configuration Dirac–Fock (MCDF) method to calculate the autoionization spectrum, energy levels, transition rates, oscillator strengths, and lifetimes of Ar^{7+}. Saloman [32] identified the energy spectra of Ar^{+} to Ar^{17+}, which he studied from the year 2006 to 2009, employing beam foil Spectroscopy (BFS), an electron beam ion trap (EBIT), laser-excited plasmas, fusion devices, astronomical observations, and ab initio calculations with quantum electrodynamic corrections.

Similarly, krypton ion spectra were detected in the interstellar medium [33,34], the galactic disc [35], and the planetary nebulae [36]. Fine structure intervals, fine structure inversions, and core-polarization study of the Kr^{7+} ion were performed by different groups [37–39] including third-order many-body perturbation theory and Møller–Pleset perturbation theory [40,41] for the energy levels.

It is found that Cu I isoelectronic sequence ions are prominent impurities at high-temperature magnetically confined plasmas [42], and their emission spectra are observed under the spark sources [43–46] of the laser-produced plasma [47,48] and in the beam-foil excitations [49–51]. The abundance of photospheric lines of trans-iron group elements in the emission spectra of the white dwarfs opens a new way of studying their radiative transfer mechanism [52]. The presence of the spectral lines of Cu-like ions motivates more accurate determination of atomic data of the radiative properties of these ions for modeling the chemical abundances. These studies are essential for deducing the stellar parameters necessary to investigate the environmental condition of the white dwarfs. There are studies of electronic properties for Xe^{7+} using various many-body methods [53–57]. Dimitrijevc et al. [24] identified the importance of Stark broadening at the spectral lines observed in extremely metal-poor halo PNH4-1 in primordial supernova [58]. However, we study Ar^{+7}, Kr^{7+}, and Xe^{7+} here again to mitigate the lack of all-order many-body calculations or precise experiments and to estimate their spectroscopic properties under a plasma environment. Recently, one of the present authors [59] studied Xe^{7+} exclusively as a single valance system without the plasma screening effect.

Unlike other noble gas ions, studies of radon ions are rare. However, there are a few many-body calculations on Rn^{+} [60] and Rn^{2+} [61]. The observation of several forbidden lines of Kr and Xe ions in the planetary nebula NGC 7027 was reported recently [62]. For Rn^{7+}, only Migdalek [63] computed a few energy levels and oscillator strengths of allowed transitions using the Dirac–Fock method corrected by the core-polarisation effect.

The aim of this paper is to estimate (a) energies of the ground and low-lying excited states, (b) the oscillator strengths of electromagnetically allowed transitions, (c) transition probabilities of the forbidden transitions, and (d) lifetimes for a few excited states of Ar^{7+}, Kr^{7+}, Xe^{7+}, and Rn^{7+} using the relativistic coupled-cluster (RCC) method [64–66]. The accuracies of the RCC calculations are well established by our group for different applications [67–76]. The all-order structure of electron correlation in the RCC theory has been elaborated in our earlier paper [77] and the review article by Bartlett [78]. Our special effort here is to study the plasma screening effect on the radiative transition parameters. It is obvious that the nuclear attraction to the bound electrons of atoms or ions immersed in plasma is screened by the neighboring ions and the free electrons. The essential feature to note is that the electron correlation of atomic systems in this environment is remarkably different from their corresponding isolated candidate. Therefore, the screening estimations on the transition parameters play a crucial role in the precise diagnostics of plasma temperature and density in the emitting region. In the plasma environment, the ionization potentials decrease gradually with the increasing strength of plasma screening [79] until they become zero at some critical parameter of plasma. Beyond these critical values of plasma, the states become a continuum state. The corresponding ionization potential beyond which instability occurs is known as ionization potential depression (IPD) according to the Stewart–Pyatt (SP) model [80]. Accurate determination of the IPD can infer much useful information about the plasma atmosphere, such as providing the proper equation of the state, estimating the radiate opacity of stellar plasma, internal confinement fusion plasma, etc. We have investigated the change in spectroscopic properties of Ar^{7+}, Kr^{7+}, Xe^{7+}, and Rn^{7+} in the plasma environment.

2. Theory

Precise generation of wave functions is important for accurately estimating the atomic properties of few-electron monovalent ions presented in this paper. Here, we employ a non-linear version of the well-known RCC theory, a many-body approach which exhaustively pools together correlations. Initially, we solve the Dirac–Coulomb Hamiltonian H, satisfying the eigenvalue equation $H|\Phi\rangle = E_0|\Phi\rangle$ to generate closed-shell atomic wave function under the potential of $(N-1)$ electrons where

$$H = \sum_i \left(c\alpha_i \cdot \mathbf{p}_i + (\beta_i - 1)c^2 + V_{nuc}(\mathbf{r}_i) + \sum_{j<i} \left(\frac{1}{\mathbf{r}_{ij}} \right) \right).$$

Here, α_i and β are the usual Dirac matrices and $V_{nuc}(\mathbf{r_i})$ is the potential at the site of the i-th electron due to the atomic nucleus. The rest mass energy of the electron is subtracted from the energy eigenvalues. The last term corresponds to the Coulomb interaction between the i-th electron and j-th electron. A single valence reference state for the RCC calculation is generated by adding a single electron in the v-th orbital following Koopman's theorem [81]. In RCC formalism, the single valence correlated state $|\Psi_v\rangle$ is connected with the single valence reference state $|\Phi_v\rangle$ as

$$|\Psi_v\rangle = e^T \{1 + S_v\}|\Phi_v\rangle, \quad \text{where } |\Phi_v\rangle = a_v^\dagger|\Phi\rangle. \tag{1}$$

The operator T deals with the excitations from core orbitals and can generate core-excited configurations from closed-shell Dirac–Fock state $|\Phi\rangle$. Whereas, S_v excites at least one electron from the valence orbital, giving rise to valence and core-valence excited configurations [64]. The operator S_v can yield the valence and core-valence excited configurations with respect to the open-shell Dirac–Fock state $|\Phi_v\rangle$ [69]. Here, we generate single- and double-excited correlated configurations from Equation (1). The amplitudes of these excitations are solved from the energy eigenvalue equations of the closed-shell and open-shell systems, which are $He^T|\Phi\rangle = Ee^T|\Phi\rangle$ and $H_v e^T|\Phi_v\rangle = E_v e^T|\Phi_v\rangle$, respectively [82]. In the present method, these amplitudes are solved following the Jacobi iteration scheme, which is considered all-ordered. The initial guesses of the single- and double-excitation amplitudes are made consistent with the first order of the perturbation theory [83]. In the present version of RCC theory, we also consider some important triple excitations and hence the abbreviation is used RCCSD(T).

The matrix elements of an arbitrary operator can be written as

$$
\begin{aligned}
O_{ki} &= \frac{\langle \Psi_k | \hat{O} | \Psi_i \rangle}{\sqrt{\langle \Psi_k | \Psi_k \rangle \langle \Psi_i | \Psi_i \rangle}} \\
&= \frac{\langle \Phi_k | \{1 + S_k^\dagger\} e^{T^\dagger} \hat{O} e^T \{1 + S_i\} | \Phi_i \rangle}{\sqrt{\langle \Phi_k | \{1 + S_k^\dagger\} e^{T^\dagger} e^T \{1 + S_k\} | \Phi_k \rangle \langle \Phi_i | \{1 + S_i^\dagger\} e^{T^\dagger} e^T \{1 + S_i\} | \Phi_i \rangle}}.
\end{aligned}
\tag{2}
$$

The detailed derivations and explanations of the matrix elements associated with electric dipole (E_1), electric quadrupole (E_2), and magnetic dipole (M_1) transitions can be found in the literature [84]. Emission transition probabilities (s^{-1}) for the E_1, E_2, and M_1 from $|\Psi_k\rangle$ to $|\Psi_i\rangle$ state are [85]

$$
A_{k \to i}^{E_1} = \frac{2.0261 \times 10^{-6}}{\lambda^3 (2J_k + 1)} S^{E_1},
\tag{3}
$$

$$
A_{k \to i}^{E_2} = \frac{1.12 \times 10^{-22}}{\lambda^5 (2J_k + 1)} S^{E_2},
\tag{4}
$$

$$
and \qquad A_{k \to i}^{M_1} = \frac{2.6971 \times 10^{-11}}{\lambda^3 (2J_k + 1)} S^{M_1}.
\tag{5}
$$

where, λ is in cm and S is the square of the transition matrix elements of O (corresponding transition operator) in atomic unit of $e^2 a_0^2$ (e is the charge of an electron and a_0 is the Bohr radius). The oscillator strength for the $E1$ transition is related to the corresponding transition probability (s^{-1}) with the following equation [86]

$$
f_{k \to i}^{osci} = 1.4992 \times 10^{-16} A_{k \to i} \frac{g_k}{g_i} \lambda^2,
\tag{6}
$$

where g_k and g_i are the degeneracies of the final and initial states, respectively. The lifetime of the k-th state is calculated by considering all transition probabilities to the lower energy states (i-th) and is given by

$$
\tau_k = \frac{1}{\sum_i A_{k \to i}}.
\tag{7}
$$

In order to incorporate the plasma screening effect on the atomic spectroscopic properties, the Dirac–Coulomb potential takes the form as

$$
H_{\text{eff}}^D = H + V_{\text{eff}}^D(\mathbf{r}_i).
\tag{8}
$$

Here, $V_{\text{eff}}^D(\mathbf{r}_i)$ is the effective potential of the nucleus on the i-th electron due to the presence of the plasma environment. The Debye–Hückle potential is considered to examine the effect of the screening of the nuclear coulomb potential due to the presence of ions and free electrons in plasma [87,88]. In the case of a weekly interacting plasma medium, the effective potential experienced by the i-th electron is given as

$$
V_{\text{eff}}^D(r_i) = \frac{Z e^{-\mu r_i}}{r_i}.
\tag{9}
$$

The nuclear charge Z and the Debye screening parameter μ are related to the ion density n_{ion} and plasma temperature T through the following relation

$$
\mu = \left[\frac{4\pi (1 + Z) n_{ion}}{K_B T} \right]^2,
\tag{10}
$$

where, k_B is the Boltzmann constant. Therefore, a given value of μ represents a range of plasma conditions with different ion densities and temperatures.

3. Results and Discussions

The single-particle Dirac–Fock (DF) wavefunctions are the building blocks of the RCC calculations yielding the many-electron correlation energies and correlated wavefunctions. We calculate the bound Dirac–Hartree–Fock orbitals as accurately as possible using a sophisticated numerical approach, GRASP92 [89]. Further, we apply the basis-set expansion technique [90] in the self-consistent field approach to obtain the Gaussian-type DF orbital (GTO) used in the RCC calculations. The radial part of each basis function has two parameters, α_0, and β, as exponents [91] to be optimized. The parameters are required to optimize due to the finite size of the basis set. The exponent parameters are optimized compared to the DF bound orbitals obtained from GRASP92, discussed in detail in our old papers [12,66]. In the basis optimization method, we consider 33, 30, 28, 25, 21, and 20 basis functions for s, p, d, f, g, and h symmetries, respectively. This basis set is considered for all the ions. However, the choice of the active orbitals in the RCC calculation relies on the convergence of the correlation contribution to the closed-shell energy with the increasing number of the orbitals [66,92]. Therefore, the active orbitals for the converged correlation contribution to the closed-shell energy are found to be distinct for different ions investigated in this work.

In this article, we calculate the ionization potential of Ar^{7+}, Kr^{7+}, Xe^{7+}, and Rn^{7+} using the RCC method and compare them in Table 1 with the results published in the National Institute of Standards and Technology (NIST) [93] wherever available. The NIST estimations are considered to have the best accuracy. We find that our calculated ground state energies of Ar^{7+}, Kr^{7+}, and Xe^{7+} are in excellent agreement with NIST results, and deviations are estimated to be -0.01%, 0.45%, and -0.03%, respectively. Table 1 presents the ionization potential of the low-lying excited states of these ions with average deviations around -0.08%, 0.42%, and 0.30%, respectively. In these cases, the maximum difference is -0.23% and occurred for the $5p_{3/2}$ state of Ar^{7+}, 0.60% for $5g_{7/2,9/2}$ of Kr^{7+}, and 1.2% for $6d_{3/2,5/2}$ of Xe^{7+}.

Table 1. Comparison of our RCC ionization potential (in cm^{-1}) with NIST data and our estimations of plasma screening effect on them. Estimations for $5g$ states of Rn^{7+} were not available in the literature (a) [63]. Plasma screening strength (μ) is in a.u. unit. Energy levels are indicated as $nL(2J+1)$. The bold values indicate that beyond which the system becomes unbound.

State	NIST	$\mu = 0$	$\mu = 0.025$	$\mu = 0.05$	$\mu = 0.075$	$\mu = 0.1$
Ar^{7+}						
3s2	1,157,056	1,157,201	1,059,866	965,330	873,513	784,345
3p2	1,016,961	1,016,995	919,704	825,299	733,697	644,828
3p4	1,014,248	1,014,184	916,898	822,505	730,924	642,084
3d4	824,447	824,210	726,923	632,529	540,956	452,148
3d6	824,302	824,027	726,741	632,349	540,779	451,974
4s2	581,098	581,069	485,070	394,397	308,793	228,054
4p2	528,815	528,528	432,658	342,356	257,356	177,460
4p4	527,813	527,428	431,565	341,287	256,323	176,473
4d4	459,524	459,475	363,709	273,714	189,240	110,134
4d6	459,435	459,386	363,620	273,627	189,156	110,055
4f6	440,204	440,190	344,137	253,339	167,618	86,891
4f8	440,181	440,159	344,107	253,309	167,590	86,865
5s2	349,750	349,752	255,504	169,770	91,961	21,742
5p2	324,795	323,193	229,236	144,332	67,862	**557**
5p4	324,307	322,543	228,611	143,769	67,376	
5d4	291,782	291,699	197,867	113,352	37,637	
5d6	291,778	291,652	197,818	113,302	37,591	
5f6	281,727	281,736	187,657	102,487	25,846	
5f8	281,707	281,720	187,642	102,473	25,833	

Table 1. *Cont.*

State	NIST	$\mu = 0$	$\mu = 0.025$	$\mu = 0.05$	$\mu = 0.075$	$\mu = 0.1$
5g8	281,051	281,015	186,389	99,684	20,685	
5g10	281,037	281,015	186,380	99,676	20,678	
Kr^{7+}						
4s2	1,014,665	1,010,099	815,902	628,237	446,944	271,892
4p2	870,969	867,027	673,080	486,149	306,059	132,689
4p4	861,189	857,253	663,348	476,541	296,652	123,557
4d4	640,619	636,965	443,698	258,762	81,891	**86,678**
4d6	639,288	635,514	442,257	257,352	80,623	
5s2	524,578	523,198	331,842	152,413	**15,482**	
5p2	467,984	465,448	274,592	96,629		
5p4	464,221	461,406	270,630	92,891		
4f6	451,900	450,180	257,831	75,655		
4f8	451,934	450,186	257,837	75,660		
5d4	373,589	371,922	182,111	7230		
5d6	373,048	371,273	181,476	6635		
6s2	322,147	321,635	133,957	**34,847**		
5f6	289,666	288,637	100,173			
5f8	289,661	288,634	100,170			
5g8	281,574	281,387	92,121			
5g10	281,572	281,379	92,114			
Xe^{7+}						
5s2	854,769	854,995	564,858	286,731	20,272	
5p2	738,302	737,059	447,391	170,643	**93,515**	
5p4	719,717	718,263	428,759	152,487		
4f6	589,608	588,730	297,599	16,588		
4f8	589,058	588,088	296,970	16,085		
5d4	544,881	543,506	255,203	**17,518**		
5d6	541,953	540,549	252,295			
6s2	459,272	455,364	170,155			
6p2	411,391	406,318	121,973			
6p4	403,996	398,801	114,717			
5f6	357,190	356,376	70,683			
5f8	356,751	355,922	70,245			
6d4	327,344	323,450	41,320			
6d6	325,975	322,176	40,102			
7s2	289,473	276,902	**451**			
5g8	284,501	283,609				
5g10	284,501	283,617				
Rn^{7+}	(a)					
6s2	834,624	839,362	377,923	**63,552**		
6p2	712,821	718,015	257,667			
6p4	661,071	665,627	205,616			
5f6	536,092	531,079	69,545			
5f8	534,720	529,291	68,106			
6d4	498,636	501,235	43,424			
6d6	491,240	493,699	36,116			
7s2	446,847	445,577	**8041**			
7p2	397,883	398,535				
7p4	377,435	377,332				
5g8		287,997				
5g10		288,252				

Our calculated energies agree well with estimations by Fischer et al. [31], who computed energy levels of Ar^{7+} using the core polarization effect on the Dirac–Hartree–Fock (CP-DHF) theory. Cheng and Kim [37] tabulated the energy levels of Kr^{7+} from the relativistic Hartree–Fock (RHF) calculations. As expected, our RCC calculated results are found to be in better agreement with the NIST values. For Rn^{7+}, we have not found any experimental measurement in the literature nor NIST compiled values. Only one theoretical calculation based on the CP-DHF method by [63] is available with the average deviation of IP being 0.66% from our calculations.

The percentage of electron correlation correction, i.e., $\frac{(\text{RCC-DF})\times 100\%}{\text{DF}}$ in IP of the ground state monotonically increases from Ar^{7+} to Rn^{7+} with the values 0.39%, 0.52%, 1.81%, and 1.87%, respectively.

Now, we investigate the impact of the plasma screening potential on the energy levels of the considered ions. Table 1 shows that IP monotonically decreases with the increase in the μ value. The bold values for each ion in the table represent the limiting case beyond which the system becomes unbound. Figure 1 presents the plasma screening contribution in IP for a few low-lying states, such as ground state $S_{1/2}$, excited $P_{1/2,3/2}$, and $D_{3/2,5/2}$ states of Ar^{7+}, Kr^{7+}, Xe^{7+}, and Ra^{7+} ions. The panels of the figure show the plasma screening contribution increases from the ground to higher excited states, as the latter states are less bound by the Coulomb attraction. For Xe^{7+} and Rn^{7+}, we could plot the effect up to a certain value of μ as most of the states become continuum states beyond that. We observe that the plasma screening effect is practically strong for fine structure levels for Ar^{7+} and Kr^{7+} ions and weak for Xe^{7+}, and Ra^{7+} ions.

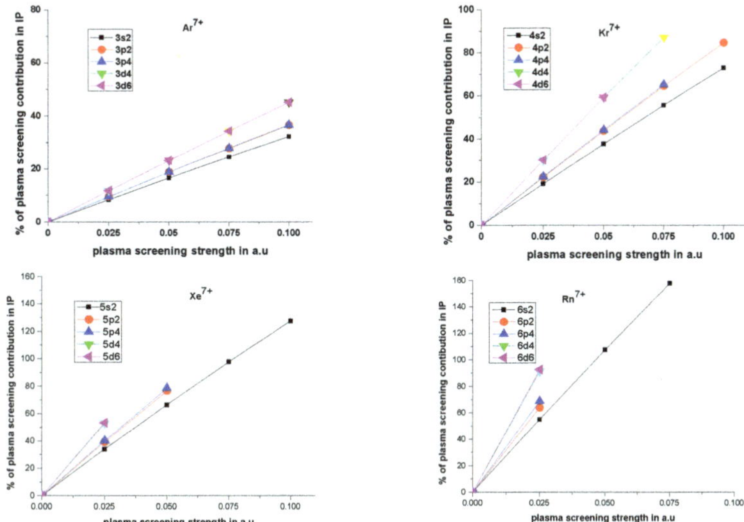

Figure 1. The plasma contribution in IP of low-lying septuple ionized atoms: % variation of IP with plasma screening strength. Energy levels are indicated as $nL(2J+1)$. Results are calculated from $\frac{\text{IP}_{\mu=0}-\text{IP}_{\mu>0}}{\text{IP}_{\mu=0}} \times 100$.

We present the electric dipole matrix elements for the ions in the plasma medium in Table 2. The table also displays our computed DF values of the matrix elements to reveal the correlation contributions. The separate presentation of the DF and RCC values over the span of the plasma screening parameter, μ, in the table is an intentional move. Here, we want to highlight that the plasma screening impacts the DF and the RCC correlation parts differently. In the case of Ar^{7+}, the average changes in the matrix element due to the increasing values of the plasma screening parameter are less than 1%. However, Figure 2 shows significant changes for $3s_{1/2} \to 4p_{1/2,3/2}$ and $3s_{1/2} \to 5p_{1/2,3/2}$ for finite values of μ, especially for $\mu = 0.05$. This is true for any $n\,{}^2S_{1/2} \to n'\,{}^2P_{1/2,3/2}$ of this ionic series. However, apart from such a few transitions, the plasma effects lie between 1% and 2% for most of the other transitions in the series. Table 2 shows that the average correlations (for the non-plasma environment) in the transition amplitudes for Ar^{7+}, Kr^{7+}, Xe^{7+}, and Rn^{7+} are 0.3858%, 4.0831%, 7.6074%, and 11.6379% apart from the $4d_{3/2} \to 5f_{5/2}$ transition where the correlation is 67.11%.

Table 2. Our DF and RCC matrix element (a.u.), in length gauge, of electric dipole ($E1$) transitions in plasma medium. Energy levels are indicated as $nL(2J+1)$.

	$\mu = 0$		$\mu = 0.025$		$\mu = 0.05$		$\mu = 0.075$		$\mu = 0.1$	

Ar^{7+}

Transitions	DF	RCC	DF	RCC	DF	RCC	DF	RCC	DF	RCC
3s2→3p2	0.9617	0.9341	0.9623	0.9346	0.9638	0.9390	0.9663	0.9405	0.9697	0.9436
3s2→3p4	1.3619	1.3228	1.3626	1.3236	1.3648	1.3299	1.3683	1.3317	1.3732	1.3355
3s2→4p2	0.2002	0.2093	0.1999	0.2089	0.1989	0.2210	0.1973	0.2181	0.1950	0.2164
3s2→4p4	0.2770	0.2898	0.2765	0.2893	0.2751	0.3059	0.2728	0.3028	0.2698	0.3012
3s2→5p2	0.1011	0.1062	0.1006	0.1057	0.0992	0.0995	0.0972	0.1105	0.0944	0.1007
3s2→5p4	0.1397	0.1470	0.1390	0.1463	0.1372	0.1215	0.1345	0.1541	0.1307	0.1528
4s2→4p2	1.8958	1.8813	1.8996	1.8851	1.9105	1.9059	1.9284	1.8987	1.9537	1.9012
4s2→4p4	2.6827	2.6622	2.6879	2.6675	2.7033	2.7006	2.7286	2.6857	2.7644	2.6844
3p2→4s2	0.3604	0.3646	0.3608	0.3649	0.3619	0.3419	0.3637	0.3437	0.3661	0.3432
3p4→4s2	0.5172	0.5231	0.5178	0.5236	0.5193	0.4898	0.5219	0.4943	0.5253	0.4952
3p2→5s2	0.1237	0.1269	0.1240	0.1271	0.1246	0.1275	0.1252	0.1144	0.1253	0.1274
3p4→5s2	0.1771	0.1816	0.1775	0.1819	0.1783	0.2021	0.1792	0.1653	0.1793	0.1659
3p2→3d4	1.3534	1.3174	1.3547	1.3186	1.3583	1.3207	1.3642	1.3218	1.3725	1.3305
3p4→3d4	0.6060	0.5899	0.6066	0.5905	0.6082	0.5916	0.6109	0.5918	0.6146	0.5956
3p4→3d6	1.8184	1.7700	1.8201	1.7720	1.8250	1.7736	1.8330	1.7750	1.8442	1.7795
3p2→4d4	0.3857	0.3957	0.3842	0.3942	0.3800	0.3925	0.3730	0.3926	0.3629	0.3551
3p4→4d4	0.1756	0.1801	0.1750	0.1794	0.1731	0.1770	0.1699	0.1791	0.1654	0.1645
3p4→4d6	0.5262	0.5394	0.5242	0.5375	0.5185	0.5321	0.5091	0.5372	0.4956	0.5357
4p2→4d4	2.7891	2.7687	2.7966	2.7761	2.8181	2.7737	2.8536	2.7737	2.9038	2.8032
4p4→4d4	1.2496	1.2405	1.2529	1.2438	1.2626	1.2470	1.2785	1.2420	1.3010	1.2498
4p4→4d6	3.7490	3.7217	3.7590	3.7316	3.7880	3.7880	3.8357	3.7255	3.9033	3.7279
3d4→4p2	0.5411	0.5434	0.5426	0.5449	0.5470	0.5616	0.5542	0.5407	0.5644	0.5701
3d4→4p4	0.2395	0.2406	0.2402	0.2412	0.2487	0.2450	0.2454	0.2395	0.2499	0.2530
3d6→4p4	0.7193	0.7225	0.7213	0.7244	0.7271	0.7193	0.7368	0.7201	0.7504	0.7184
3d4→4f6	1.7707	1.7378	1.7705	1.7377	1.7702	1.7384	1.7691	1.7401	1.7663	1.7778
3d6→4f6	0.4734	0.4671	0.4734	0.4646	0.4733	0.4649	0.4730	0.4653	0.4722	0.4658
3d6→4f8	2.1172	2.078	2.1171	2.0779	2.1166	2.0793	2.1154	2.0768	2.1121	2.0799
4d4→4f6	3.1546	3.1495	3.1686	3.1634	3.2100	3.1508	3.2796	3.1504	3.3822	3.1414
4d6→4f6	0.8430	0.8416	0.8468	0.8454	0.8578	0.8420	0.8764	0.8419	0.9087	0.8423
4d6→4f8	3.7706	3.7644	3.7874	3.7812	3.7652	3.7644	3.9201	3.7778	4.0427	3.7790

Kr^{7+}

Transitions	DF	RCC	DF	RCC	DF	RCC	DF	RCC	DF	RCC
4s2→4p2	1.1348	1.0794	1.1363	1.1017	1.1405	1.1223	1.1475	1.1088	1.1575	1.1245
4s2→4p4	1.6095	1.5314	1.6115	1.5632	1.6175	1.5900	1.6275	1.5724	1.6416	1.6085
4s2→5p2	0.1488	0.1655	0.1479	0.1584	0.1451	0.1537				
4s2→5p4	0.1812	0.2057	0.1799	0.1944	0.1761	0.1839				
4p2→4d4	1.7539	1.6812	1.7569	1.7412	1.7657	1.7506	1.7804	1.7422		
4p4→4d4	0.7940	0.7616	0.7954	0.7734	0.7995	0.7836	0.8064	0.7771		
4p4→4d6	2.3818	2.2847	2.3860	2.3218	2.3984	2.3511	2.4190	2.3301		
4p2→5s2	0.4884	0.4934	0.4894	0.4892	0.4925	0.4838				
4p4→5s2	0.7316	0.7379	0.7331	0.7325	0.7377	0.7272				
4p2→6s2	0.1655	0.1686	0.1654	0.1664	0.1646	0.1624				
4p4→6s2	0.2447	0.2486	0.2445	0.2460	0.2427	0.2424				
4d4→4f6	2.8243	2.7466	2.8340	2.7759	2.8623	2.7985				
4d6→4f6	0.7565	0.7358	0.7591	0.7436	0.7668	0.7499				
4d4→5p2	1.1740	1.1630	1.1809	1.1776	1.2015	1.1742				
4d4→5p4	0.5090	0.5124	0.5120	0.5107	0.5212	0.5168				
4d6→5p4	1.5415	1.5522	1.5507	1.5464	1.5786	1.5435				
4d6→4f8	3.3836	3.2916	3.3953	3.3257	3.4297	3.3529				
4d4→5f6	0.0596	0.0427	0.0582	0.0445						
4d6→5f6	0.0133	0.0091	0.0175	0.0089						
4d6→5f8	0.0592	0.0418	0.0779	0.0394						
4f6→5g8	3.8576	3.7993	3.8688	3.8112						
4f8→5g8	0.7422	0.7311	0.7444	0.7340						
4f8→5g10	4.3914	4.3256	4.4042	4.3380						
4f6→5d4	2.0421	2.0387	2.0728	2.0318	2.1683	2.0704				
4f6→5d6	0.5430	0.5420	0.5512	0.5401	0.5766	0.5498				
4f8→5d6	2.4274	2.4232	2.4641	2.4137	2.5780	2.4567				
5p2→5d4	3.1683	3.1275	3.1867	3.1373	3.2401	3.1494				
5p4→5d4	1.4407	1.4226	1.4489	1.4271	1.4728	1.4325				
5p4→5d6	4.3172	4.2629	4.3417	4.2781	4.4131	4.2912				
5p2→6s2	1.0220	1.0239	1.0277	1.0215						
5p4→6s2	1.5278	1.5310	1.5347	1.5276						

Table 2. *Cont.*

	$\mu = 0$		$\mu = 0.025$		$\mu = 0.05$		$\mu = 0.075$		$\mu = 0.1$	
Xe^{7+}	DF	RCC	DF	RCC	DF	RCC	DF	RCC	DF	RCC
5s2→5p2	1.3758	1.1736	1.3791	1.1769	1.3889	1.1866				
5s2→5p4	1.9516	1.6705	1.9562	1.6752	1.9700	1.6890				
5p2→5d4	2.1135	1.8629	2.1197	1.8697						
5p4→5d4	0.9768	0.8646	0.9799	0.8679						
5p4→5d6	2.9252	2.5911	2.9344	2.6011						
5p2→6s2	0.5866	0.6026	0.5893	0.6046						
5p4→6s2	0.9449	0.9582	0.9494	0.9619						
4f6→5d4	1.7952	1.5958	1.8177	1.6155						
4f6→5d6	0.4761	0.4237	0.4822	0.4291						
4f8→5d6	2.1392	1.9075	2.1663	1.9313						
4f6→5g8	1.8079	1.5869	1.7885	1.5731						
4f8→5g8	0.3495	0.3073	0.3457	0.3046						
4f8→5g10	2.0684	1.819	2.0462	1.8032						
5d4→5f6	2.9732	2.8328	2.9687	2.8347						
5d6→5f6	0.8050	0.7670	0.8040	0.7676						
5d6→5f8	3.5922	3.4233	3.5876	3.4252						
6s2→6p2	2.7436	2.5326	2.5640	2.5931						
6s2→6p4	3.3346	3.5052	3.6133	3.6569						
6p2→6d4	3.7253	3.5930	3.7520	3.6089						
6p4→6d4	1.7277	1.6725	1.7399	1.6762						
6p4→6d6	5.1635	4.9966	5.1994	5.0147						
5f6→5g8	5.5926	5.3626								
5f6→5g8	1.0772	1.0330								
5f8→5g10	6.3725	6.1114								
Rn^{7+}	DF	RCC	DF	RCC	DF	RCC	DF	RCC	DF	RCC
6s2→6p2	1.4159	1.1344	1.4210	1.1414						
6s2→6p4	1.9902	1.6147	1.9969	1.6264						
6p2→7s2	0.6184	0.6284	0.6189	0.6678						
6p4→7s2	1.2512	1.2162	1.2657	1.2024						
6p2→6d4	2.0887	1.7562	2.0978	1.7629						
6p4→6d4	1.0493	0.8953	1.0556	0.9008						
6p4→6d6	3.1200	2.6596	3.1380	2.6691						
5f6→6d4	2.3362	2.0900	2.3794	2.0911						
5f6→6d6	0.6118	0.5500	0.6236	0.5478						
5f8→6d6	2.7709	2.4962	2.8234	2.4931						
5f6→5g8	2.7042	2.3829								
5f8→5g8	0.5263	0.4650								
5f8→5g10	3.1174	2.7455								
7s2→7p2	2.5750	2.4195								
7s2→7p4	3.5648	3.3636								
6d4→7p2	2.0098	1.9415								
6d4→7p4	0.7525	0.7381								
6d6→7p4	2.4112	2.3331								

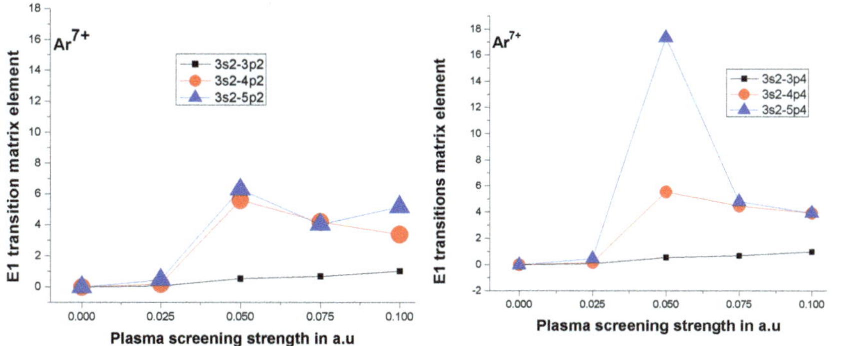

Figure 2. The contribution of plasma screening in E1 matrix elements of Ar^{7+} of transitions $^{2}S_{1/2} \rightarrow {}^{3}P_{1/2,3/2}$. The figures display % variation of the E1 matrix elements vs. plasma screening strength.

For observational astronomy and laboratory spectroscopy, we present a tabulation of a list of our computed oscillator strengths ($f_{R\,CC}$) of $E1$ transitions along with their previously reported theoretical and experimental values in Table 3. Most of the transitions fall in the far and mid-UV regions of the electromagnetic spectrum. f_{RCC} is calculated using the RCC transition amplitudes in length gauge [94] form presented in Table 2 and the NIST [93] wavelengths, wherever available (in the case of Rn^{7+}, our computed RCC wavelengths are used). The ratios between the length and velocity gauge amplitudes of our calculated $E1$ transitions are also displayed in table to show the accuracy of our RCC wavefunctions, which is close to unity for all the cases, confirming the accuracy of our correlated atomic wavefunctions. However, we find that the ratio is almost two for $^2F \rightarrow {}^2D$ transitions of Xe^{7+} and Rn^{7+}. A point to note is that this disagreement is also available in the ratio at the DF level, where we also employed the numerically accurate GRASP92 Code [89]. One of the reasons for this outcome is due to the strong correlation effect from the d- and f-states, as so in similar alkali systems [95]. In addition, the consistency of the accuracy of our calculations can be drawn from the approximate consistency of the ratios 3:2:1 among the transition matrix estimations of $^2P_{3/2} \rightarrow {}^2D_{5/2} : {}^2P_{1/2} \rightarrow {}^2D_{3/2} : {}^2P_{3/2} \rightarrow {}^2D_{3/2}$ [96].

The $E1$ Oscillator strengths for Ar^{7+} are well studied in the literature [31,97–105], and they are in good agreement with our estimations based on the correlation exhaustive RCC method. Table 3 shows that the same is true for Kr^{7+}. For Ar^{7+}, our calculations for f_{RCC} are almost as accurate as those found from other sophisticated theoretical approaches, such as the relativistic many-body perturbation theory [102], and for the most latest theoretical results employing the multiconfigurational Dirac–Hartree–Fock approximation [31]. To the best of our knowledge, in the case of Kr^{7+}, we could not find any correlation-exhaustive many-body result of $E1$ transition. There have been experiments, mostly using beam-foil experiments, on the $E1$ transition from the ground state to the first excited states of Ar^{7+} [97], and Kr^{7+} [49–51,106,107]. Our estimations are well within the uncertainty limit of the latest experiments. We also see that some of the old calculations either underestimate or overestimate the oscillator strength values due to non-appropriate considerations of correlations and relativistic effects.

Table 3. Our RCC oscillator strengths of electric dipole transitions. We compare our results with other estimations available in the recent literature (experimental endeavors are highlighted with "exp" subscript). Our results ("RCC") are obtained using the RCC calculations, except NIST wavelengths are used for $\mu = 0$ wherever available. Transition states are designated with the outermost orbital followed by $(2J + 1)$ of the state. Values at the parenthesis in the second column are ratios between length- and velocity-gauged dipole matrix elements.

	$\mu = 0$		$\mu = 0.025$	$\mu = 0.050$	$\mu = 0.075$	$\mu = 0.1$
Transition	RCC	Other				
Ar^{7+}						
$3s2 \rightarrow 3p2$	0.1857 (0.99)	0.183 (4)$_{exp}^{a_1}$, 0.188 [b1], 0.193 [c1], 0.186 [d1], 0.1864 [e1], 0.185 [f1,i1], 0.187 [g1], 0.196 [h1]	0.1859	0.1875	0.1878	0.1887
$3s2 \rightarrow 3p4$	0.3795 (0.99)	0.398(10)$_{exp}^{a_1}$, 0.385 [b1], 0.394 [c1], 0.381 [d1,g1], 0.3811 [e1], 0.379 [f1], 0.401 [h1], 0.378 [i1]	0.3804	0.3837	0.3840	0.3854
$3s2 \rightarrow 4p2$	0.0418 (1.00)	0.0414 [b1], 0.0401 [c1], 0.0376 [d1], 0.0415 [e1], 0.0385 [h1], 0.0432 [i1]	0.0416	0.0462	0.0445	0.0432
$3s2 \rightarrow 4p4$	0.0803 (1.00)	0.0829 [b1], 0.0766 [c1], 0.0751 [d1], 0.0798 [e1], 0.0739 [h1], 0.0836 [i1]	0.0793	0.0877	0.0859	0.0838
$3s2 \rightarrow 5p2$	0.0143 (0.98)	0.0146 [b1,e1]	0.0141	0.0123	0.0149	0.0121
$3s2 \rightarrow 5p4$	0.0271 (1.00)	0.0292 [b1], 0.0284 [e1]	0.0270	0.0184	0.0291	0.0278
$4s2 \rightarrow 4p2$	0.2810 (1.01)	0.2821 [e1], 0.2824 [i1]	0.2829	0.2871	0.2816	0.2777
$4s2 \rightarrow 4p4$	0.5736 (1.01)	0.5755 [e1], 0.5759 [i1]	0.5782	0.5883	0.5748	0.5645
$3p2 \rightarrow 4s2$	0.0880 (1.00)	0.0876 [e1], 0.0884 [i1]	0.0879	0.0765	0.0763	0.0760
$3p4 \rightarrow 4s2$	0.0900 (1.01)	0.0896 [e1], 0.08947 [i1]	0.0899	0.0780	0.0783	0.0783
$3p2 \rightarrow 5s2$	0.0163 (1.01)	0.0161 [e1]	0.0163	0.0162	0.0128	0.0154
$3p4 \rightarrow 5s2$	0.0166 (1.01)	0.0164 [e1]	0.0166	0.0202	0.0133	0.0130

Table 3. *Cont.*

Transition	$\mu = 0$ RCC	$\mu = 0$ Other	$\mu = 0.025$	$\mu = 0.050$	$\mu = 0.075$	$\mu = 0.1$
3p2 → 3d4	0.5074 (0.96)	0.532 [c1], 0.5097 [e1], 0.5074 [g1] 0.508 [i1], 0.47 [j1]	0.5091	0.5107	0.5115	0.5180
3p4 → 3d4	0.0502 (0.96)	0.0527 [c1], 0.0504 [e1], 0.0501 [g1] 0.0502 [i1], 0.046 [j1]	0.0503	0.0505	0.0505	0.0512
3p4 → 3d6	0.4556 (0.96)	0.475 [c1], 0.4539 [e1], 0.4517 [g1] 0.452 [i1], 0.42 [j1]	0.4534	0.4542	0.4549	0.4571
3p2 → 4d4	0.1326 (1.05)	0.1310 [e1], 0.1344 [i1]	0.1312	0.1291	0.1275	0.1024
3p4 → 4d4	0.0137 (1.03)	0.0135 [e1], 0.0136 [i1]	0.0135	0.0131	0.0132	0.0109
3p4 → 4d6	0.1226 (1.03)	0.1212 [e1], 0.1228 [i1]	0.1214	0.1180	0.1187	0.1159
4p2 → 4d4	0.8067 (0.95)	0.8085 [i1]	0.8070	0.8020	0.7930	0.8035
4p4 → 4d4	0.0798 (0.98)	0.0796 [i1]	0.0797	0.0798	0.0786	0.0787
4p4 → 4d6	0.7192 (0.98)	0.7177 [i1]	0.7185	0.7215	0.7079	0.7009
3d4 → 4p2	0.0663 (0.98)	0.0657 [e1], 0.0663 [i1]	0.0663	0.0695	0.0630	0.0678
3d4 → 4p4	0.0130 (1.00)	0.0129 [e1], 0.0131 [i1]	0.0130	0.0133	0.0124	0.0134
3d6 → 4p4	0.0784 (0.97)	0.0778 [e1], 0.0786 [i1]	0.0784	0.0762	0.0747	0.0720
3d4 → 4f6	0.8812 (1.00)	0.8776 [i1]	0.8777	0.8702	0.8584	0.8762
3d6 → 4f6	0.0424 (1.00)	0.0418 [i1]	0.0418	0.0415	0.0409	0.0401
3d6 → 4f8	0.8397 (1.00)	0.8360 [i1]	0.8364	0.8297	0.8149	0.7796
Kr^{7+}						
4s2 ⟶ 4p2	0.2543 (1.00)	0.25(1) [a2,f2]$_{exp}$, 0.24(2) [b2]$_{exp}$, 0.246 [c2], 0.2781 [d2], 0.278 [e2], 0.28 [g2], 0.2578 [h2], 0.220 [i2], 0.2448 [j2]	0.2633	0.2718	0.2705	0.2673
4s2 ⟶ 4p4	0.5466 (0.97)	0.53(2) [a2]$_{exp}$, 0.47(4) [b2]$_{exp}$, 0.526 [c2], 0.5965 [d2], 0.60 [e2], 0.59 (9) [f2]$_{exp}$, 0.59 [g2], 0.554 [h2], 0.473 [i2] 0.5265 [j2]	0.5661	0.5825	0.5833	0.5829
4s2 ⟶ 5p2	0.0227 (1.01)	0.0176 [d2]	0.0206	0.0191		
4s2 ⟶ 5p4	0.0354 (1.00)	0.0265 [d2]	0.0313	0.0275		
4p2 → 5s2	0.1281 (1.02)	0.1212 [d2]	0.1261	0.1187		
4p4 → 5s2	0.1392 (1.02)	0.1321 [d2]	0.1351	0.1302		
4p2 → 4d4	0.9888 (1.00)	1.057 [d2]	1.0562	1.0583	1.1068	
4p4 → 4d4	0.0972 (1.00)	0.1038 [d2]	0.0998	0.1015	0.1045	
4p4 → 4d6	0.8796 (1.00)	0.9395 [d2]	0.9051	0.9200	0.9480	
4p2 → 6s2	0.0237 (1.06)	0.0208 [d2]	0.0227	0.0181		
4p4 → 6s2	0.0253 (1.07)	0.0220 [d2]	0.0243	0.0197		
4d4 → 4f6	1.0811 (0.98)	1.126 [d2]	1.0876	1.0890		
4d6 → 4f6	0.0514 (0.98)	0.0535 [d2]	0.0508	0.0509		
4d6 → 4f8	1.0277 (0.98)	1.0700 [d2]	1.0326	1.0340		
4d4 → 5p2	0.1828 (1.00)	0.1784 [d2]	0.1781	0.1697		
4d4 → 5p4	0.0352 (1.03)	0.0344 [d2]	0.0345	0.0329		
4d6 → 5p4	0.2135 (1.00)	0.2088 [d2]	0.2078	0.1984		
4d4 → 5f6	0.0001 (1.09)	0.0001 [d2]	0.0005			
4f6 → 5d4	0.1647 (0.98)	0.1638 [d2]	0.1582	0.1485		
4f6 → 5d6	0.0117 (0.98)	0.0117 [d2]	0.0113	0.0106		
4f8 → 5d6	0.1759 (0.98)	0.1750 [d2]	0.1689	0.1582		
4f6 → 5g8	1.2447 (1.00)	1.261 [d2]	1.2185			
4f8 → 5g8	0.0346 (1.00)	0.0350 [d2]	0.0339			
4f8 → 5g10	1.2103 (1.00)	1.226 [d2]	1.1841			
5p2 → 5d4	1.4023 (1.00)	1.401 [d2]	1.3825	1.3467		
5p4 → 5d4	0.1393 (1.01)	0.1383 [d2]	0.1369	0.1335		
5p4 → 5d6	1.2582 (1.01)	1.249 [d2]	1.2391	1.2062		
4d6 → 5f6	0.00002 (0.28)	0.00003 [d2]	0.00001			
4d6 → 5f8	0.0003 (0.29)	0.0007 [d2]	0.0003			
5p2 → 6s2	0.2322 (1.07)	0.2023 [d2]	0.2229			
5p4 → 6s2	0.2529 (1.08)	0.2179 [d2]	0.2422			
Xe^{7+}						
5s2 → 5p2	0.2436 (1.01)	0.294 [a3], 0.234 [b3], 0.242 [c3] 0.253 [d3], 0.237 [e3], 0.237 [f3] 0.232 [g3], 0.223 [h3], 0.232 [i3]	0.2471	0.2482		

Table 3. *Cont.*

Transition	RCC	Other ($\mu = 0$)	$\mu = 0.025$	$\mu = 0.050$	$\mu = 0.075$	$\mu = 0.1$
5s2 → 5p4	0.5724 (1.01)	0.697 [a3], 0.550 [b3], 0.569 [c3] 0.596 [d3], 0.560 [e3], 0.563 [f3], 0.543 [g3], 0.522 [h3], 0.537 [i3]	0.5801	0.5816		
5p2 →5d4	1.0195 (1.02)	1.189 [a3], 0.977 [b3], 1.020 [c3] 1.025 [d3], 1.003 [e3], 1.000 [f3] 1.057 [i]	1.0204			
5p4 →5d4	0.0992 (1.02)	0.095 [b3], 0.089 [c3], 0.099 [d3] 0.097 [e3], 0.097 [f3], 0.095 [i3]	0.0993			
5p4 →5d6	0.9064 (1.02)	0.523 [a3], 0.868 [b3], 0.904 [c3] 0.907 [d3], 0.889 [e3], 0.886 [f3] 0.875 [i3]	0.9066			
5p2 → 6s2	0.1539 (1.05)	0.160 [c3], 0.156 [d3], 0.155 [e3] 0.153 [f3], 0.199 [i3]	0.1539			
5p4 → 6s2	0.1816 (1.05)	0.188 [c3], 0.186 [d3], 0.184 [e3] 0.182 [f3], 0.186 [i3]	0.1817			
4f6 → 5d4	0.0577 (2.35)	0.130 [a3], 0.058 [b3], 0.060 [i3]	0.0560			
4f6 → 5d6	0.0043 (2.12)	0.0044 [b3]	0.0042			
4f8 → 5d6	0.0651 (2.19)	0.075 [a3], 0.065 [b3], 0.068 [i3]	0.0633			
4f6 → 5g8	0.3890 (1.02)	0.3646 [b3], 0.354 [i3]				
4f8 → 5g8	0.0109 (1.02)	0.0102 [b3]				
4f8 → 5g10	0.3826 (1.02)	0.3595 [b3], 0.343 [i3]				
5d4 → 5f6	1.1437 (1.04)	1.099 [i3]	1.1259			
5d6 → 5f6	0.0550 (1.04)	0.052 [i3]	0.0813			
5d6 → 5f8	1.0981 (1.04)	1.032 [i3]	1.0812			
5f6 → 5g8	1.0583 (0.99)	1.071 [i3]				
5f8 → 5g8	0.0293 (0.99)	0.030 [i3]				
5f8 → 5g10	1.0246 (0.99)	1.035 [i3]				
6s2 → 6p2	0.4778 (1.13)		0.4921			
6s2 → 6p4	1.0555 (1.13)		1.1260			
6p2 → 6d4	1.6248 (1.11)		1.5953			
6p4 → 6d6	1.4527 (1.11)		1.4249			
6p4 → 6d4	0.1601 (1.11)		0.1566			
Rn^{7+}						
6s2 → 6p2	0.2372 (1.00)	0.234 [a4]	0.2379			
6s2 → 6p4	0.6880 (1.02)	0.689 [a4]	0.6922			
6p2 → 7s2	0.1634 (1.06)	0.173 [a4]				
6p4 → 7s2	0.2472 (1.07)	0.259 [a4]				
6p2 →6d4	1.0154 (1.01)	1.059 [a4]	1.0113			
6p4 →6d4	0.1001 (1.01)	0.103 [a4]	0.0999			
6p4 →6d6	0.9234 (1.01)	0.956 [a4]	0.9170			
5f6 → 6d4	0.0660 (2.13)	0.0753 [a4]	0.0578			
5f8 → 6d6	0.0842 (1.83)	0.0981 [a4]	0.0755			
5f6 → 6d6	0.0057 (1.73)	0.0069 [a4]	0.0051			
5f6 → 5g8	0.6982 (1.00)					
7s2 → 7p2	0.4182 (1.11)	0.414 [a4]				
7s2 → 7p4	1.1727 (1.12)	1.126 [a4]				
6d4 → 7p2	0.2940 (1.05)	0.292 [a4]				
6d4 → 7p4	0.0513 (1.04)	0.052 [a4]				
6d6 → 7p4	0.3207 (1.05)	0.33 [a4]				

$a_1 \implies$ Beam-foil technique [97]; $b_1 \implies$ calculations are based on high level methods such as the R-matrix method and asymptotic techniques developed by Seaton [98]; $c_1 \implies$ Single Configuration Interaction Hartree–Fock method using a pseudopotential [99]; $d_1 \implies$ non-relativistic WKB approaches (Klein–Gordon dipole matrix) [100]; $e_1 \implies$ single configuration Dirac–Fock method [101]; $f_1 \implies$ relativistic many-body perturbation theory [102]; $g_1 \implies$ realistic model potential [103]; $h_1 \implies$ relativistic Hartree–Fock method [104]; $i_1 \implies$ multiconfiguration Dirac–Hartee-Fock approximation [31]; $j_1 \implies$ relativistic effective orbital quantum number [105]; $a_2 \implies$ jointly analyzed decay curves: beam-foil [106]; $b_2 \implies$ multiexponential fits: beam-foil [106]; $c_2 \implies$ Non-Relativistic Multi Configuration Hartree–Fock approximation [108]; $d_2 \implies$ relativistic Hartree–Fock [37]; $e_2 \implies$ Hartree-Fock oscillator strength using the Dirac correction factor [109]; $f_2 \implies$ Arbitrarily Normalized Decay curve method for cascade-correction in beam-foil [107]; $g_2 \implies$ Hartree–Fock with relativistic correction [110]; $h_2 \implies$ semi-empirical Coulomb approximation [111]; $i_2 \implies$ model potential [112]; $j_2 \implies$ Hartree-Slater method [23]; $a_3 \implies$ RPTMP [57]; $b_3 \implies$ RMBPT(3) [55]; $c_3 \implies$ DF+CP [54]; $d_3 \implies$ DX+CP method with SCE model potential [54]; $e_3 \implies$ DX+CP method with CAFEGE model potential [54]; $f_3 \implies$ DX+CP method with HFEGE model potential [54]; $g_3 \implies$ CIDF method with integer occupation number [56]; $h_3 \implies$ CIDF(q) method with non-occupation number [56]; $i_3 \implies$ HFR+CP method [53]; $a_4 \implies$ relativistic core-polarization corrected Dirac–Fock method (DF+CP) [63].

Over the last two decades, a few of the low-lying $E1$ transitions of Xe^{7+} are estimated using core-polarization or model potential as an effective means of correlation calculations, apart from third-order perturbation calculations [55]. It is known that our RCC method is an all-order extension of many-body perturbation theory [83]. Further, it includes most of the correlation features, including core correlation, pair correlation, and higher order correlation effects [69] for a given level of excitation. For Rn^{7+}, we find only one theoretical endeavor [63] using the model potential. The presence of d- and f-orbitals for Xe^{7+} and Rn^{7+} ions in the core makes these two ions highly correlated. Because of the large atomic number and highly stripped configurations, we expect a strong relativistic effect in their spectroscopy. Therefore, it is necessary to do relativistic ab initio correlation exhaustive calculations for them and our computations exactly mitigate that requirement. In Table 3, we also present the effect of the plasma atmosphere on the oscillator strengths for observational and laboratory spectroscopy. The oscillator strengths for $\mu > 0$ are calculated using the $E1$ matrix elements presented in Table 2 and the corresponding transition wavelengths computed from RCC theory. The table exhibits the significant effects of plasma screening parameters on the oscillator strengths.

Tables 4 and 5 present transition probabilities for the relatively strong forbidden transitions governed by the electric quadruple ($E2$) and the magnetic dipole ($M1$) moments. Similar to oscillator strength in Table 3, here we use NIST wavelengths for the transition probability wherever available. For Rn^{7+}, we use the RCC calculated transition wavelengths. We do not find any estimation of the forbidden transitions in the literature of this ionic series which fall either in the ultraviolet or in the near infra-red regions of the electromagnetic spectrum. Transitions falling in the ultraviolet region are significant in astronomical observation and plasma research [113–116]. In comparison, the infra-red transitions have applications in astronomy using space-based telescopes ([117]). Moreover, infrared spectroscopy provides major information about cool astronomical regions in space, such as interstellar medium [118] and planetary nebulae [119]. It is found that $5p_{1/2} \rightarrow 4f_{5/2}$ of Kr^{7+} and $5p_{1/2} \rightarrow 5p_{3/2}$ of Xe^{7+} emit orange and green lights, respectively, which can be used in laser spectroscopy [120–124].

It is found from Table 5 that the $M1$ transition probability is stronger among fine-structure levels than the $E2$ transition. Table 4 reveals that the maximum $M1$ transition probability, A_{RCC}^{M1}, occurs for the transition $3p_{1/2} \rightarrow 3p_{3/2}$ of Ar^{7+}, $4p_{1/2} \rightarrow 4p_{3/2}$ of Kr^{7+}, $4f_{7/2} \rightarrow 5f_{7/2}$ of Xe^{7+} and $6p_{1/2} \rightarrow 6p_{3/2}$ of Rn^{7+}, and they have values of 0.17949, 8.4000, 158.33 and 1137.7, respectively. Moreover, our estimations of $M1$ transition probability for the $4f_{5/2} \rightarrow 4f_{7/2}$ transition of Xe^{7+} has excellent agreement with the calculations using the multi-configuration Dirac–Hartree–Fock method [125,126]. Table 6 presents the lifetime of the low-lying states of this series. We compare our results with other experimental and theoretical estimations wherever available and find good agreement with the recent endeavors. We provide lifetimes of many excited states calculated for the first time in the literature to our knowledge.

The comparisons of our computed results with the other estimations obtained from correlation exhaustive ab initio theoretical computations or precise experiments are one of the measures of accuracy of our calculations. Further, the differences between the calculated matrix elements in the length and velocity gauge forms are characteristic of the preciseness of our calculations. A recent piece of literature [95] also claims that the difference in length gauge and velocity gauge is a measure of accuracy. Another factor of accuracy in ab initio calculations arises from the DF wavefunctions used for correlation calculations. In addition, we should consider the uncertainty that arises from the other correlation terms (which we did not consider in this article) and the quantum electrodynamics effect, which is at most 2% in total. Taking all these into account, the maximum calculated uncertainties for Ar^{7+}, Kr^{7+}, Xe^{7+}, and Rn^{7+} are about 5.6%, 5.37%, 5% and 5.01%, respectively.

To understand the critical effect of the plasma atmosphere on the ionization potential of the ions, we highlight the IPD values in bold fonts in Table 1 for different values of screening length, μ. These IPD values reflect critical electron or plasma density at a particular temperature for the ionic system when a few of the bound ionic states are elevated to continuum states.

Table 1 also reveals that the fine structure splittings (FSS) are suppressed as the screening strength increases from $\mu = 0$ to 1.0. For example, the energy differences between $4p_{3/2}$ and $4p_{1/2}$ of Kr^{7+} are evaluated as 9774 a.u, 9732 a.u, 9608 a.u., and 9407 a.u. for $\mu = 0$, 0.025, 0.05, and 0.075 a.u., respectively. This phenomenon is consistent with earlier calculations for sodium D line [127] and hydrogen-like atoms [128]. The suppression of the transition rate among the fine-structure levels is mainly arising from the energy quench.

From Figure 3, we pictorially estimate the critical values of plasma screening strength (μ_c) where the ionization potential becomes zero for a particular atomic state. We also tabulate these values in Table 7. The critical screening strength is essential in photo-ionization cross-section, which increases with increasing μ until $\mu = \mu_c$. This increment is obvious due to the decrease in bound state energy leading to the increase in radial expansion of the bound state wavefunction [128]. This phase shift of bound state to continuum state is induced by the plasma atmosphere, and the ionization threshold decreases with the Debye screening length (μ^{-1}). In terms of the photo-ionization cross-section [129], the plasma decreases the threshold cross-section, and the discrete bound wavefunctions become diffused. Therefore, critical screening strength plays an important role in atomic structure. However, we have not found any spectroscopic data in the literature for these ions in plasma medium to compare with our results.

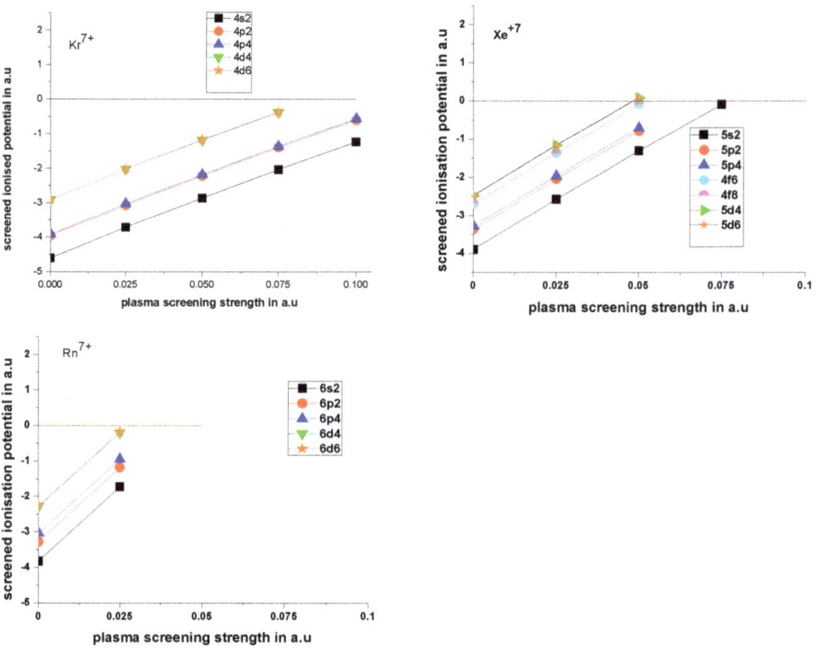

Figure 3. Determination of the value of critical plasma screening strength.

Table 4. Transition rate (in s^{-1}) of E2 (A_{RCC}^{E2}) in plasma screened and unscreened medium. Here, we have used our RCC matrix element (in a.u) and RCC wavelength (in Å). Note that the notation $P(Q)$ in the case of transition rates means $P \times 10^Q$.

Transitions	$\mu = 0$	$\mu = 0.025$	$\mu = 0.050$	$\mu = 0.075$	$\mu = 0.1$
Ar^{7+}					
3p4 → 4f6	1.4501 (+06)	1.4349 (+06)	1.3927 (+06)	1.3250 (+06)	1.2333 (+06)
3p4 → 4f8	6.5288 (+06)	6.4605 (+06)	6.2705 (+06)	5.9658 (+06)	5.5526 (+06)
3s2→ 3d4	2.1840 (+05)	2.2032 (+05)	2.2179 (+05)	2.2417 (+05)	2.2742 (+05)
3s2⟶ 3d6	2.1919 (+05)	2.2126 (+05)	2.2256 (+05)	2.2494 (+05)	2.2819 (+05)
3d4 → 5g8	2.7225 (+06)	2.6258 (+06)	2.3685 (+06)	1.9874 (+06)	
3d6 → 5g8	3.0252 (+05)	2.9179 (+05)	2.6352 (+05)	2.2077 (+05)	
3d6 → 5g10	3.0267 (+06)	2.9190 (+06)	2.6362 (+06)	2.2085 (+06)	
Kr^{7+}					
4s2⟶ 4d4	8.9514 (+05)	8.7868 (+05)	8.6208 (+05)	8.3473 (+05)	
4s2⟶ 4d6	9.1180 (+05)	8.9640 (+05)	8.7912 (+05)	8.5066 (+05)	
5p2 → 4f6	7.0537 (−01)	8.9779 (−01)	3.0592 (00)		
5p4 → 4f8	2.3536 (−01)	2.9840 (−01)	1.4702 (00)		
5d6 → 5g8	2.7840 (+03)	2.6100 (+03)			
5d6 → 5g10	2.9352 (+04)	2.6114 (+04)			
Xe^{7+}					
5s2→ 5d4	6.6714 (+05)	6.7116 (+05)			
5s2→ 5d6	6.9674 (+05)	7.0069 (+05)			
5p2 → 5p4	6.0045 (−01)	6.1555 (−01)	5.6058 (−01)		
5p2 → 4f6	8.5467 (+03)	2.7288 (+04)	3.3098 (+04)		
5p4 → 4f6	1.2882 (+03)	2.0657 (+03)	2.6049 (+03)		
5p4 → 4f8	5.9866 (+03)	1.2836 (+04)	1.6523 (+04)		
5p2→ 5f6	1.6471 (+06)	1.5379 (+06)			
5p4→ 5f6	4.3300 (+05)	4.0395 (+05)			
5p4→ 5f8	1.9479 (+06)	1.8169 (+06)			
4f6 → 5f6	5.9296 (+04)	5.4103 (+04)			
4f6 → 5f8	7.4168 (+03)	6.7680 (+03)			
4f8 → 5f6	9.9224 (+03)	9.0350 (+03)			
4f8 → 5f8	6.2129 (+04)	5.6576 (+04)			
5d4→ 6s2	5.3813 (+03)				
5d4 → 5g8	1.1670 (+06)				
5d6 → 5g8	1.2597 (+05)				
5d6→ 5g10	1.2601 (+06)				
5d6→ 6s2	7.0259 (+03)				
Rn^{7+}					
6s2 → 6d4	1.0243 (+06)	9.8377 (+05)			
6s2 → 6d6	1.0371 (+06)	9.9749 (+05)			
6p2 → 5f6	5.0885 (+04)	5.4520 (+04)			
6p4 → 5f6	3.2019 (+03)	3.5290 (+03)			
6p4 → 5f8	1.5773 (+04)	1.7127 (+04)			
6d4 → 5g8	6.3782 (+05)				
6d6 → 5g8	6.2707 (+04)				
6d6 → 5g10	6.2238 (+05)				
6p2 → 6p4	1.2855 (+02)	1.2763 (+02)			
7p2 → 7p4	1.5299 (+01)				
6p4 → 7p4	2.3245 (+05)				
5f6 → 7p2	1.4611 (+04)				
5f6 → 7p4	3.0317 (+03)				
5f8 → 7p4	1.7756 (+04)				

Table 5. Magnetic dipole transition rate (in s^{-1}) in plasma screened and unscreened medium. Note: the notation $P(Q)$ in the case of transition rates means $P \times 10^Q$. For $4f_{5/2} \to 4f_{7/2}$ (Xe^{7+}), transition rates 1.9227 (−03) and 1.9277 (−03) are available in the literature (a) using the multiconfiguration Dirac–Fock method without and with Breit interaction plus the quantum electrodynamics effect, respectively.

Transitions	$\mu = 0$	$\mu = 0.025$	$\mu = 0.050$	$\mu = 0.075$	$\mu = 0.1$
Ar^{7+}					
3p2 → 3p4	1.7951 (−01)	1.9857 (−01)	1.9603 (−01)	1.9165 (−01)	1.8570 (−01)
3p4 → 4f6	3.7845 (−02)	3.7248 (−02)	3.5575 (−02)	3.2958 (−02)	2.9561 (−02)
3d6 → 5g8	1.0634 (−02)	1.0124 (−02)	8.8719 (−03)	7.0637 (−03)	
Kr^{7+}					
4s2→ 4d4	9.0675 (−03)	7.2173 (−03)	8.2040 (−03)	6.7390 (−03)	
4p2 → 4p4	8.4015 (00)	8.2789 (00)	7.9664 (00)	7.4776 (00)	6.8397 (00)
4d4 → 4d6	2.5436 (−02)	3.2234 (−02)	3.0177 (−02)	2.7000 (−02)	
5p2 → 5p4	4.7841 (−01)	5.3517 (−01)	4.6911 (−01)		
5d4 → 5d6	1.7080 (−03)	1.0904 (−03)	2.2682 (−03)		
Xe^{7+}					
5p2 → 5p4	5.7437 (+01)	5.7875 (+01)	5.3500 (+01)		
5p4 → 5f6	6.3588 (−03)	5.7557 (−03)			
4f6 → 5f6	5.0459 (+01)	4.9605 (+01)			
4f6 → 5f8	4.9212 (−01)	5.3389 (−01)			
4f8 → 5f6	1.1142 (+01)	1.0620 (+01)			
4f8 → 5f8	1.5833 (+02)	1.5517 (+02)			
5d4 → 5d6	2.7070 (−01)	2.6538 (−01)	2.2900 (−01)		

Table 5. *Cont.*

Transitions	$\mu = 0$	$\mu = 0.025$	$\mu = 0.050$	$\mu = 0.075$	$\mu = 0.1$
Rn^{7+}					
6p2 → 6p4	1.2442 (+03)	1.2212 (+03)			
6d4 → 6d6	4.5873 (00)	4.1847 (00)			
5f6 → 5f8	6.6129 (−02)	3.4422 (−02)			
6s2 → 7s2	1.0113 (00)				
7p2 → 7p4	8.1938 (+01)				
6p2 → 7p2	7.2065 (−02)				
6p4 → 7p4	2.4557 (00)				
5f6 → 7p4	2.4281 (−03)				

[125,126].

Table 6. Lifetimes in ns of few low-lying states.

Level	Present Work	Other Work (Experiment)	Other Work (Theory)
Ar^{7+}			
3p2	0.411	0.417 ± 0.010 [a], 0.423 ± 0.040 [b], 0.48 ± 0.05 [g], 0.49 ± 0.05 [h] 0.55 ± 0.03 [l], 0.53 ± 0.11 [m]	0.413 [c], 0.407 [d], 0.397 [e], 0.409 [f] 0.389 [i], 0.408 [j], 0.4121 [k]
3p4	0.387	0.389 ± 0.010 [a], 0.421 ± 0.030 [b], 0.428 ± 0.027 [g], 0.48 ± 0.06 [h] 0.54 ± 0.02 [l], 0.527 ± 0.018 [m]	0.389 [c], 0.382 [d], 0.373 [e], 0.386 [f] 0.366 [i], 0.388 [j], 0.3872 [k]
3d4	0.132	0.170 ± 0.010 [a], 0.130 ± 0.005 [b], 0.158 ± 0.008 [g],	0.127 [e], 0.134 [f], 0.133 [j], 0.1318 [k]
3d6	0.137	0.166 ± 0.008 [a], 0.131 ± 0.005 [b], 0.160 ± 0.008 [g]	0.131 [e], 0.138 [f,j], 0.1361 [k]
4f6	0.003		
4f8	0.002		
Kr^{7+}			
4p2	0.293	0.41 ± 0.04 [n], 0.291 ± 0.012 [o] 0.290 ± 0.015 [g], 0.401 ± 0.018 [l]	0.282 [p], 0.29653 [q]
4p4	0.235	0.33 ± 0.03 [n], 0.243 ± 0.01 [o] 0.218 ± 0.033 [g], 0.331 ± 0.011 [l]	0.230 [p], 0.24176 [q]
4d4	0.048		0.05019 [q]
4d6	0.052	0.048 ± 0.004 [o]	0.05388 [q]
4f6	0.055		
4f8	0.055		
Xe^{7+}			
5p2	0.45	0.52 (3) [r], 0.50 ± 0.05 [s], 0.380 ± 0.040 [g]	0.37 [t], 0.47 [u], 0.48 [v], 0.53 [w]
5p4	0.29	0.35 (2) [r], 0.33 ± 0.03 [s], 0.272 ± 0.037 [g]	0.23 [t], 0.30 [u], 0.31 [v], 0.33 [w]
5d4	0.08	0.10 (2) [r]	0.07 [t], 0.07 [v], 0.06 [w]
5d6	0.08	0.14 (2) [r]	0.14 [t], 0.08 [v], 0.07 [w]
Rn^{7+}			
6p2	0.429		
6p4	0.144		
6d4	0.056		
6d6	0.082		

$a \implies$ Beam-foil technique [97]; $b \implies$ [130]; $c \implies$ third-order many-body perturbation theory [131]; $d \implies$ R-matrix theory [98]; $e \implies$ single Configuration interaction Hartree–Fock method using a pseudo potential [99]; $f \implies$ relastic model potential [103]; $g \implies$ Arbitrarily Normalized Decay curve method for cascade-correction in beam-foil [107]; $h \implies$ beam-foil technique in the vacuum u.v [132]; $i \implies$ Multiconfiguration Dirac–Fock method [104]; $j \implies$ charge expansion technique [133]; $k \implies$ multiconfiguration Dirac–Hartree–Fock theory including core polarization [31]; $l \implies$ beam-foil [50]; $m \implies$ beam-foil [134]; $n \implies$ beam-foil [49]; $o \implies$ foil excitation [106]; $p \implies$ Coulomb approximation [111]; $q \implies$ Hartree–Slater method [23]; $r \implies$ beam-foil spectroscopy [53]; $s \implies$ relativistic Hartree–Fock method [135]; $t \implies$ relativistic perturbation theory with a zero approximation model potential [57]; $u \implies$ relativistic many-dody perturbation theory(RMBPT(3)) [55]; $v \implies$ relativistic HFR+CP [53]; $w \implies$ relativistic MCDF [53].

Table 7. Critical values of plasma screening strength (μ_c) in a.u. for the following ions.

Kr^{7+}		Xe^{7+}		Rn^{7+}	
State	μ_c	State	μ_c	State	μ_c
4d4	0.087152	5s2	0.076987	5s2	0.046401
4d6	0.086956	5p2	0.066149	6p2	0.039652
		5p4	0.064474	6p4	0.036763
		4f6	0.051526	6d4	0.027521
		4f8	0.051486	6d6	0.027100
		5d4	0.048395		
		5d6	0.048139		

4. Conclusions

The continuous progress in astrophysical and astronomical observations demands accurate theoretical transition data in a realistic environment. In many cases, the experiment is difficult to extract the data used to estimate the abundance of the ions in the stellar chemical composition. Here, the highly correlated relativistic coupled-cluster theory is applied to precisely determine the excitation energies of a few low-lying states of astrophysically relevance such as Ar^{7+}, Kr^{7+} and Xe^{7+}, and Rn^{7+}. Furthermore, we calculate various properties of allowed and forbidden transitions, such as transition probabilities, oscillator strengths, and lifetimes, and compare them with previously reported data in the literature. We found an overall good agreement between our results with the other theoretical and experimental results. Moreover, the concurrence between the length and velocity gauge allowed transition amplitudes signifies the exact calculations of our correlated wavefunctions. We found that most of the transitions shown here fall in the ultraviolet region of the electromagnetic spectrum, useful for astrophysical plasma research and telescope-based astronomy. A few transitions, such as $4d_{3/2} - 4f_{5/2}$, $4d_{5/2} - 4f_{5/2}$ and $4d_{5/2} - 4f_{7/2}$ of Ar^{7+}, $5p_{1/2} - 4f_{5/2}$ of Kr^{7+} and $5p_{1/2} - 5p_{3/2}$ of Xe^{7+} emit the visible light, which can have application in laser spectroscopy. Our presented transition line parameters of Rn^{7+} may help the astronomer identify the ion's unknown lines. To the best of our knowledge, some of the oscillator strengths of allowed transitions and most of the transition rates of the forbidden transitions are reported here for the first time in the literature.

The main focus of this paper is to evaluate the above spectroscopic properties under a realistic astronomical atmosphere. We showed that the variation of our results for different values of Debye screening lengths and ionization potential depression values for each atomic state are useful for atomic structure characterization.

Author Contributions: S.B. and S.M. have visualize and conceptualize the problem, S.B. has augmented the code for plasma environment, did former analysis, and data curation. S.B., A.D., A.B., R.R.P. and S.M. worked on the manuscript writing. S.M. has supervised the total work. All authors have read and agreed to the published version of the manuscript.

Funding: This research received no external funding.

Data Availability Statement: Not applicable.

Conflicts of Interest: The authors declare no conflict of interest.

References

1. Barlow, M.J.; Swinyard, B.M.; Owen, P.J.; Cernicharo, J.; Gomez, H.L.; Ivison, R.J.; Krause, O.; Lim, T.L.; Matsuura, M.; Miller, S.; et al. Detection of a Noble Gas Molecular Ion $^{36}ArH^+$. *Science* **2013**, *342*, 1343–1345. [CrossRef]
2. Güsten, R.; Wiesemeyer, H.; Neufeld, D.; Menten, K.; Graf, U.; Jacobs, K.; Klein, B.; Ricken, O.; Risacher, C.; Stutzki, J. Astrophysical detection of the helium hydride ion HeH^+. *Nature* **2019**, *568*, 357–359. [CrossRef] [PubMed]
3. Benna, M.; Mahaffy, P.R.; Halekas, J.S.; Elphic, R.C.; Delory, G.T. Variability of helium, neon, and Argom in the lunar exosphere as observed by the LADEE NMS instrument. *Geophys. Res. Lett.* **2015**, *42*, 3723. [CrossRef]
4. Müller, H.S.P.; Muller, S.; Schilke, P.; Bergin, E.A.; Black, J.H.; Gerin, M.; Lis, D.C.; Neufeld, D.A.; Suri, S. Detection of extragalatic argonium ArH^+, towards PKS 1830-211. *Astron. Astrophys.* **2015**, *582*, L4. [CrossRef]
5. Schilke, P.; Neufeld, D.; Mueller, H.; Comito, C.; Bergin, E.; Lis, D.; Gerin, M.; Black, J.H.; Wolfire, M.; Indriolo, N.; et al. Ubiquitous argonium (ArH^+) in the diffuse interstellar medium: A molecular tracer of almost purely atomic gas. *Astron. Astrophys.* **2014**, *566*, A29. [CrossRef]
6. Taresch, G.; Kudritzki, R.P.; Hurwitz, M.; Bowyer, S.; Pauldrach, A.W.A.; Puls, J.; Butler, K.; Lennon, D.J.; Haser, S.M. Quantitative analysis of the FUV, UV and optical spectrum of the O3 star HD 93129A. *Astron. Astrophys.* **1997**, *321*, 531–548.
7. Pérez-Montero, E.; Hägele, G.F.; Contini, T.; Díaz, Á.I. Neon and Argon optical emission lines in ionized gaseous nebulea: Implications and aplications. *Mon. Not. R. Astron. Soc.* **2007**, *381*, 125–135. [CrossRef]
8. Panigrahy, S.N.; Dougherty, R.W.; Das, T.P.; Andriessen, J. Theory of hyperfine interactions in lithiumlike systems. *Phys. Rev. A* **1989**, *40*, 1765. [CrossRef]
9. Owusu, A.; Yuan, X.; Panigrahy, S.N.; Doughetry, R.W.; Das, T.P.; Andriessen, J. Theory of hyperfine interactions in potassium and Sc^{2+} Sion: Trends in system isoelectronic with Potassium. *Phys. Rev. A* **1977**, *55*, 2644. [CrossRef]
10. Xiao, L.W.; Kai-Zhi, Y.; Bing-Cong, G.; Meng, Z. Calculation on the hyperfine constants of the ground states for Lithium-like system. *Chin. Phys.* **2007**, *16*, 2389. [CrossRef]

11. Roy, S.N.; Rodgers, J.E.; Das, T.P. Hyperfine interactions in stripped atoms isoelectronic with alkali atoms. *Phys. Rev. A* **1976**, *13*, 1983. [CrossRef]
12. Dutta, N.N.; Majumder, S. Electron-correlation trends in the hyperfine A and B constant of Na isoelectronic sequence. *Phys. Rev. A* **2013**, *88*, 062507. [CrossRef]
13. Herrnstein, J.R.; Moran, J.M.; Greenhill, L.J.; Dimond, P.J.; Inoue, M.; Nakai, N.; Miyoshi, M.; Henkel, C.; Riess, A. A geometric distance to the Galaxy NGC4258 from orbital motions in a nuclear gas disc. *Nature* **1999**, *400*, 539–541. [CrossRef]
14. Dimitrijević, M.S. Stark broadening in astrophysics. *Astron. Astrophys. Trans.* **2003**, *22*, 389–412. [CrossRef]
15. Eldén, B. Die deutung der Emissionslinien im spektrum der sonnenkorona. *Z. Astrophys.* **1942**, *22*, 30.
16. Eidelsberg, M.; Crifo-Magnant, F.; Zeippen, C.J. Forbidden lines in hot astronomical sources. *Astron. Astrophys. Suppl. Ser.* **1981**, *43*, 455–471.
17. Mcwhirter, R.W.P.; Summers, H.P. Atomic radiation from low density plasma. *Appl. At. Collission Phys.* **1983**, *2*, 51.
18. Eldén, B. Forbidden lines in hot plasma. *Phys. Scr.* **1984**, *26*, 71.
19. Sterling, N.C.; Dinerstein, L.H.; Charles, B.W. Discovery of enhanced Germanium abundances in planetary nebulae with the *far ultraviolet spectroscopic explorer*. *Astrophys. J.* **2002**, *578*, L55. [CrossRef]
20. O'Toole, S.J. beyond the iron group: Heavy metals in hot subdwarfs. *Astron. Atrophys.* **2004**, *423*, L25–L28. [CrossRef]
21. Chayer, P.; Vennes, S.; Dupuis, J.; Kruk, J.W. Abundance of elements beyond the Iron group in cool DO white Dwarfs. *Astrophys. J.* **2005**, *630*, L169. [CrossRef]
22. Vennes, S.; Chayer, P.; Dupuis, J. Discovery of photospheric Germanium in Hot DA White Dwarfs. *Astrophys. J.* **2005**, *622*, L121. [CrossRef]
23. Curtis, L.J.; Theodosiou, C.E. Comprehensive calculations of 4p and 4d lifetimes for the Cu sequence. *Phys. Rev. A* **1989**, *39*, 605. [CrossRef]
24. Dimitrijević, M.S.; Simić, Z.; Kovačević, A.; Valjarević, A.; Sahal-Bréchot, S. Stark broadening of Xe VIII spectral lines. *Mon. Not. R. Astron. Soc.* **2015**, *454*, 1736–1741. [CrossRef]
25. Beer, H.; Käppeler, F.; Reffo, G.; Venturini, G. Neutron capture cross-sections of stable xenon isotopes and their application in stellar nucleosynthesis. *Astrophys. Space Sci.* **1983**, *97*, 95. [CrossRef]
26. Clayton, D.D.; Ward, R.A. S process studies: Xenon and krypton isotopic abundances. *Astrophys. J.* **1978**, *224*, 1000. [CrossRef]
27. Berry, H.G.; Hallin, R.; Sjödin, R.; Gaillard, M. Beam -foil observations of Na 1 doubly-excited states. *Phys. Lett.* **1974**, *50A*, 191. [CrossRef]
28. Striganov, A.R.; Odintsova, G.A. *Tables of Spectral Lines of Atom and Ions*; Energy: Moscow, Russia, 1982.
29. Jupén, C.; Engström, L.; Hutton, R.; Reistad, N.; Westrlind, M. Analysis of core-excited n = 3 configurations in S VI, Cl VII, Ar VIII, and Ti XII. *Phys. Scr.* **1990**, *42*, 44. [CrossRef]
30. Biémont, E.; Frémat, Y.; Quinet, P. Ionization potentials of atoms and ions from Lithium to Tin (z = 50). *At. Nucl. Data Tables* **1999**, *71*, 117. [CrossRef]
31. Fischer, C.F.; Tachiev, G.; Irimia, A. Relativistic energy levels, lifetimes, and transition probabilities for the sodium-like to argon-like sequences. *At. Nucl. Data Tables* **2006**, *92*, 607. [CrossRef]
32. Saloman, E.B. Energy levels and observed spectral lines of ionized Argon, Ar II through Ar XVIII. *J. Phys. Chem. Ref. Data* **2010**, *39*, 033101. [CrossRef]
33. Cardelli, J.A.; Mayer, D.M. The Abundance of Interstellar krypton. *Astrophys. J.* **1997**, *477*, L57. [CrossRef]
34. Cardelli, J.A.; Sarage, B.D.; Ebbets, D.C. Interstellar Gas phase Abundance Carbon, Oxygen, Nitrogen, Copper, Gallium, Germanium, and Krypton toward zeta ophiuchi. *Astrophys. J. Lett.* **1991**, *383*, L23. [CrossRef]
35. Cartledge, S.I.B.; Meyer, D.M.; Lauroesch, J.T. The Homogeneity of Interstellar Krypton in the Galactic Disk. *Astrophys. J.* **2003**, *597*, 408–413. [CrossRef]
36. Dinerstein, H.L. Neutron-Capture Elements in Planetary Nebulae: Identification of two near-Infrared Emission Lines as [Kr III]. *Astrophy. J.* **2001**, *550*, L223–L226. [CrossRef]
37. Cheng, K.-T.; Kim, Y.-K. Energy levels, wavelengths, and transition probabilities of Cu-like ions. *At. Data Nucl. Data Tables* **1978**, *22*, 547–563. [CrossRef]
38. Reader, J.; Acquista, N.; Kaufman, V. Spectrum and energy levels of seven-times-ionized Krypton (Kr VIII) and resonance lines of eight-times-ionized Krypton (Kr VIII). *J. Opt. Soc. Am. B* **1991**, *8*, 538–547. [CrossRef]
39. Boduch, P.; Chantepie, M.; Hennecart, D.; Husson, X.; Lecler, D.; Druetta, M.; Wilson, M. Spectroscopic analysis of visible and near UV light emitted in 120-kev Kr^{8+}-He and Kr^{8+}-H_2 collisions. *Phys. Scr.* **1992**, *46*, 337. [CrossRef]
40. Johnson, W.R.; Blundell, S.A.; Sapirstein, J. Many-body perturbation-theory calculations of energy levels along the copper isoelectronic sequence. *Phys. Rev. A* **1990**, *42*, 1087. [CrossRef]
41. Vilkas, M.J.; Ishikawa, Y.; Hirao, K. Ionization energies and fine structure splitting of highly correlated systems: Zn, zinc-like ions and Copper-like ions. *Chem. Phys. Lett.* **2000**, *321*, 243–252. [CrossRef]
42. Hinnov, E. Highly ionized atoms in tokamak discharges. *Phys. Rev. A* **1976**, *14*, 1533. [CrossRef]
43. Alexander, E.; Even-Zohar, M.; Fraenkel, B.S.; Goldsmith, S. Classification of transitions in the EUV spectra of Y ix-xiii, Zn x-xiv, Nb xi-xv and Mo xii-xvi. *J. Opt. Soc. Am.* **1971**, *61*, 508. [CrossRef]
44. Reader, J.; Acquista, N. 4s-4p resonance transitions in highly charged Cu- and Zn-like ions. *Phys. Rev. Lett.* **1977**, *39*, 184. [CrossRef]

45. Curtis, L.J.; Lindgård, A.; Eldén, B.; Martinson, I.; Nielsen, S.E. Energy levels and transition probabilities in Mo XIV. *Phys. Scr.* **1977** *16*, 72. [CrossRef]
46. Joshi, Y.N.; Van Kleef, T.A.M. The sixth spectrum of selenium: Se VI. *Physica B+C* **1978**, *94*, 270–274. [CrossRef]
47. Reader, J.; Luther, G.; Acquista, N. Spectrum and energy levels of thirteen-times ionized Molybdenum Mo XIV. *J. Opt. Soc. Am.* **1979**, *69*, 144–149. [CrossRef]
48. Mansfield, M.W.D.; Peacock, N.J.; Smith, C.C.; Hobby, M.G.; Cowan, R.D. The XUV spectra of highly ionized Molybdenum. *J. Phys. B Atom. Mol. Phys.* **1978**, *11*, 1521–1544. [CrossRef]
49. Druetta, M.; Buchet, J.P. Beam-foil study of krypton between 400 and 800 Å. *J. Opt. Soc. Am.* **1976**, *66*, 433. [CrossRef]
50. Irwin, D.J.G.; Karnahan, J.A.; Pinnington, E.H.; Livingston, A.E. Beam-foil mean-life measurements in krypton. *J. Opt. Soc. Am.* **1976**, *66*, 1396–1400. [CrossRef]
51. Knystautas, E.J.; Drouin, R. Oscillator strength of resonance lines in Br(VI), (VII) and Kr(VII), (VIII). *J. Quant. Spectrosc. Radiat. Transfer.* **1977**, *17*, 551–553. [CrossRef]
52. Werner, K.; Rauch, T.; Ringat, E.; Kruk, J.W. First detection of Krypton and Xenon in a white dwarf. *Astrophys. J. Lett.* **2012**, *753*, L7. [CrossRef]
53. Biémont, E.; Clar, M.; Fivet, V.; Garnir, H.-P.; Palmeri, P.; Quinet, P.; Rostohar, D. Lifetime and transition probability determination in xenon ions. *Eur. Phys. J. D* **2007**, *44*, 23–33. [CrossRef]
54. Migdalek, J.; Garmulewicz, M. The relativistic ab initio model potential versus Dirac–Fock oscillator strengths for silver and gold isoelectronic sequences. *J. Phys. B At. Mol. Opt. Phys.* **2000**, *33*, 1735. [CrossRef]
55. Safronova, U.I.; Savukov, I.M.; Safronova, M.S.; Johnson, W.R. Third-order relativistic many-body calculations of energies and lifetimes of levels along the silver isoelectronic sequence. *Phys. Rev. A* **2003**, *68*, 062505. [CrossRef]
56. Glowacki, L.; Migdałek, J. Relativistic configuration-interaction Oscillator strength for lowest E1 transitions in silver and gold isoelectronic sequences. *Phys. Rev. A* **2009**, *80*, 042505. [CrossRef]
57. Ivanova, E.P. Transitions probabilities for 5s-5p, 5p-5d, 4f-5d, and 5d-5f transitions in Ag-like ions with Z = 50–86. *At. Data Nucl. Data Tables* **2011**, *97*, 1–22. [CrossRef]
58. Otsuka, M.; Tajitsu, A. Chemical abundances in the Carbon-rich and Xenon-rich halo planetary nebula H4-1. *Astrophys. J.* **2013**, *778*, 146. [CrossRef]
59. Bhowmik, A.; Dutta, N.N.; Roy, S. Precise calculations of Astrophysically Important Allowed and Forbidden Transitions of Xe VIII. *Astrophys. J.* **2017**, *836*, 125. [CrossRef]
60. Pernpointner, M.; Zobel, J.P.; Kryzhevoi, N.V. Strong configuration interaction in the double ionization spectra of noble gases studied by the relativistic propagator method. *Phys. Rev. A* **2012**, *85*, 012505. [CrossRef]
61. Eser, S.; Özdemir, L. Energies and lifetimes of levels for doubly ionized Xenon and Radon. *Phys. Pol. A* **2018**, *133*, 1324. [CrossRef]
62. Pequignot, D.; Baluteau, J.-P. The identification of krypton, xenon and other elements of rows 4, 5, and 6 of the periodic table in the planetary nebula NGC 7027. *Astron. Astrophys.* **1994**, *283*, 593–625.
63. Migdalek, J. Core-polarisation corrected relativistic energy and radiative data for one- electron spectra of Bi v through U XIV. *Can. J. Phys.* **2018**, *96*, 610. [CrossRef]
64. Lindgren, I.; Mukherjee, D. On the connectivity criteria in the open shell coupled-cluster theory for general model spaces. *Phys. Rep.* **1987**, *151*, 93–127. [CrossRef]
65. Bishope, R.F. An overview of coupled cluster theory and it's applications in physics. *Theor. Chim. Acta* **1991**, *80*, 95–148. [CrossRef]
66. Roy, S.; Majumder, S. Ab initio estimations of the isotope shift for the first three elements of the K isoelectronic sequence. *Phys. Rev. A* **2015**, *92*, 012508. [CrossRef]
67. Dutta, N.N.; Majumder, S. Accurate estimations of stellar and interstellar transition lines of triply ionized germanium. *Astrophys. J.* **2011**, *737*, 25. [CrossRef]
68. Dutta, N.N.; Majumder, S. Ab initio studies of electron correlation and Gaunt interaction effects in the boron isoelectronic sequence using coupled cluster theory. *Phys. Rev. A* **2012**, *85*, 032512. [CrossRef]
69. Dutta, N.N.; Majumder, S. *Ab initio* calculations of spectroscopic properties of Cr^{5+} using coupled cluster-theory. *Indian J. Phy.* **2016**, *90*, 373. [CrossRef]
70. Bhowmik, A.; Roy, S.; Dutta, N.N.; Majumder, S. Study of Coupled-Cluster correlations on electromagnetic transitions and hyperfine structure constants of W VI. *J. Phys. B At. Mol. Opt. Phys.* **2017**, *50*, 125005. [CrossRef]
71. Bhowmik, A.; Dutta, N.N.; Majumder, S. Magic wavelengths for trapping with focused Laguerre-Gaussian beams. *Phy. Rev. A* **2018**, *97*, 022511. [CrossRef]
72. Bhowmik, A.; Dutta, N.N.; Majumder, S. Vector polarizability of an atomic state induced by a linearly polarized vortex beam. External control of magic tune-out wavelengths and heteronuclear spin oscillations. *Phy. Rev. A* **2020**, *102*, 063116. [CrossRef]
73. Bhowmik, A.; Dutta, N.N.; Das, S. Role of vector polarizability induced by a linearly polarized focused Lagueree-Gaussian light:application in optical trapping and ultracold spinor mixture. *Eur. Phys. J. D* **2022**, *76*, 139. [CrossRef]
74. Biswas, S.; Das, A.; Bhowmik, A.; Majumder, S. Accurate estimations of electromagnetic transitions of Sn IV for stellar and interstellar media. *Mon. Not. R. Astron. Soc.* **2018**, *477*, 5605. [CrossRef]
75. Das, A.; Bhowmik, A.; Dutta, N.N.; Majumder, S. Many-body calculations and hyperfine-interaction effect on dynamic polarizabilities at low-lying energy levels of Y^{2+}. *Phys. Rev. A* **2020**, *102*, 012801. [CrossRef]

76. Das, A.; Bhowmik, A.; Dutta, N.N.; Majumder, S. Two-Photon Polarizability of Ba^+ ion: Control of Spin-Mixing Process in an Ultracold $^{137}Ba^+$. *Atoms* **2022**, *10*, 109. [CrossRef]

77. Sahoo, B.K.; Majumder, S.; Chaudhuri, R.K.; Das, B.P.; Mukherjee, D. Ab initio determination of the lifetime of the $6^2P_{3/2}$ state for $^{207}Pb^+$ by relativistic many-body theory. *J. Phys. B* **2004**, *37*, 3409. [CrossRef]

78. Bartlett, R.J. *Modern Electronic Structure Theory*; Yarkony, D.R., Ed.; World Scientific: Singapore, 1995; Volume II, p. 1047.

79. Mondal, P.K.; Dutta, N.N.; Majumder, S. Effect of screening on spectroscopic properties of Li-like ions in a plasma medium. *Phys. Rev. A* **2013**, *87*, 062502. [CrossRef]

80. Stewart, J.C.; Pyatt, K.D., Jr. Lowering of ionization potential in plasmas. *Astrophys. J.* **1966**, *144*, 1203. [CrossRef]

81. Szabo, A.; Ostlund, N.S. *Modern Quantum Chemistry: Int roduction to Advanced Elec-tronic Structure Theory*; Courier Corporation: Dover, Mineola, 1996.

82. Dixit, G.; Sahoo, B.K.; Chaudhuri, R.K.; Majumder, S. Ab initio calculations of forbidden transition amplitudes and lifetimes of the low-lying states in V^{4+}. *Phy. Rev. A* **2007**, *76*, 059901. [CrossRef]

83. Lindgren, I.; Morrison, J. *Atomic Many-Body Theory*; Springer: Berlin/Heidelberg, Germany, 1986; Volume 3. [CrossRef]

84. Savukov, I.M.; Johnson, W.R.; Safronova, U.I. Multipole(E1, M1, E2, M2) transition wavelength and rates between states with $n \leq 6$ in He-like Carbon, Nitrogen, Oxygen, Neon, Silicon and Argon. *At. Data Nucl. Data Tables* **2003**, *85*, 83–167. [CrossRef]

85. Shore, B.W.; Menzel, D.H. *Principles of ATOMIC SPECTRA*; John Wiley and Sons: New York, NY, USA, 1968.

86. Kelleher, D.E.; Podobedova, L.I. Atomic transition probabilities of Sodium and Magnetium. A critical compilation. *J. Phys. Chem. Ref. Data* **2008**, *37*, 267. [CrossRef]

87. Ichimaru, S. Strongly coupled plasmas: High-density classical plasmas and degenerate electron liquids. *Rev. Mod. Phys.* **1982**, *54*, 1017. [CrossRef]

88. Akhiezer, A.I.; Akhiezer, I.A.; Polovin, R.V.; Sitenko, A.G.; Stepano, K.N. *Plasma Electrodynamics. Volume 1. Linear Theory*; Pergamon Press: New York, NY, USA, 1975; 431p.

89. Parpia, F.A.; Fischer, C.F.; Grant, I.P. GRASP92 a package for large-scale relativistic atomic structure calculations. *Comput. Phys. Commun.* **2006**, *175*, 745. [CrossRef]

90. Clementi, E. (Ed.) *Modern Techniques in Computational Chemistry: MOTECC-90*; Springer: Amsterdam, The Netherlands, 1990.

91. Huzinaga, S.; Klobukowski, M. Well-tempered gaussian bassis set for the calculation of matrix Hartree-fock wave function. *Chem. Phys. Letts.* **1993**, *212*, 260–264. [CrossRef]

92. Roy, S.; Dutta, N.N.; Majumder, S. Relativistic coupled-cluster calculations on hyperfine structures and electromagnetic transition amplitudes of In III. *Phys. Rev. A* **2014**, *89*, 042511. [CrossRef]

93. Kramida, A.; Ralchenko, Y.; Reader, J.; NIST ASD Team. *NIST Atomic Spectra Database (Ver. 5.10)*; National Institude of Standards and Technology: Gaithersburg, MD, USA, 2022. Available online: http://physics.nist.gov/asd (accessed on 10 March 2023).

94. Dutta, N.N.; Roy, S.; Dixit, G.; Majumder, S. Ab initio calculations of transition amplitudes and hyperfine A and B constants of Ga III. *Phys. Rev. A* **2013**, *87*, 012501. [CrossRef]

95. Jyoti, K.H.; Arora, B.; Sahoo, B.K. Radiative properties of Cu-isoelectronic As, Se, and Br ions for astrophysical applications. *Mon. Not. R. Astron. Soc.* **2022**, *511*, 1999–2007. [CrossRef]

96. Cowan, R.D. *The Theory of Atomic Structure and Spectra*; University of California Press: Berkeley, CA, USA, 1981.

97. Reistad, N.; Engström, L.; Berry, H.G. Oscillator strength measurements of resonance transitions in Sodium- and Magnesium-like Argon. *Phys. Scr.* **1986**, *34*, 158. [CrossRef]

98. Verner, D.A.; Verner, E.M.; Ferl, G.J. Atomic data for permitted resonance lines of atoms and ions from H to Si and S, Ar, Ca, and Fe. *At. Data Nucl. Data Tables* **1996**, *64*, 180. [CrossRef]

99. Féret, L.; Pascale, J. Study of the multi charged ion Ar^{6+} by a configuration interaction Hartee Fock method using a pseudopotential. *J. Phys. Rev. B At. Mol. Opt. Phys.* **1999**, *32*, 4175. [CrossRef]

100. Lagmago, K.G.; Nana Engo, S.G.; Kwato Njock, M.G.; Oumarou, B. Consistent description Klein-Gordon dipole matrix elements. *J. Phys. B At. Mol. Opt. Phys.* **1998**, *31*, 963. [CrossRef]

101. Siegel, W.; Migdalek, J.; Kim, Y.K. Dirac–Fock oscillator strengths for E1 transition in the sodium isoelectronic sequence(Na I-Ca X). *At. Data Nucl. Data Tables* **1988**, *68*, 303–322. [CrossRef]

102. Guet, C.; Blundell, S.A.; Johnson, W.R. Relativistic Many-body calculations of oscillator strength for sodium-like ions. *Phys. Lett. A* **1990**, *143*, 384. [CrossRef]

103. Theodosiou, C.E.; Curtis, L.J. Accurate calculations of 3p and 3d lifetimes in the Na sequence. *Phys. Rev. A* **1988**, *38*, 4435. [CrossRef] [PubMed]

104. Kim, Y.-K.; Cheng, K.-T. Transition probabilities for the resonance transitions of Na-like ions. *J. Opt. Soc. Am.* **1978**, *68*, 836. [CrossRef]

105. Gruzdev, P.F.; Sherstyuk, A.I. Relativistic generalization of the effective orbital quantum number method. *Opt. Spectrosc.* **1979**, *46*, 353–355.

106. Livingston, A.E.; Curtis, L.J.; Schectman, R.M.; Berry, H.G. Energies and lifetimes of excited states in Copper like Kr VIII. *Phys. Rev. A* **1980**, *21*, 771. [CrossRef]

107. Pinnington, E.H.; Gosselin, R.N.; O'Neill, J.A.; Kernahan, J.A.; Donnely, K.E.; Brooks, R.L. Beam-foil lifetime measurements using ANDC analysis for the resonance doublet in A VIII, Kr VIII and Xe VIII. *Phys. Scr.* **1979**, *20*, 151. [CrossRef]

108. Fischer, C.F. Oscillator strength of $^2S - {}^2P$ transitions in the copper sequence. *J. Phys. B* **1977**, *10*, 1241. [CrossRef]

109. Weiss, A.W. Hartree–Fock line strength for the Lithium, Sodium, and Copper isoelectronic sequences. *J. Quant. Spectrosc. Radiat. Transf.* **1977**, *18*, 481–490. [CrossRef]
110. Cowan, R.D. *Los Alamos Report No. LA-6679, MS, National Information Service*; Spring: London, UK, 1977.
111. Lindgård, A.; Curtis, L.J.; Matrinson, I.; Nielsen, S.E. Semi-Empirical Oscillator Strengths for Cu I Isoelectronic sequence. *Phys. Scr.* **1980**, *21*, 47. [CrossRef]
112. Migdalek, J.; Baylis, W.E. Influence of core polarization on oscillator strength along the Copper isoelectronic sequence. *J. Phys. B Atom. Mol. Phys.* **1979**, *12*, 1113. [CrossRef]
113. Saloman, E.B. Energy levels and observed spectral lines of Xenon, Xe I through Xe LIV. *J. Phys. Chem. Ref. Data* **2004**, *33*, 765. [CrossRef]
114. Morita, S.; Goto, M.; Katai, R.; Dong, C.; Sakaue, H.; Zhou, H. Observation of Magnetic dipole forbidden transition in LHD and its application to burning plasma diagnostics. *Plasma Sci. Technol.* **2010**, *12*, 341. [CrossRef]
115. Fahy, K.; Sokell, E.; O'Sullivan, G.; Aguilar, A.; Pomeroy J.M.; Tan, J.N.; Gillaspy, J.D. Extreme-ultraviolet spectroscopy of highly charged xenon created using an electron-beam ions trap. *Phys. Rev. A* **2007**, *75*, 032520. [CrossRef]
116. Morgan, C.A.; Serpa, F.G.; Takacs, E.; Meyer, E.S.; Gillaspy, J.D.; Sugar, J.; Roberts, J.R.; Brown, C.M.; Feldman, U. Observation of visible and uv magnetic dipole transition in highly charged Xenon and Barium. *Phys. Rev. Lett.* **1995**, *74*, 1716. [CrossRef]
117. Kessler, M.F.; Steinz, J.A.; Anderegg, M.E.; Clavel, J.; Drechsel, G.; Estaria, P.; Faelker, J.; Riedinger, J.R.; Robson, A.; Taylor, B.G.; et al. The Infrared Space Observatory (ISO) Mission. *Astron. Astrophys.* **1996**, *315*, L27–L31. [CrossRef]
118. Feuchtgruber, H.; Lutz, D.; Beintema, D.A.; Valentijn, E.A.; Bauer, O.H.; Boxhoorn, D.R.; De Graauw, T.; Haser, T.N.; Haerendel, G.; Heras, A.M. New wavelength determination of mid-infrared fine structure lines by *infrared space observatory* short wavelength spectrometer. *Astrophys. J.* **1997**, *487*, 962. [CrossRef]
119. Liu, X.-W.; Barlow, M.J.; Cohen, M.; Danziger, I.J.; Luo, S.-G.; Baluteau, J.P.; Cox, P.; Emery, R.J.; Lim, T.; Péquignot, D. *ISO LWS* observations of planetary nebula fine-structure lines. *Mon. Not. R. Astron. Soc.* **2001**, *323*, 343–361. [CrossRef]
120. Ponciano-Ojeda, F.; Hernández-Gómez, S.; López-Hernández, O.; Mojica-Casique, C.; Colín-Rodríguez, R.; Ramírez-Martínez, F.; Flores-Mijangos, J.; Sahagún, D.; Jáuregui, R.; Jiménez-Mier, J. Observation of the $5p3/2 \rightarrow 6p3/2$ electric-dipole-forbidden transition in atomic rubidium using optical-optical double-resonance spectroscopy. *Phys. Rev. A* **2015**, *92*, 042511. [CrossRef]
121. Ponciano-Ojeda, F.; Hernández-Gómez, S.; López-Hernández, O.; Mojica-Casique, C.; Colín-Rodríguez, R.; Ramírez-Martínez, F.; Flores-Mijangos, J.; Sahagún, D.; Jáuregui, R.; Jiménez-Mier, J. Laser spectroscopy of the $5P3/2 \rightarrow 6Pj$ (j = 1/2 and 3/2) electric dipole forbidden transitions in atomic rubidium. *AIP Conf. Proc.* **2018**, *1950*, 030001.
122. Ponciano-Ojeda, F.; Hernández-Gómez, S.; López-Hernández, O.; Mojica-Casique, C.; Colín-Rodríguez, R.; Ramírez-Martínez, F.; Flores-Mijangos, J.; Sahagún, D.; Jáuregui, R.; Jiménez-Mier, J. One step beyond the electric dipole approximation: An experiment to observe the $5p \rightarrow 6p$ forbidden transition in atomic rubidium. *Am. J. Phys.* **2018**, *86*, 7. [CrossRef]
123. Bhattacharya, M.; Haimberger, C.; Bigelow, N.P. Forbidden Transitions in a Magneto-Optical Trap. *Phys. Rev. Letts.* **2003**, *91*, 213004. [CrossRef] [PubMed]
124. Preston, D.W. Doppler-free saturated absorption: Laser spectroscopy. *Am. J. Phys.* **1996**, *64*, 1432. [CrossRef]
125. Grumer, J.; Zhao, R.; Brage, T.; Li, W.; Huldt, S.; Hutton, R.; Zou, Y. Coronal lines and the importance of deep-core-valence correlation in Ag-like ions. *Phys. Rev. A* **2014**, *89*, 062511. [CrossRef]
126. Ding, X.B.; Koike, F.; Murakami, I. M1 transition energies and probabilities between the multiplets of the ground states of Ag-like ions with Z = 47–92. *J. Phys. B* **2012**, *45*, 035003. [CrossRef]
127. Basu, J.; Ray, D. Suppression of fine-structure splitting and oscillator strength of sodium D-line in a Debye plasma. *Phys. Plasmas* **2014**, *21*, 013301. [CrossRef]
128. Xie, L.Y.; Wang, J.G.; Janev, R.K. Relativistic effects in the photoionization of hydrozen-like ions with screened coulomb interaction. *Phys. Plasmas* **2014**, *21*, 063304. [CrossRef]
129. Sahoo, S.; Ho, Y.K. On the appearance of a Cooper Minimum in the Photoionization Cross Section of the Plasma-Embedded Li Atom. *Res. Lett. Phys.* **2009**, *2009*, 832413. [CrossRef]
130. Buchet-Poulizac, M.C.; Buchet, J.P.; Ceyzeriat, P. Spectroscopic et durees de vie dand Ar VI-VII. *Nucl. Instrum. Methods* **1982**, *202*, 13–18. [CrossRef]
131. Johnson, W.R.; Liu, Z.W.; Sapirstein, J. Transition rates for lithium-like ions, sodium-like ions, and neutral alkali-metal atoms. *At. Data Nucl. Data Tables* **1996**, *64*, 279–300. [CrossRef]
132. Knystautas, E.J.; Drouin, R.; Druetta, M. Nouvelles identification et mesures de durees de vie dans l'argon hautement ionise. *J. Phys. Colloques* **1979**, *40*, C1-186–C1-189. [CrossRef]
133. Crossley, R.J.S.; Dalgarno, A. An expansion method for calculating atomic properties, V. Transition probabilities of M shell electrons. *Proc. R. Soc. A* **1965**, *286*, 510–518. [CrossRef]

134. Livingston, A.E.; Irwin, D.J.G.; Pinnington, E.H. Lifetime measurement in Ar II-Ar VIII. *J. Opt. Soc. Am.* **1972**, *62*, 1303–1308. [CrossRef]
135. Cheng, K.-T.; Kim, Y.-K. Excitation energy and oscillator strength in the silver isoelectronic sequence. *J. Opt. Soc. Am.* **1979**, *69*, 125. [CrossRef]

Article

Attosecond Time Delay Trends across the Isoelectronic Noble Gas Sequence

Brock Grafstrom * and Alexandra S. Landsman

Department of Physics, The Ohio State University, Columbus, OH 43210, USA; landsman.7@osu.edu
* Correspondence: grafstrom.1@osu.edu

Abstract: The analysis and measurement of Wigner time delays can provide detailed information about the electronic environment within and around atomic and molecular systems, with one the key differences being the lack of a long-range potential after a halogen ion undergoes photoionization. In this work, we use relativistic random-phase approximation to calculate the average Wigner delay from the highest occupied subshells of the atomic pairings (2p, 2s in Fluorine, Neon), (3p, 3s in Chlorine, Argon), (4p, 4s, 3d, in Bromine, Krypton), and (5p, 5s, 4d in Iodine, Xenon). The qualitative behaviors of the Wigner delays between the isoelectronic pairings were found to be similar in nature, with the only large differences occurring at photoelectron energies less than 20 eV and around Cooper minima. Interestingly, the relative shift in Wigner time delays between negatively charged halogens and noble gases decreases as atomic mass increases. All atomic pairings show large differences at low energies, with noble gas atoms showing large positive Wigner delays, while negatively charged halogen ions show negative delays. The implications for photoionization studies in halide-containing molecules is also discussed.

Keywords: attosecond time delay; noble gas; halogen atoms; relativistic random-phase approximation

Citation: Grafstrom, B.; Landsman, A.S. Attosecond Time Delay Trends across the Isoelectronic Noble Gas Sequence. *Atoms* **2023**, *11*, 84. https://doi.org/10.3390/atoms11050084

Academic Editors: Himadri S. Chakraborty, Hari R. Varma and Yew Kam Ho

Received: 28 February 2023
Revised: 10 May 2023
Accepted: 12 May 2023
Published: 15 May 2023

1. Introduction

The recent advancement of attosecond extreme ultraviolet infrared (XUV-IR) laser metrology over the past decade [1–6] has enabled access for observing ultrafast phenomena across a variety of atomic and molecular systems at the natural time scale of their electronic motion. One common experimental technique utilizes an XUV-IR pump–probe process [7,8], where an electron is first ionized through the absorption of an XUV photon and subsequently streaked by the few-cycle IR laser field, which imprints itself on the photoelectron's final energy and momentum. By varying the time delay between the XUV and IR pulses, it is possible to measure the photoionization time delay relative to a reference. Another common technique is reconstruction of attosecond beating by interference of two-photon transitions (RABBITT) [9,10], where the target is first ionized by an XUV attosecond pulse train of high-order harmonics, and the photoelectron can then either absorb or emit a secondary IR photon in the continuum. By adjusting the delay between the IR laser field and the high-order harmonics, it is possible to extract the time delay for a particular transition.

These techniques have been used to measure photoionization delays in atoms [11–15], molecules [16–20], and solid-state systems [21–23], thereby providing new information about their electronic structure. The total time delay τ, also known as the streaking delay, given by a pump-probe experiment is frequently separated into two components, $\tau = \tau_w + \tau_{CLC}$. This convention of separating the total delay into two separate terms is also followed in traditional RABBITT experiments, with the only difference being that τ_{CLC} is replaced by the continuum–continuum delay τ_{CC} [10]. The first contribution is the Wigner time delay [24,25] τ_w, which describes the group delay of the ionized photoelectron due to the absorption of an XUV photon and depends on the nature of the target. The second component is the Coulomb-laser coupling time delay [26] τ_{CLC}, as it describes the delay caused by the coupling between the IR field in the continuum and the long-range potential of the remaining ion. Unlike the Wigner time delay, which requires an accurate description of the target potential, the Coulomb-laser coupling delay can be computed with an analytical formula [26–28] and does not rely on the precise nature of the target species. It does, however, depend on the photoelectron's kinetic energy, the energy of the photons from the laser, and the charge of the residual ion. For example, the photoionization of a neutral atom will create a positively charged ion of $+1$ and, therefore, τ_{CLC} will be finite, but if a negatively charged halogen undergoes photoionization, the remaining ion will have a neutral charge and τ_{CLC} will vanish. This implies that it is possible to directly measure the Wigner delay of negatively charged halogen ions.

The process of removing an electron from a neutral atom is called photoionization, while the removal of an electron from a negatively charged ion is called photodetachment. Many theoretical and experimental studies have been conducted to accurately describe and predict various aspects of the photoionization process in noble gas atoms (with some of the key features being photoemission angle dependence [29–32], autoionization resonances [33], the 4d giant dipole resonance in Xenon [34], photorecombination [35], and relativistic effects [36–38]) but only recently has work been conducted on Wigner time delays in negatively charged species [39–45].

This understanding of the time delays of negatively charged halogens is important for two key reasons. First, the halides F^-, Cl^-, Br^-, and I^- are isoelectronic to the well-studied systems of Ne, Ar, Kr, and Xe, respectively, and, hence, they allow for an interesting comparison of two systems where the initial electronic configurations are identical and yet the binding energies significantly differ. Second, it has been shown that in the case of iodine-containing molecules, such as methyl iodide [46], the 4d orbitals of iodide are non-bonding and reasonably agree with the predicted cross-section data of an isolated iodine atom [47,48]. Therefore, comparisons between the Wigner delays in halogen ions and noble gases should help to inform future molecular photoionization studies while also helping to confirm the driving mechanisms behind Wigner time delay phenomena.

In this paper, we utilize the relativistic random-phase approximation (RRPA) formalism of Johnson and Lin [49] to calculate the average phases and average Wigner time delays of the highest occupied s, p, and d shells of the noble gas series and their halide counterparts. The theoretical details of RRPA are given in Section 2, along with a description of the methodology used to calculate the average Wigner time delays, the results of which are illustrated in Section 3, with the time delays of each halide–noble gas pairing being plotted with respect to photoelectron kinetic energy. Section 3 also includes a discussion regarding the similar qualitative behavior in the known regions of the Cooper minimum, while also acknowledging the sharp contrast in time delays at photoelectron energies below $\sim 20\,\text{eV}$ for the highest occupied p and s orbitals. The first part of Section 3 briefly describes the methodology used to calculate the Dirac–Hartree–Fock orbitals and their associated binding energies, along with specific details regarding the number of photoionization channels used for the RRPA calculations of each atom. Section 4 provides a summary of the results and a brief discussion regarding applications for future molecular ionization studies.

2. Theoretical Overview of the Relativistic Random-Phase Approximation (RRPA)

The same theoretical formalism was applied to negative halogen ions and neutral noble gas atoms in order to calculate the photoionization dipole transition matrix amplitudes and phases for an initial bound state. This overview follows the same outline as the previous RRPA photoionization studies of Kheifets and Deshmukh [50] and the original multi-channel paper of Johnson and Lin [49]. Atomic units ($\hbar = e = m_e = 1$) are used throughout, unless stated otherwise.

For a time-dependent perturbation of the form $v_+ e^{-i\omega t} + v_- e^{i\omega t}$, the probability amplitude for a transition from the ground state u_i to an excited state $w_{i\pm}$, stimulated by said perturbation, is given by

$$T = \sum_{i=1}^{N} \int d^3 r \left(w_{i+}^\dagger v_+ u_i + w_{i-}^\dagger v_- u_i \right) \tag{1}$$

For an electromagnetic interaction described in the Coulomb gauge, it is possible to rewrite the transition amplitude T as a function of the vector potential $\vec{A}$, where the perturbations $v_{\pm\backslash}$ are described by, $v_+ = \vec{\alpha} \cdot \vec{A}$, $v_- = v_+^\dagger$ with $\vec{\alpha} = \begin{pmatrix} 0 & \vec{\sigma} \\ \vec{\sigma} & 0 \end{pmatrix}$.

$$T = \sum_{i=1}^{N} \int d^3 r \left(w_{i+}^\dagger \vec{\alpha} \cdot \vec{A} u_i + u_i^\dagger \vec{\alpha} \cdot \vec{A} w_{i-} \right) \tag{2}$$

Therefore, a photon of frequency ω (or equivalently of wavevector $\vec{k}$) and polarization $\hat{e}$ can be described by the vector potential in the Coulomb gauge as $\vec{A} = \hat{e} e^{-i \vec{k} \cdot \vec{r}}$, which can then be expanded in terms of its multipole components, $\vec{a}_{JM}^{(\lambda)}$. Note that an upper index of $\lambda = 1, 0$ corresponds to the electric and magnetic multipoles, respectively. In the single active electron approximation, the transition amplitude describe by Equation (2) can be reduced even further to

$$T_{JM}^{(\lambda)} = \int d^3 r w_{i+}^\dagger \vec{\alpha} \cdot \vec{a}_{JM}^{(\lambda)} u_i \tag{3}$$

where J and M are the angular momentum quantum numbers describing the incoming photon. It is common to label the initial bound state (u_i) of the electron by the quantum numbers ljm and the final continuum state (w_i) by the numbers $\bar{l} \bar{j} \bar{m}$. The spin of the electron is given by the spinor χ_ν, where $\nu = \pm 1/2$. Any final state can be uniquely described by the index $\bar{\kappa} = \mp \left(\bar{j} + 1/2 \right)$, where $\bar{j} = \bar{l} \pm 1/2$ is the total angular momentum of the outgoing electron in the continuum. The final state can also be written as a partial wave expansion, which is given explicitly in Equation (40) of [49]. Inserting this expansion into Equation (3) results in

$$T_{JM}^{(\lambda)} = i \sqrt{\frac{2\pi^2}{Ep}} \sqrt{\frac{(2J+1)(J+1)}{J}} \frac{\omega^J}{(2J+1)!!} \sum_{\bar{\kappa}\bar{m}} \left[\chi_\nu^\dagger \Omega_{\bar{\kappa}\bar{m}}(\hat{p}) \right] (-1)^{\bar{j}-\bar{m}} \begin{pmatrix} \bar{j} & J & j \\ -\bar{m} & M & m \end{pmatrix} D_{lj \to \bar{l}\bar{j}} (-1)^{\bar{j}+j+J} \tag{4}$$

The ionized photoelectron's energy and momentum are represented by E and p, respectively, while $\Omega_{\bar{\kappa}\bar{m}}(\hat{p})$ is described in terms of Clebsch–Gordan coefficients and spherical harmonics.

$$\Omega_{\bar{\kappa}\bar{m}}(\hat{p}) = \sum_\nu C_{\bar{l},\bar{m}-\nu,1/2\nu}^{\bar{j},\bar{m}} Y_{\bar{l},\bar{m}-\nu}(\hat{p}) \chi_\nu \tag{5}$$

The six-term bracket to the right of the $\Omega_{\bar{\kappa}\bar{m}}(\hat{p})$ in Equation (4) corresponds to the Wigner 3-j symbol, and $D_{lj \to \bar{l}\bar{j}}$ represents the reduced matrix element describing the transition from the initial state $a = n\kappa$ to the final state $\bar{a} = \left(E, \bar{\kappa} \right)$ multiplied by the phase shift of the continuum photoelectron wave $\delta_{\bar{\kappa}}$.

$$D_{lj \to \bar{l}\bar{j}} = i^{1-\bar{l}} e^{i\delta_{\bar{\kappa}}} \left\langle \bar{a} \left\| Q_J^{(\lambda)} \right\| a \right\rangle \tag{6}$$

One should note that the electric (or magnetic) multipole operator $Q_J^{(\lambda)}$ in the reduced matrix element is the only component that changes in Equation (4) for different values of λ, as the entire matrix element can be deconstructed as

$$\langle \bar{a} || Q_J^{(\lambda)} || a \rangle = (-1)^{j+1/2} [\bar{j}][j] \begin{pmatrix} \bar{j} & j & J \\ -1/2 & 1/2 & 0 \end{pmatrix} \pi\left(\bar{l}, l, J - \lambda + 1 \right) R_J^{(\lambda)}\left(\bar{a}, a \right) \tag{7}$$

where $R_J^{(\lambda)}\left(\bar{a}, a \right)$ is a radial integral listed in [49] and $\pi\left(\bar{l}, l, J - \lambda + 1 \right)$ is simply the parity factor responsible for imposing the necessary selection rules for a given transition.

$$\pi\left(\bar{l}, l, J - \lambda + 1 \right) = \begin{cases} 1, \bar{l} + l + J - \lambda + 1 = \text{even} \\ 0, \bar{l} + l + J - \lambda + 1 = \text{odd} \end{cases} \tag{8}$$

Despite its use for describing single-electron transitions, Equation (4) is also valid for any closed-shell atomic species, as the only change required in Equations (4) and (6) is that the single-electron reduced matrix element is modified slightly to include multi-electron RRPA effects in both the initial and final states. An explanation of this modification is given in Appendix A.

$$\langle \bar{a} || Q_J^{(\lambda)} || a \rangle \rightarrow \langle \bar{a} || Q_J^{(\lambda)} || a \rangle_{RRPA} \tag{9}$$

In this work, we restrict ourselves to electric dipole transitions ($\lambda = 1$ and $J = 1$) with linearly polarized light oriented along the $\hat{z}$ axis ($M = 0$). From these restrictions, Equation (4) simplifies to

$$T_{10}^{1\pm} \equiv \left[T_{10}^{(1\nu)} \right]^m_{nlj} = \sum_{\kappa m} C^{\bar{j}\bar{m}}_{\bar{l},\bar{m}-\nu,1/2\nu} \, Y_{\bar{l},\bar{m}-\nu}(\hat{p})(-1)^{2\bar{j}+j+1-\bar{m}} \begin{pmatrix} \bar{j} & 1 & j \\ -\bar{m} & 0 & m \end{pmatrix} D_{\bar{l}\bar{j}\rightarrow l j} \tag{10}$$

Here, we have followed the convention of ref. [51] by defining separate transition amplitudes for the spin-up and spin-down cases and by omitting the scaling factor of $\frac{2\pi i}{\sqrt{3Ep}}\omega$. By choosing the linear polarization of the incoming photon to be purely along $\hat{z}$, it is possible to take advantage of the axial symmetry of the system. Therefore, we will introduce the shorthand $Y_{lm} \equiv Y_{lm}(\hat{p}) = Y_{lm}(\theta, \phi = 0)$, where $\theta = 0$ corresponds to a photoelectron emission parallel to the direction of the initial photon.

It useful to observe how Equation (10) directly reduces in the simple case of electric dipole transitions for the $np_{1/2}$ and $np_{3/2}$ states. The expressions for the $nd_{3/2}$ and $nd_{5/2}$ transition amplitudes can be found in ref. [50].

$$\left[T_{10}^{(1+)} \right]^{m=1/2}_{np_{1/2}} = \frac{1}{\sqrt{6}} D_{np_{1/2}\rightarrow \epsilon s_{1/2}} Y_{00} + \frac{1}{\sqrt{15}} D_{np_{1/2}\rightarrow \epsilon d_{3/2}} Y_{20} \tag{11a}$$

$$\left[T_{10}^{(1-)} \right]^{m=1/2}_{np_{1/2}} = -\frac{1}{\sqrt{10}} D_{np_{1/2}\rightarrow \epsilon d_{3/2}} Y_{21} \tag{11b}$$

$$\left[T_{10}^{(1+)} \right]^{m=1/2}_{np_{3/2}} = \frac{1}{\sqrt{6}} D_{np_{3/2}\rightarrow \epsilon s_{1/2}} Y_{00} - \frac{1}{5}\left(\frac{1}{\sqrt{6}} D_{np_{3/2}\rightarrow \epsilon d_{3/2}} + \sqrt{\frac{3}{2}} D_{np_{3/2}\rightarrow \epsilon d_{5/2}} \right) Y_{20} \tag{12a}$$

$$\left[T_{10}^{(1-)} \right]^{m=1/2}_{np_{3/2}} = \frac{1}{10}\left(D_{np_{3/2}\rightarrow \epsilon d_{3/2}} - 2 D_{np_{3/2}\rightarrow \epsilon d_{5/2}} \right) Y_{21} \tag{12b}$$

$$\left[T_{10}^{(1+)} \right]^{m=3/2}_{np_{3/2}} = -\left(\frac{3}{\sqrt{10}} D_{np_{3/2}\rightarrow \epsilon d_{3/2}} + \frac{2\sqrt{3}}{15} D_{np_{3/2}\rightarrow \epsilon d_{5/2}} \right) Y_{21} \tag{12c}$$

$$\left[T^{(1-)}_{10}\right]^{m=3/2}_{np_{3/2}} = \frac{\sqrt{3}}{5}\left(D_{np_{3/2}\to\epsilon d_{3/2}} - \frac{1}{3}D_{np_{3/2}\to\epsilon d_{5/2}}\right)Y_{22} \tag{12d}$$

The reduced matrix elements $D_{lj\to l\,j}$ can be evaluated numerically for both its real and imaginary components, as doing so allows for the phase η and Wigner time delay to be calculated using the standard formulation of

$$\tau_w = \frac{d\eta}{dE}, \eta = \arctan\left[\frac{\mathrm{Im}\left\{T^{1\pm}_{10}\right\}}{\mathrm{Re}\left\{T^{1\pm}_{10}\right\}}\right] \tag{13}$$

For a given subshell nl_j, the angle-dependent time delay can be calculated as a weighted average over all possible transition amplitudes and spin states.

$$\tau_{nl_j}(\theta) = \frac{\sum\limits_{m,v}\tau_{nl_j,m,v}(\theta)\left|\left[T^{(1v)}_{10}\right]^m_{nl_j}\right|^2}{\sum\limits_{m,v}\left|\left[T^{(1v)}_{10}\right]^m_{nl_j}\right|^2} \tag{14}$$

For the purposes of this work, we will only consider the case of $\theta = 0$, as it is commonly the most dominant direction of photoelectron emission. As with Equation (10), it is informative to see how Equation (14) simplifies into the simple weighted average of the spin-up and spin-down Wigner delays for the case of an $np_{1/2}$ state.

$$\tau_{np_{1/2}} = \frac{\tau^+_{np_{1/2}}\left|T^{1+}_{10}\right|^2 + \tau^-_{np_{1/2}}\left|T^{1-}_{10}\right|^2}{\left|T^{1+}_{10}\right|^2 + \left|T^{1-}_{10}\right|^2} \tag{15}$$

This process of averaging Wigner delays was performed for every subshell listed in Table 1 below. By taking the average Wigner delays $\tau_{np_{1/2}}$ and $\tau_{np_{3/2}}$ and weighting them by their respective differential cross-sections, the total average time delay τ_{np} was also calculated. An analogous process was also used to compute τ_{ns} and τ_{nd} for every halogen and noble gas.

It should be noted that the accuracy of this averaging process depends not only on the values of the Wigner delays, but also with regard to the accuracy of the transition amplitudes themselves. Because the photoionization cross-section can be computed from the transition amplitudes $\left[T^{(1v)}_{10}\right]^m_{nl_j}$, it is possible to estimate the accuracy of the RRPA calculations by simply comparing the predicted cross-sections with those of experimental measurements. Reference [38] lists the predicted RRPA cross-sections for all of the noble gases being studied in this work, and found there to be a good agreement with experimental values. This result implies that the calculated RRPA transition amplitudes given by Equation (10) should also be quite accurate.

3. Results and Discussion

In this section, we present the calculated binding energies and average Wigner delays for each of the highest subshells of the halide ions and noble gas atoms. It should be noted that RRPA often produces autoionization resonances in the time delay spectra for any given noble gas. However, they are not the focus of this paper and, therefore, have been filtered out to prevent the obfuscation of more general time delay features. The locations of autoionization resonances in noble gases are well documented [38], but they generally occur at photon energies close to the binding energies of orbital subshells. In Argon, for example, the autoionization resonances produced by RRPA occur around photon energies of 35 eV, which corresponds to the ionization threshold of the $3s_{1/2}$ subshell.

3.1. Dirac–Hartree–Fock (DHF) Orbital Subshell Ionization Calculations

RRPA requires the use of Dirac–Hartree–Fock orbitals in order to account for ab initio relativistic effects and to obtain accurate subshell ionization potential energies. It is important to note that DHF calculations are typically the most accurate for the highest occupied shells regardless of the atom being studied; however, they can reasonably predict the binding energy of lower-lying subshells as atomic mass increases. Table 1 confirms this trend, as the predicted value of the 4d orbitals in xenon are within ~4 eV of experimental measurements. The binding energies of halide ions are not well known, yet it is possible to approximate their highest experimental binding energies with electron affinity measurements. The electron affinity values were found to closely match the calculated DHF energies, with the approximation being increasingly valid for the higher-mass ions of bromide and iodide. These trends and absolute energies were also found to agree with the calculated values of Saha et al. [38,41] and Lindroth and Dahlstrom [39]. The congruence between our calculated DHF binding energy for F^- and the $2p_{3/2}$ energy reported in [39], which utilized a non-relativistic HF theory with exchange, is of particular interest as it implies that despite not being necessary, relativistic effects do not negatively impact Wigner time delay calculations for lighter-mass atomic systems.

Table 1. Calculated and experimental binding energies in eV.

	DHF	Exp. [52,53]
F^- (9 Channels)	$2p_{3/2} = 4.889$ $2p_{1/2} = 4.968$ $2s_{1/2} = 29.334$	3.401 *
Ne (9 Channels)	$2p_{3/2} = 23.083$ $2p_{1/2} = 23.207$ $2s_{1/2} = 52.677$	21.565 21.627 48.365
Cl^- (14 Channels)	$3p_{3/2} = 4.027$ $3p_{1/2} = 4.169$ $3s_{1/2} = 20.132$	3.613 *
Ar (14 Channels)	$3p_{3/2} = 15.995$ $3p_{1/2} = 16.201$ $3s_{1/2} = 35.010$	15.760 15.946 29.307
Br^- (20 Channels)	$4p_{3/2} = 3.565$ $4p_{1/2} = 4.122$ $4s_{1/2} = 19.393$ $3d_{5/2} = 76.152$ $3d_{3/2} = 77.317$	3.364 *
Kr (20 Channels)	$4p_{3/2} = 13.996$ $4p_{1/2} = 14.735$ $4s_{1/2} = 32.320$ $3d_{5/2} = 101.411$ $3d_{3/2} = 102.795$	13.999 14.627 27.464 93.788 [a] 95.038 [a]
I^- (33 Channels)	$5p_{3/2} = 3.089$ $5p_{1/2} = 4.207$ $5s_{1/2} = 16.555$ $4d_{5/2} = 53.897$ $4d_{3/2} = 55.734$	3.059 *
Xe (33 Channels)	$5p_{3/2} = 11.968$ $5p_{1/2} = 13.404$ $5s_{1/2} = 27.487$ $4d_{5/2} = 71.668$ $4d_{3/2} = 73.779$	12.130 13.436 23.361 67.548 [a] 69.537 [a]

* Electron affinity NIST data [54], [a] Krypton and Xenon d-shell data [55].

Table 1 also lists the number of coupled photoionization channels used for the subsequent calculations of the reduced matrix elements. For Neon and Fluorine, all nine possible channels ($2p, 2s, 1s$) were coupled. Argon and Chlorine used 14 channels ($3p, 3s, 2p, 2s$) and omitted the $1s$ channels. Krypton and Bromine used 20 channels ($4p, 4s, 3d, 3p, 3s$) and omitted the $2p, 2s,$ and $1s$ channels. Finally, Xenon and Iodine used 33 channels ($5p, 5s, 4d, 4p, 4s, 3d, 3s$) and omitted the core $2p, 2s,$ and $1s$ channels. The omitted channels could be ignored due to the fact that they are significantly farther away in energy from the other channels and do not have substantial impact at the photon energies of this study.

3.2. Individual Photoionization Channel Phases and Wigner Time Delays of Neon and F$^-$

While a previous study compared the Wigner delays of individual ionization channels in Argon and Chlorine [41], to the best of our knowledge, no similar study has been performed for the lighter pair of Neon and Fluorine. The phase η of each channel was calculated directly by computing the argument of the reduced matrix element associated with each channel (i.e., via Equations (6) and (13)). By simply taking the energy derivative of the phase, the Wigner time delays could also be determined, as seen in Figure 1. Due to their low atomic mass and relatively small number of electrons, relativistic effects do not play a crucial role in time delays in Neon and Fluorine. This is reflected in the behavior of both the phase and Wigner time delays for any given channel, since the j and $\bar{j}$ values of a particular $l \to l+1$ or $l \to l-1$ transition have a negligible impact (e.g., the $2p_{1/2} \to \epsilon s_{1/2}$ and $2p_{3/2} \to \epsilon s_{1/2}$ both produce the same η and Wigner delay). This fact also holds true for the $3p_{1/2}$ and $3p_{3/2}$ channels in both Chlorine and Argon [41], but begins to break down for the lower-lying $3s_{1/2} \to \epsilon p_{1/2}$ and $3s_{1/2} \to \epsilon p_{3/2}$ channels. While not the primary focus of this work, the study of time delays in individual channels is often useful for analyzing the results of the average orbital delays, since it is possible to determine where a given channel dominates in a particular energy region and instructive to see how the average delay results from Wigner delays of individual transitions. For the sake of brevity, the individual channel phases and time delays for Cl$^-$, Ar, Br$^-$, Kr, I$^-$, and Xe are omitted.

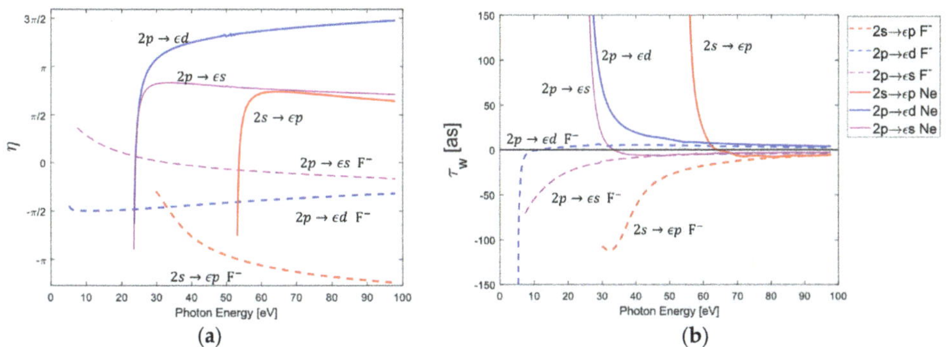

Figure 1. Individual 2s and 2p channel phases (**a**) and Wigner time delays (**b**) for Neon and Fluorine. The individual labels are a descriptive shorthand to describe the identical behavior of multiple ionization channels. For instance, the label $2p \to \epsilon s$ relates to the two ionization channels ($2p_{1/2} \to \epsilon s_{1/2}$ and $2p_{3/2} \to \epsilon s_{1/2}$), while the label $2p \to \epsilon d$ corresponds to the following three channels: ($2p_{1/2} \to \epsilon d_{3/2}, 2p_{3/2} \to \epsilon d_{3/2}$, and $2p_{3/2} \to \epsilon d_{5/2}$). The behavior of $2s \to \epsilon p$ is equivalent to the behavior of the ($2s_{1/2} \to \epsilon p_{1/2}$ and $2s_{1/2} \to \epsilon p_{3/2}$) channels. In the case of Ne, a small autoionization resonance was removed near 48 eV for the $2p \to \epsilon d$ and $2p \to \epsilon s$ channels.

By comparing Figures 1b and 2b, it is apparent that the average $2p$ delay in Ne is dominated by the $2p \to \epsilon d$ channels across all photon energies, whereas the average $2p$ delay in F$^-$ primarily corresponds to $2p \to \epsilon d$ transitions at higher energies and $2p \to \epsilon s$ channels below photon energies near 10 eV. This agrees with the fact that for typical photoionization in noble gases, the $l \rightarrow l + 1$ photoionization channels are known to dominate regardless of energy, with the only exception being Cooper minima. However, in photodetachment, the $l \rightarrow l - 1$ channels dominate near the threshold and the $l \rightarrow l + 1$ channels only begin to dominate once photon energy increases [41,45]. A cursory comparison of the time delays between Neon and Fluorine shows a strong agreement for photon energies greater than 50 eV for any given $2p$ transition. The same can also be said of $2s \rightarrow \epsilon p$ transitions at photon energies above 75 eV. There is, however, a sharp contrast between the time delays at lower energies.

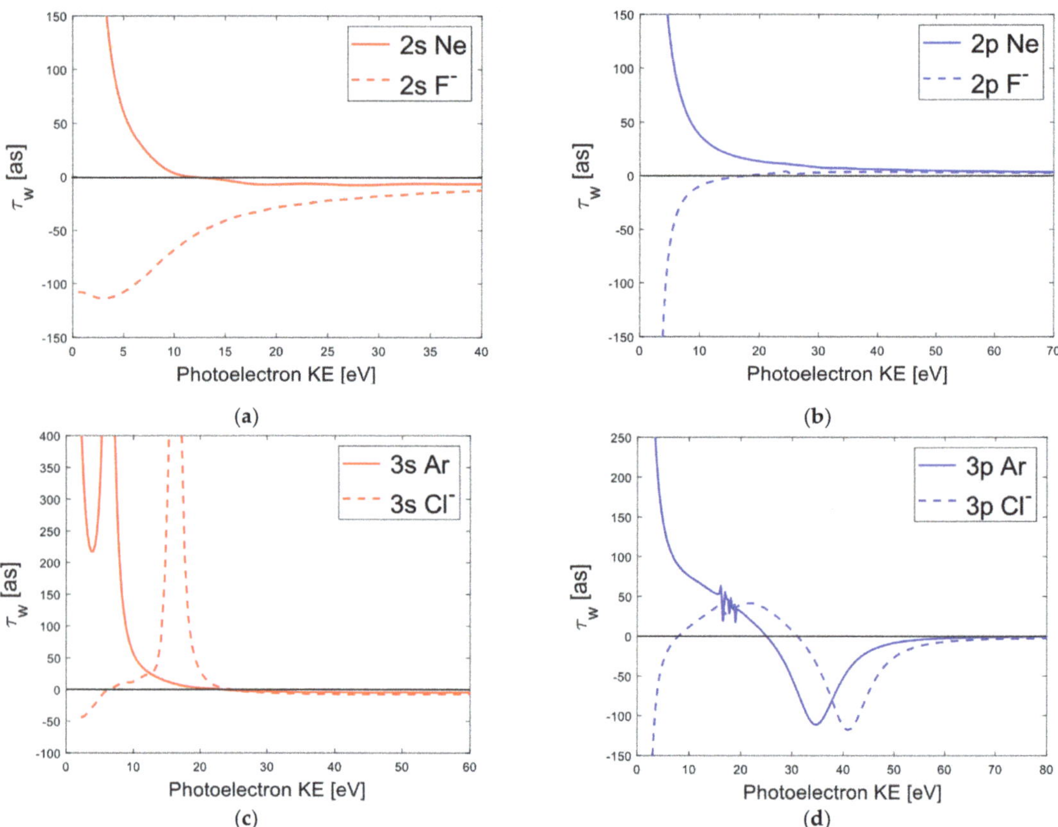

Figure 2. All four subfigures plot the average Wigner time delay with respect to the kinetic energy of the ionized electron. The first row compares the average Wigner time delays of Neon and Fluoride for the $2s$ (**a**) and the $2p$ (**b**) orbitals. The bottom row compares the average Wigner delays for Argon and Chloride with regard to their $3s$ (**c**) and $3p$ (**d**) orbitals. The resonant peaks in the Argon $3p$ delay spectra between 15 and 20 eV correspond to autoionization resonances that were not entirely removed. The average 3p Wigner delay for Cl^- displays a deep minimum of -900 as near 6 eV. This minimum matches that of [41] although it is not shown due to the scale of the figure. The average $2s$ and $3s$ Wigner delays of F^- and Cl^-, respectively, originally displayed oscillations at photoelectron energies below 15 eV. These oscillations were determined to be caused by small variations in the average phase data and were subsequently removed by taking a best fit of the phase data and then computing the energy derivative of that best fit.

3.3. Comparison of Average Wigner Delays for Halogen Ions and Noble Gases

We observe in Figure 2a,b that Neon exhibits the familiar time delay behavior of having a large positive delay near the threshold energies of the $2s$ and $2p$ subshells, which slowly vanishes as photon energy increases. Fluorine, by contrast, exhibits a strong negative delay on the order of -100 as near the $2p$ threshold. This difference can be explained by comparing the differences in the calculated phases for both Fluorine and Neon illustrated in Figure 1a, as each of Neon's ionization channels displays a dramatic increase near their respective threshold energies, while Fluorine's $2s \rightarrow \epsilon p$ and $2p \rightarrow \epsilon s$ channels tend to decrease more gradually over a longer energy range. For noble gas atoms, the Wigner time delay of any given orbital will trend towards positive infinity at energies near the threshold due to the drastic increase in the Coulomb phase, which is known to dominate [40,41], and the individual channels of Ne confirm this trend. The Coulomb phase for negatively charged atoms is essentially nonexistent due to the lack of a long-range Coulomb potential and, therefore, the Wigner delays corresponding to photodetachment do not possess the same behavior of trending towards infinity at low energies. In the case of photodetachment, however, it is still possible for a short-range potential to play a role at energies extremely close to the threshold. For example, in Figure 1a, the phase of the $2p \rightarrow \epsilon d$ ionization channels in Fluorine rapidly decreases in an energy region of ~ 1 eV near threshold and then slowly increases with photon energy. This also explains the sharp increase in the average $2p$ Wigner delays in Fluorine at low photoelectron energies below ~ 10 eV and small positive delay times in the higher photoelectron energy region.

Just as with the individual channel analysis in Fluorine and Neon, we find that the average $2s$ and $2p$ Wigner delays also diverge at lower photoelectron energies and converge towards zero in the large photoelectron kinetic energy limit. The physical picture underlying this vanishing time delay is quite clear, as a photoelectron with high kinetic energy will spend less time near the perturbative effects of the remaining ion and instead behave more similarly to a free electron wave. Conversely, at low kinetic energies, the photoelectron will spend more time near the remaining ion and be more susceptible to collective electron effects. We also observe the same general feature of diverging Wigner delays between the $3s$ and $3p$ states in Chlorine and Argon at low energy (see Figure 2c,d), with the only difference being the introduction of Cooper minima. A Cooper minimum occurs when the transition matrix element changes sign and undergoes a phase shift of π. Typically, this happens when the initial state radial wavefunction contains at least one node and overlaps with the continuum wavefunctions. This is the process responsible for the well-known $3p$ Cooper minima illustrated for Argon and Chlorine in Figure 2d. However, a different mechanism is responsible for the observed minimum in the average $3s$ Wigner delay of Argon and Chlorine. Instead of being the result of a radial node in the initial wavefunction, the behavior of the Cooper minimum in the 3s channel is caused by a π shift in the phases of the $3s \rightarrow \epsilon p$ channels, which occurs due to the result of significant interchannel coupling with the $3p$ ionization channels [38,56,57]. Without the effect of interchannel coupling, this minimum does not appear in the $3s$ ionization channels, which is why the $3s$ minimum can be deemed an "induced Cooper minimum" as it is still the result of a π shift but its origin differs from that of the $3p$ Cooper minimum. However, the location of this induced Cooper minimum in Chlorine appears to be shifted ~ 10 eV higher than the induced Cooper minimum in Argon. A similar shift occurs for the Cooper minimum [58] in the $3p$ spectra, with Chlorine again displaying a higher shift of ~ 6 eV. Because Figure 2 plots both time delays with respect to the kinetic energy of the ionized electron, any difference in the location of the two Cooper minima must be the result of properties not related to the threshold energies of the atomic systems. If the shift was solely caused by a difference in binding energies, the two Wigner delay spectra would overlap.

For the higher-mass systems of Bromine, Krypton, Iodine, and Xenon illustrated in Figure 3, the average Wigner time delays for each orbital were also found to diverge at low photoelectron energies, as Br^- and I^- both display large negative time delays (again on the order of -100 as) in the low energy region. This appears to indicate a universal time delay trend between negative charged halogens and noble gas atoms at photoelectron energies below 20 eV. While it is possible to explain the negative delay times as the result of a negative energy derivative of the phase, a physical explanation is less obvious. If one interprets a positive delay time as the retardation of the photoelectron wave with respect to a free electron wave of the same kinetic energy, then a negative Wigner delay can be interpreted as an acceleration in the outgoing electron wave packet. In the case of the induced Cooper minimum in the average 3s state for Argon at low energy, the "dip" in the delay time can be thought to be the result of induced oscillations in the outer $3p$ subshells that screen the $3s$ electrons [57]. This only can occur for systems where the ns and np states are strongly coupled, which explains the absence of a similar feature in the $2s$ time delay spectra of Neon due to its $2s$ and $2p$ interchannel coupling being much weaker.

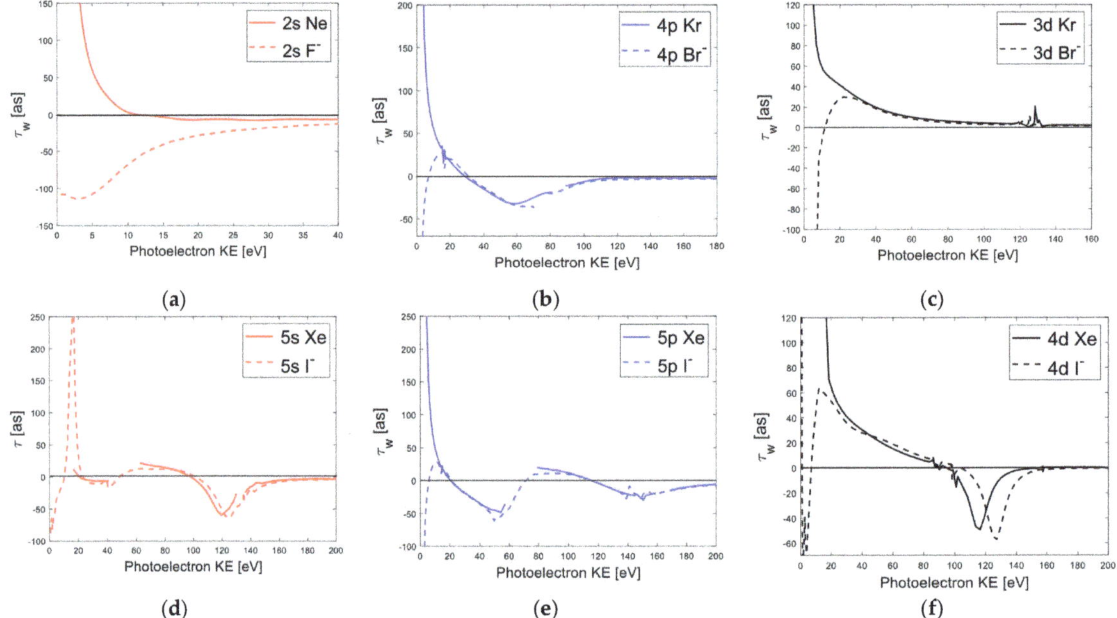

Figure 3. Average Wigner time delays with respect to the kinetic energy of the ionized electron for Krypton and Bromide (**a–c**). Average Wigner time delays for the $5s, 5p,$ and $4d$ states of Xenon and Iodide (**d–f**). The autoionization resonances in Br$^-$ and Kr were removed at the approximate photon energies of $10\,\text{eV}, 76\,\text{eV}, 194\,\text{eV},$ and $201\,\text{eV}$. For I$^-$ and Xe, autoionization resonances were removed near $16\,\text{eV}, 65 - 70\,\text{eV}, 80 - 90\,\text{eV}, 159\,\text{eV}, 176\,\text{eV},$ and $229\,\text{eV}$.

The Cooper minima in the average $5s$ Wigner delay for Xenon have also been explained to be the result of interchannel couplings, with the first Cooper minimum near 35 eV being the result of couplings with the $5p_{1/2}$ and $5p_{3/2}$ states and the second minimum at ~ 120 eV being the result of couplings with the $4d_{3/2}$ and $4d_{5/2}$ states [59,60]. In the case of I$^-$, we also find the same $5s$ Cooper minima at energies approximately equal to those of Xenon. A similar equivalence was observed between the average $5p$ Wigner delays of Iodine and Xenon, as well as for the $4s, 4p,$ and $3d$ delays in Br$^-$ and Kr. However, the minimum in the $4d$ time delay spectra for I$^-$ was found to lie ~ 11 eV higher than for Xenon.

Comparing the relative shifts between Cooper minima for the average Wigner delays of each halogen and noble gas pairing, we find a qualitative trend where the difference decreases as atomic mass increases. The origin of this trend is still not entirely clear, although it could be related to the relative difference in the electronegativities and polarizabilities of the halogen and noble gas atoms, as relative shifts have been observed in the photodetachment cross-sections for different isotopes of Chlorine [61]. A previous theoretical study [62] also reported shifts in the respective $3s$ and $3p$ cross-sections for Ar and Cl$^-$ and other high-Z isoelectronic species that were positively charged (i.e., Sc^{3+}, Mn^{7+}) and concluded that interchannel coupling and initial-state correlation effects can impact the location of Cooper minima, although they do not directly account for the difference in binding energies between Ar and Cl$^-$. Despite not focusing on the relative shift in photoionization cross-sections, our work regarding the 3s Wigner delays between Ar and Cl$^-$ seems to agree with the conclusions of [62], since the location and appearance of an "induced Cooper minimum" in Figure 2c was found to be the result of significant coupling to the $3p$ ionization channels in Argon. However, more work must be carried out to determine the origin for the relative shift in the Cooper minima between isoelectronic systems.

4. Conclusions

In this work, we performed RRPA Wigner time delay calculations for the halogen ion and noble gas pairings of $(F^-, Ne), (Cl^-, Ar), (Br^-, Kr),$ and (I^-, Xe). The individual photoionization channels were then averaged to obtain the average delay times for the highest occupied $s, p,$ and d states. The Wigner delays were plotted with respect to photoelectron energy in order to account for energy shifts due to differences in binding energies. For photoelectron energies below 20 eV, negatively charged halogen ions exhibit large negative Wigner delays that sharply increase. This qualitative difference is due to the absence of a large Coulomb phase, which is known to dominate the Wigner delay behavior of noble gases at low energies. Overall, the qualitative time delay behaviors of halogens and noble gases were found to be similar, with each pairing displaying the same general features and Cooper minima. The location of the $3p$ Cooper minimum in Cl$^-$ was found to be shifted ~6 eV higher than Ar. A similar shift of ~10 eV was noticed in the 3s Wigner delays between Ar and Cl$^-$. As atomic mass increased, this relative shift between Cooper minima was found to decrease, except in the case of the $4d$ Cooper minima of I$^-$, which displayed an ~11 eV shift above that of Xe. The physical process underlying this relative shift is still not clear, indicating the need for more detailed analysis of negatively charged ions and noble gas atoms in the regions of Cooper minima. Molecular photoionization studies are becoming of greater interest to the attosecond community, with Iodine-containing molecules having already been favored due to iodine's $4d$ orbitals. Our findings indicate that the average p and s Wigner delays of Br$^-$ and I$^-$ are the most similar to the average d orbital Wigner delays of their noble gas counterparts Kr and Xe. This implies that Bromine and Iodine are the best halogens for studying time delay phenomena in molecular systems.

Author Contributions: A.S.L. and B.G. conceptualized the study. B.G. performed the calculations and prepared the original draft. All authors have read and agreed to the published version of the manuscript.

Funding: This work was supported by the U.S. Department of Energy, Office of Basic Energy Sciences, Atomic, Molecular and Optical Sciences Program under Award No. DE-SC0022093.

Institutional Review Board Statement: Not applicable.

Informed Consent Statement: Not applicable.

Data Availability Statement: Simulation data can be provided upon reasonable request.

Acknowledgments: We thank Subhasish Saha for valuable discussions.

Conflicts of Interest: The authors declare no conflict of interest.

Appendix A Determining $\langle \bar{a} || Q_J^{(\lambda)} || a \rangle_{RRPA}$

From Equation (3) in Section 2, it was possible to write the transition amplitude in terms of a single-electron reduced matrix element, $\langle \bar{a} || Q_J^{(\lambda)} || a \rangle$, describing the transition between an initial state, $a = n\kappa$, and final state, $\bar{a} = \left(E, \bar{\kappa} \right)$. In order to generalize this to the multi-electron case, one must start with Equation (2) and essentially repeat the same process that was carried out with the single-electron case (i.e., perform a multipole expansion of $\vec{A}$ for both the positive frequency perturbations and negative frequency perturbation terms $w_{i\pm}$).

$$T = \sum_{i=1}^{N} \int d^3r \left(w_{i+}^{\dagger} \vec{\alpha} \cdot \vec{A} u_i + u_i^{\dagger} \vec{\alpha} \cdot \vec{A} w_{i-} \right) \rightarrow T = \sum_{i=1}^{N} \int d^3r \left(w_{i+}^{\dagger} \vec{\alpha} \cdot \vec{a}_{JM}^{(\lambda)} u_i + u_i^{\dagger} \vec{\alpha} \cdot \vec{a}_{JM}^{(\lambda)} w_{i-} \right) \quad \text{(A1)}$$

It was the partial wave expansion given by Equation (40) in Johnson and Lin's original paper [49] that, when substituted into Equation (3), resulted in the single-electron transition amplitude described by Equation (4). By following the same process of describing the final continuum state as a partial wave expansion, Equation (A1) leads to a generalized formula for the multi-electronic transition amplitude, which is identical to Equation (4) in Section 2, with the only necessary modification occurring to $D_{lj \to \bar{l}\bar{j}}$ so that it now includes the multi-electron reduced matrix element.

$$D_{lj \to \bar{l}\bar{j}} = i^{1-\bar{l}} e^{i\delta_{\bar{\kappa}}} \langle \bar{a} || Q_J^{(\lambda)} || a \rangle \rightarrow D_{lj \to \bar{l}\bar{j}} = i^{1-\bar{l}} e^{i\delta_{\bar{\kappa}}} \langle \bar{a} || Q_J^{(\lambda)} || a \rangle_{RRPA} \quad \text{(A2)}$$

where the reduced matrix element is now described by

$$\langle \bar{a} || Q_J^{(\lambda)} || a \rangle_{RRPA} = \sum_{\bar{b}} \left(\langle \bar{b}_+ || Q_J^{(\lambda)} || b \rangle + \langle \bar{b}_- || Q_J^{(\lambda)} || b \rangle \right) \quad \text{(A3)}$$

Therefore, instead of considering only the transition between one set of waves a and $\bar{a}$ as was the case before with Equations (4) and (6), we are now summing all of the possible transitions between the remaining waves b and $\bar{b}$ under the condition that the waves $b \to \bar{b}$ vanish in the asymptotic limit $r \to \infty$. The reduced matrix elements are identical to the single-electron matrix elements in every way except in terms of the radial integral $R_J^{(\lambda)} \left(\bar{b}_\pm, b \right)$, which is given as Equation (46) in [49].

$$\langle \bar{b}_\pm || Q_J^{(\lambda)} || b \rangle = (-1)^{j+1/2} [\bar{j}][j] \begin{pmatrix} j & \bar{j} & J \\ -1/2 & 1/2 & 0 \end{pmatrix} \pi \left(\bar{l}, l, J - \lambda + 1 \right) R_J^{(\lambda)} \left(\bar{b}_\pm, b \right) \quad \text{(A4)}$$

References

1. Hentschel, M.; Kienberger, R.; Spielmann, C.; Reider, G.A.; Milosevic, N.; Brabec, T.; Corkum, P.; Heinzmann, U.; Drescher, M.; Krausz, F. Attosecond Metrology. *Nature* **2001**, *414*, 509–513. [CrossRef] [PubMed]
2. Krausz, F. From Femtochemistry to Attophysics. *Phys. World* **2001**, *14*, 41. [CrossRef]
3. Drescher, M.; Hentschel, M.; Kienberger, R.; Tempea, G.; Spielmann, C.; Reider, G.A.; Corkum, P.B.; Krausz, F. X-Ray Pulses Approaching the Attosecond Frontier. *Science* **2001**, *291*, 1923–1927. [CrossRef] [PubMed]
4. Varjú, K.; Johnsson, P.; Mauritsson, J.; L'Huillier, A.; López-Martens, R. Physics of Attosecond Pulses Produced via High Harmonic Generation. *Am. J. Phys.* **2009**, *77*, 389–395. [CrossRef]
5. Orfanos, I.; Makos, I.; Liontos, I.; Skantzakis, E.; Förg, B.; Charalambidis, D.; Tzallas, P. Attosecond Pulse Metrology. *APL Photonics* **2019**, *4*, 080901. [CrossRef]
6. Frank, F.; Arrell, C.; Witting, T.; Okell, W.A.; McKenna, J.; Robinson, J.S.; Haworth, C.A.; Austin, D.; Teng, H.; Walmsley, I.A.; et al. Invited Review Article: Technology for Attosecond Science. *Rev. Sci. Instrum.* **2012**, *83*, 071101. [CrossRef]
7. Itatani, J.; Quéré, F.; Yudin, G.L.; Ivanov, M.Y.; Krausz, F.; Corkum, P.B. Attosecond Streak Camera. *Phys. Rev. Lett.* **2002**, *88*, 173903. [CrossRef]
8. Zaïr, A.; Mével, E.; Cormier, E.; Constant, E. Ultrastable Collinear Delay Control Setup for Attosecond IR-XUV Pump–Probe Experiment. *J. Opt. Soc. Am. B JOSAB* **2018**, *35*, A110–A115. [CrossRef]
9. Klünder, K.; Dahlström, J.M.; Gisselbrecht, M.; Fordell, T.; Swoboda, M.; Guénot, D.; Johnsson, P.; Caillat, J.; Mauritsson, J.; Maquet, A.; et al. Probing Single-Photon Ionization on the Attosecond Time Scale. *Phys. Rev. Lett.* **2011**, *106*, 143002. [CrossRef]
10. Vos, J.; Cattaneo, L.; Patchkovskii, S.; Zimmermann, T.; Cirelli, C.; Lucchini, M.; Kheifets, A.; Landsman, A.S.; Keller, U. Orientation-Dependent Stereo Wigner Time Delay and Electron Localization in a Small Molecule. *Science* **2018**, *360*, 1326–1330. [CrossRef]

11. Kheifets, A.S. Wigner Time Delay in Atomic Photoionization. *J. Phys. B At. Mol. Opt. Phys.* **2023**, *56*, 022001. [CrossRef]
12. Schultze, M.; Fieß, M.; Karpowicz, N.; Gagnon, J.; Korbman, M.; Hofstetter, M.; Neppl, S.; Cavalieri, A.L.; Komninos, Y.; Mercouris, T.; et al. Delay in Photoemission. *Science* **2010**, *328*, 1658–1662. [CrossRef]
13. Palatchi, C.; Dahlström, J.M.; Kheifets, A.S.; Ivanov, I.A.; Canaday, D.M.; Agostini, P.; DiMauro, L.F. Atomic Delay in Helium, Neon, Argon and Krypton*. *J. Phys. B At. Mol. Opt. Phys.* **2014**, *47*, 245003. [CrossRef]
14. Guénot, D.; Klünder, K.; Arnold, C.L.; Kroon, D.; Dahlström, J.M.; Miranda, M.; Fordell, T.; Gisselbrecht, M.; Johnsson, P.; Mauritsson, J.; et al. Photoemission-Time-Delay Measurements and Calculations Close to the 3s-Ionization-Cross-Section Minimum in Ar. *Phys. Rev. A* **2012**, *85*, 053424. [CrossRef]
15. Kheifets, A.S.; Ivanov, I.A. Delay in Atomic Photoionization. *Phys. Rev. Lett.* **2010**, *105*, 233002. [CrossRef]
16. Huppert, M.; Jordan, I.; Baykusheva, D.; von Conta, A.; Wörner, H.J. Attosecond Delays in Molecular Photoionization. *Phys. Rev. Lett.* **2016**, *117*, 093001. [CrossRef]
17. Caillat, J.; Maquet, A.; Haessler, S.; Fabre, B.; Ruchon, T.; Salières, P.; Mairesse, Y.; Taïeb, R. Attosecond Resolved Electron Release in Two-Color Near-Threshold Photoionization of N_2. *Phys. Rev. Lett.* **2011**, *106*, 093002. [CrossRef]
18. Serov, V.V.; Kheifets, A.S. Time Delay in XUV/IR Photoionization of H_2O. *J. Chem. Phys.* **2017**, *147*, 204303. [CrossRef]
19. Baykusheva, D.; Wörner, H.J. Theory of Attosecond Delays in Molecular Photoionization. *J. Chem. Phys.* **2017**, *146*, 124306. [CrossRef]
20. Haessler, S.; Fabre, B.; Higuet, J.; Caillat, J.; Ruchon, T.; Breger, P.; Carré, B.; Constant, E.; Maquet, A.; Mével, E.; et al. Phase-Resolved Attosecond near-Threshold Photoionization of Molecular Nitrogen. *Phys. Rev. A* **2009**, *80*, 011404. [CrossRef]
21. Zhang, C.-H.; Thumm, U. Streaking and Wigner Time Delays in Photoemission from Atoms and Surfaces. *Phys. Rev. A* **2011**, *84*, 033401. [CrossRef]
22. Borrego-Varillas, R.; Lucchini, M.; Nisoli, M. Attosecond Spectroscopy for the Investigation of Ultrafast Dynamics in Atomic, Molecular and Solid-State Physics. *Rep. Prog. Phys.* **2022**, *85*, 066401. [CrossRef] [PubMed]
23. Tao, Z.; Chen, C.; Szilvási, T.; Keller, M.; Mavrikakis, M.; Kapteyn, H.; Murnane, M. Direct Time-Domain Observation of Attosecond Final-State Lifetimes in Photoemission from Solids. *Science* **2016**, *353*, 62–67. [CrossRef] [PubMed]
24. Wigner, E.P. Lower Limit for the Energy Derivative of the Scattering Phase Shift. *Phys. Rev.* **1955**, *98*, 145–147. [CrossRef]
25. Smith, F.T. Lifetime Matrix in Collision Theory. *Phys. Rev.* **1960**, *118*, 349–356. [CrossRef]
26. Dahlström, J.M.; Guénot, D.; Klünder, K.; Gisselbrecht, M.; Mauritsson, J.; L'Huillier, A.; Maquet, A.; Taïeb, R. Theory of Attosecond Delays in Laser-Assisted Photoionization. *Chem. Phys.* **2013**, *414*, 53–64. [CrossRef]
27. Dahlström, J.M.; Carette, T.; Lindroth, E. Diagrammatic Approach to Attosecond Delays in Photoionization. *Phys. Rev. A* **2012**, *86*, 061402. [CrossRef]
28. Pazourek, R.; Nagele, S.; Burgdörfer, J. Time-Resolved Photoemission on the Attosecond Scale: Opportunities and Challenges. *Faraday Discuss.* **2013**, *163*, 353–376. [CrossRef]
29. Wätzel, J.; Moskalenko, A.S.; Pavlyukh, Y.; Berakdar, J. Angular Resolved Time Delay in Photoemission. *J. Phys. B At. Mol. Opt. Phys.* **2014**, *48*, 025602. [CrossRef]
30. Heuser, S.; Galán, Á.J.; Cirelli, C.; Marante, C.; Sabbar, M.; Boge, R.; Lucchini, M.; Gallmann, L.; Ivanov, I.; Kheifets, A.S.; et al. Angular Dependence of Photoemission Time Delay in Helium. *Phys. Rev. A* **2016**, *94*, 063409. [CrossRef]
31. Cirelli, C.; Marante, C.; Heuser, S.; Petersson, C.L.M.; Galán, Á.J.; Argenti, L.; Zhong, S.; Busto, D.; Isinger, M.; Nandi, S.; et al. Anisotropic Photoemission Time Delays Close to a Fano Resonance. *Nat. Commun.* **2018**, *9*, 955. [CrossRef]
32. Ivanov, I.A.; Kheifets, A.S. Angle-Dependent Time Delay in Two-Color XUV+IR Photoemission of He and Ne. *Phys. Rev. A* **2017**, *96*, 013408. [CrossRef]
33. George, J.; Pradhan, G.B.; Rundhe, M.; Jose, J.; Aravind, G.; Deshmukh, P.C. Autoionization Resonances in the Argon Iso-Electronic Sequence. *Can. J. Phys.* **2012**, *90*, 547–555. [CrossRef]
34. Magrakvelidze, M.; Madjet, M.E.-A.; Chakraborty, H.S. Attosecond Delay of Xenon 4d Photoionization at the Giant Resonance and Cooper Minimum. *Phys. Rev. A* **2016**, *94*, 013429. [CrossRef]
35. Magrakvelidze, M.; Madjet, M.E.-A.; Dixit, G.; Ivanov, M.; Chakraborty, H.S. Attosecond Time Delay in Valence Photoionization and Photorecombination of Argon: A Time-Dependent Local-Density-Approximation Study. *Phys. Rev. A* **2015**, *91*, 063415. [CrossRef]
36. Jordan, I.; Huppert, M.; Pabst, S.; Kheifets, A.S.; Baykusheva, D.; Wörner, H.J. Spin-Orbit Delays in Photoemission. *Phys. Rev. A* **2017**, *95*, 013404. [CrossRef]
37. Keating, D.A.; Manson, S.T.; Dolmatov, V.K.; Mandal, A.; Deshmukh, P.C.; Naseem, F.; Kheifets, A.S. Intershell-Correlation-Induced Time Delay in Atomic Photoionization. *Phys. Rev. A* **2018**, *98*, 013420. [CrossRef]
38. Saha, S.; Mandal, A.; Jose, J.; Varma, H.R.; Deshmukh, P.C.; Kheifets, A.S.; Dolmatov, V.K.; Manson, S.T. Relativistic Effects in Photoionization Time Delay near the Cooper Minimum of Noble-Gas Atoms. *Phys. Rev. A* **2014**, *90*, 053406. [CrossRef]
39. Lindroth, E.; Dahlström, J.M. Attosecond Delays in Laser-Assisted Photodetachment from Closed-Shell Negative Ions. *Phys. Rev. A* **2017**, *96*, 013420. [CrossRef]
40. Pi, L.-W.; Landsman, A.S. Attosecond Time Delay in Photoionization of Noble-Gas and Halogen Atoms. *Appl. Sci.* **2018**, *8*, 322. [CrossRef]
41. Saha, S.; Jose, J.; Deshmukh, P.C.; Aravind, G.; Dolmatov, V.K.; Kheifets, A.S.; Manson, S.T. Wigner Time Delay in Photodetachment. *Phys. Rev. A* **2019**, *99*, 043407. [CrossRef]

42. Saha, S.; Deshmukh, P.C.; Kheifets, A.S.; Manson, S.T. Dominance of Correlation and Relativistic Effects on Photodetachment Time Delay Well above Threshold. *Phys. Rev. A* **2019**, *99*, 063413. [CrossRef]
43. Banerjee, S.; Deshmukh, P.C.; Kheifets, A.S.; Manson, S.T. Effects of Spin-Orbit-Interaction-Activated Interchannel Coupling on Photoemission Time Delay. *Phys. Rev. A* **2020**, *101*, 043411. [CrossRef]
44. Saha, S.; Jose, J.; Deshmukh, P.C.; Kheifets, A.S.; Dolmatov, V.K.; Manson, S.T. Effects of Relativistic Interactions in Photodetachment Time Delay of Br^{-}. *J. Phys. Conf. Ser.* **2020**, *1412*, 092013. [CrossRef]
45. Banerjee, S.; Aarthi, G.; Saha, S.; Aravind, G.; Deshmukh, P.C. Time Delay in Negative Ion Photodetachment. *Phys. Scr.* **2021**, *96*, 114005. [CrossRef]
46. Lindle, D.W.; Kobrin, P.H.; Truesdale, C.M.; Ferrett, T.A.; Heimann, P.A.; Kerkhoff, H.G.; Becker, U.; Shirley, D.A. Inner-Shell Photoemission from the Iodine Atom in CH_3I. *Phys. Rev. A* **1984**, *30*, 239–244. [CrossRef]
47. Magrakvelidze, M.; Chakraborty, H. Attosecond Time Delays in the Valence Photoionization of Xenon and Iodine at Energies Degenerate with Core Emissions. *J. Phys. Conf. Ser.* **2017**, *875*, 022015. [CrossRef]
48. Biswas, S.; Förg, B.; Ortmann, L.; Schötz, J.; Schweinberger, W.; Zimmermann, T.; Pi, L.; Baykusheva, D.; Masood, H.A.; Liontos, I.; et al. Probing Molecular Environment through Photoemission Delays. *Nat. Phys.* **2020**, *16*, 778–783. [CrossRef]
49. Johnson, W.R.; Lin, C.D. Multichannel Relativistic Random-Phase Approximation for the Photoionization of Atoms. *Phys. Rev. A* **1979**, *20*, 964–977. [CrossRef]
50. Kheifets, A.; Mandal, A.; Deshmukh, P.C.; Dolmatov, V.K.; Keating, D.A.; Manson, S.T. Relativistic Calculations of Angle-Dependent Photoemission Time Delay. *Phys. Rev. A* **2016**, *94*, 013423. [CrossRef]
51. Mandal, A.; Deshmukh, P.C.; Kheifets, A.S.; Dolmatov, V.K.; Manson, S.T. Angle-Resolved Wigner Time Delay in Atomic Photoionization: The 4d Subshell of Free and Confined Xe. *Phys. Rev. A* **2017**, *96*, 053407. [CrossRef]
52. Kramida, A.; Ralchenko, Y.; Reader, J.; NIST ASD Team. *NIST Atomic Spectra Database (Version 5.10)*, National Institute of Standards and Technology: Gaithersburg, MD, USA, 2022. Available online: https://physics.nist.gov/asd(accessed on 22 February 2023).
53. NIST: Atomic Spectra Database—Energy Levels Form. Available online: https://physics.nist.gov/PhysRefData/ASD/levels_form.html (accessed on 22 February 2023).
54. CCCBDB Electron Affinities. Available online: https://cccbdb.nist.gov/elecaff1x.asp (accessed on 22 February 2023).
55. King, G.C.; Tronc, M.; Read, F.H.; Bradford, R.C. An Investigation of the Structure near the L2,3 Edges of Argon, the M4,5 Edges of Krypton and the N4,5 Edges of Xenon, Using Electron Impact with High Resolution. *J. Phys. B At. Mol. Phys.* **1977**, *10*, 2479. [CrossRef]
56. Amusia, M.Y.; Ivanov, V.K.; Cherepkov, N.A.; Chernysheva, L.V. Interference Effects in Photoionization of Noble Gas Atoms Outer S-Subshells. *Phys. Lett. A* **1972**, *40*, 361–362. [CrossRef]
57. Hammerland, D.; Zhang, P.; Bray, A.; Perry, C.F.; Kuehn, S.; Jojart, P.; Seres, I.; Zuba, V.; Varallyay, Z.; Osvay, K.; et al. Effect of Electron Correlations on Attosecond Photoionization Delays in the Vicinity of the Cooper Minima of Argon. *arXiv* **2019**. Available online: https://arxiv.org/pdf/1907.01219 (accessed on 27 February 2023).
58. Fano, U.; Cooper, J.W. Spectral Distribution of Atomic Oscillator Strengths. *Rev. Mod. Phys.* **1968**, *40*, 441–507. [CrossRef]
59. Ganesan, A.; Banerjee, S.; Deshmukh, P.C.; Manson, S.T. Photoionization of Xe 5s: Angular Distribution and Wigner Time Delay in the Vicinity of the Second Cooper Minimum. *J. Phys. B At. Mol. Opt. Phys.* **2020**, *53*, 225206. [CrossRef]
60. Aarthi, G.; Jose, J.; Deshmukh, S.; Radojevic, V.; Deshmukh, P.C.; Manson, S.T. Photoionization Study of Xe 5s: Ionization Cross Sections and Photoelectron Angular Distributions. *J. Phys. B At. Mol. Opt. Phys.* **2014**, *47*, 025004. [CrossRef]
61. Berzinsh, U.; Gustafsson, M.; Hanstorp, D.; Klinkmüller, A.; Ljungblad, U.; Mårtensson-Pendrill, A.-M. Isotope Shift in the Electron Affinity of Chlorine. *Phys. Rev. A* **1995**, *51*, 231–238. [CrossRef]
62. Jose, J.; Pradhan, G.B.; Radojević, V.; Manson, S.T.; Deshmukh, P.C. Electron Correlation Effects near the Photoionization Threshold: The Ar Isoelectronic Sequence. *J. Phys. B At. Mol. Opt. Phys.* **2011**, *44*, 195008. [CrossRef]

 *atoms*

Article

Projectile Coherence Effects in Twisted Electron Ionization of Helium

A. L. Harris

Physics Department, Illinois State University, Normal, IL 61790, USA; alharri@ilstu.edu

Abstract: Over the last decade, it has become clear that for heavy ion projectiles, the projectile's transverse coherence length must be considered in theoretical models. While traditional scattering theory often assumes that the projectile has an infinite coherence length, many studies have demonstrated that the effect of projectile coherence cannot be ignored, even when the projectile-target interaction is within the perturbative regime. This has led to a surge in studies that examine the effects of the projectile's coherence length. Heavy-ion collisions are particularly well-suited to this because the projectile's momentum can be large, leading to a small deBroglie wavelength. In contrast, electron projectiles that have larger deBroglie wavelengths and coherence effects can usually be safely ignored. However, the recent demonstration of sculpted electron wave packets opens the door to studying projectile coherence effects in electron-impact collisions. We report here theoretical triple differential cross-sections (TDCSs) for the electron-impact ionization of helium using Bessel and Laguerre-Gauss projectiles. We show that the projectile's transverse coherence length affects the shape and magnitude of the TDCSs and that the atomic target's position within the projectile beam plays a significant role in the probability of ionization. We also demonstrate that projectiles with large coherence lengths result in cross-sections that more closely resemble their fully coherent counterparts.

Keywords: ionization; coherence; twisted electrons; Laguerre-Gauss beam

1. Introduction

In traditional charged particle scattering theory, the incident projectile is typically considered to be delocalized with an infinitely large coherence. However, in the last decade, it has been shown for heavy ion projectiles that a finite projectile coherence length can significantly alter the collision cross-sections and must be considered when comparing theoretical results with experimental data [1–10]. In these cases, the width of the impinging particle wave packet can be similar in size or smaller than the target width.

For heavy-ion collisions, the effect of the projectile's finite coherence length went unnoticed for many decades. During this time, experimental measurements were only possible for total or single differential cross-sections, and theoretical models were limited to collisions with small perturbation parameters (ratio of projectile charge to speed). In many cases, the agreement between experiment and theory for less differential cross-sections under small perturbation parameters was quite satisfactory (e.g., [11,12]).

In more recent years, it became possible to perform fully differential cross-section measurements, which opened the door to more rigorous theory tests [11,13–19]. In some of the initial studies of fully differential cross-sections for the ionization of helium by heavy-ion impact, significant discrepancies were observed between experiment and theory, even at small perturbation parameters where theory was expected to perform well [13,14,20–23]. Many possible explanations were suggested [16,20,23–31], but it was not until the projectile's coherence properties were considered that a satisfactory explanation was found [1,2,6,7,10]. These experiments demonstrated that projectile coherence cannot be ignored in heavy-ion collisions. Since that time, numerous theoretical and experimental studies have demonstrated the effects of coherence length on collision cross-sections, as well as the ability to control the projectile coherence length [1–7,10,32–39].

Citation: Harris, A.L. Projectile Coherence Effects in Twisted Electron Ionization of Helium. *Atoms* **2023**, *11*, 79. https://doi.org/10.3390/atoms11050079

Academic Editors: Himadri S. Chakraborty and Hari R. Varma

Received: 28 February 2023
Revised: 25 April 2023
Accepted: 26 April 2023
Published: 3 May 2023

A projectile's transverse coherence length is proportional to its deBroglie wavelength, which is inversely proportional to projectile momentum [40,41]. Thus, one technique for controlling projectile coherence is through the alteration of the projectile's momentum. This control can be readily achieved with heavy ion projectiles by changing either the projectile's energy or ion type (i.e., mass) [1–4,32].

For electron projectiles, the control of coherence length through momentum is more challenging due to their small mass. Even at high energies, the electron's wavelength is large, leading to a coherence length that is generally larger than the target width. It is, however, still possible to control an electron projectile's coherence length through wave packet sculpting. In particular, electron vortex projectiles have been experimentally demonstrated [42–45], and these sculpted wave packets offer a method to control projectile coherence in electrons. To date, electron vortex projectiles have been demonstrated in the form of Bessel and Airy electrons. These sculpted (or twisted) electron wave packets differ from their traditional plane wave counterparts in several ways. They can have quantized non-zero orbital angular momentum, which, during a collision, can be transferred to the target or ionized electrons [46–49]. This leads to possible applications for the orientation and rotation control of individual atoms and molecules through electron vortex collisions [50–53]. Twisted electrons also have non-zero transverse linear momentum, which has been shown to alter the distribution of electrons in ionization collisions [54,55].

Several studies on electron-atom and electron-molecule collision cross-sections have been performed for Bessel projectiles, and they have shown that the use of an electron vortex projectile alters the collision cross-section [9,46–48,54–64]. For elastic scattering [58], it was shown that the projectile maintains its vortex properties throughout the collision process. For excitation collisions [47], orbital angular momentum was transferred from the projectile to the target atom, resulting in the alteration of the selection rules. For ionization collisions [46,54,64], it was shown that the orbital angular momentum of the projectile can be transferred to the ionized electron and that the projectile's momentum uncertainty alters the angular distribution of ejected electrons. For the ionization of helium by electron vortex projectile, it was also shown that the projectile's transverse momentum could result in the out-of-plane emission of the ejected electron, an outcome that is not possible with plane wave electrons [55]. For the ionization of H_2, the angular distribution of ionized electrons was shown to depend on the orbital angular momentum of the projectile [56].

Here, we present theoretical triple differential cross-sections (TDCSs) for the electron-impact ionization of helium using Laguerre-Gauss and Bessel projectiles. We show that the localized nature of the LG projectile causes the binary peak to shift to larger ejected electron angles and enhances the recoil peak. As the projectile becomes less localized, the cross-sections more closely resemble their delocalized counterparts. We also show that the atomic target's transverse position within the projectile beam can significantly alter the magnitude of the cross-section. Our results demonstrate that LG projectiles can be used to control the coherence length for electron projectiles and that changing the coherence length has observable effects on the collision cross-section.

The remainder of the paper is organized as follows. Section 2 contains details of the theoretical treatment. Section 3 presents the results for the LG and Bessel projectiles. Section 4 contains a summary of the work.

2. Theory

To calculate the TDCSs, we used the perturbative first Born approximation (FBA) [46,54]. For the projectile energies and scattering geometries used here, this level of approximation contains the relevant physics and captures the qualitative features of the TDCS. Within the FBA, the TDCS is proportional to the square of the transition matrix T_{fi}^V

$$\frac{\mathrm{d}^3\sigma}{d\Omega_1 d\Omega_2 dE_2} = \mu_{pa}^2 \mu_{ie} \frac{k_f k_e}{k_i} \left| T_{fi}^V \right|^2 \tag{1}$$

with

$$T_{fi}^V = -(2\pi)^{3/2} < \Psi_f | V_i | \Psi_i^V >$$

(2)

Here, μ_{ie} is the reduced mass of the He$^+$ ion and the ionized electron, μ_{pa} is the reduced mass of the projectile and target atom, $\vec{k}_i$ is the momentum of the incident projectile, $\vec{k}_f$ is the momentum of the scattered projectile, and $\vec{k}_e$ is the momentum of the ionized electron. Equation (2) can be written as an integral over all of position space for each of the particles in the collision by inserting complete sets of position states. Cylindrical coordinates (ρ_1, φ_1, z_1) are used to represent the projectile wave functions, and spherical coordinates $(r_2, \theta_2, \varphi_2)$ are used for the atomic electron. With this geometry, the projectile momenta can be written in terms of their respective longitudinal and transverse components as $\vec{k}_i = k_{i\perp}\hat{\rho}_{1i} + k_{iz}\hat{z}_1$ and $\vec{k}_f = k_{f\perp}\hat{\rho}_{1f} + k_{fz}\hat{z}_1$. We consider here the coplanar scattering geometry, in which the incident projectile, final projectile, and ionized electron momentum lie in the same plane. The incident projectile propagates along the Z-axis, and the scattered projectile lies in the x–z plane with its transverse momentum along the positive X-axis.

The initial state wave function is expressed as a product of the incident vortex wave function $\chi_{\vec{k}_i}^V \left(\vec{r}_1 \right)$ and the target atom wave function $\Phi\left(\vec{r}_2 \right)$

$$\Psi_i^V = \chi_{\vec{k}_i}^V \left(\vec{r}_1 \right) \Phi\left(\vec{r}_2 \right)$$

(3)

The incident projectile vortex beam may be either a delocalized Bessel beam or a localized Laguerre-Gauss beam. One unique feature of both Bessel and LG vortex beams is that they are non-uniform in the transverse direction with a well-defined center of symmetry. Therefore, their transverse alignment relative to the atomic target must be considered. To account for this alignment, an offset vector $\vec{b}$ (i.e., impact parameter) is introduced such that $\vec{b}$ points transversely from the atomic scattering center to the symmetry center of the impinging vortex projectile (Figure 1).

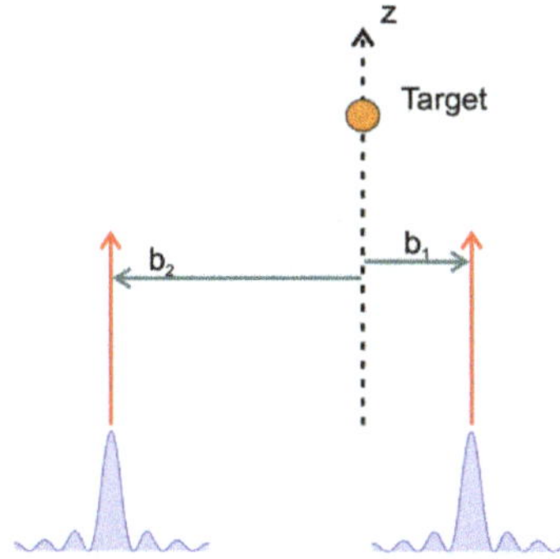

Figure 1. Schematic of incident Bessel projectile impinging on a target atom. Because the Bessel wave function is not uniform in the transverse direction and has a well-defined center of symmetry, an offset vector (or impact parameter) $\vec{b}$ must be defined. Projectiles with two possible values of $\vec{b}_1, \vec{b}_2$ are shown ($\vec{b}_1, \vec{b}_2$). The shaded blue region is the transverse profile of an incident Bessel projectile. The red arrows indicate the propagation direction of the incident projectile.

The wave function for the Bessel projectile with $\vec{b} = 0$ is given by

$$\chi_{\vec{k}_i}^V \left(\vec{r}_1, \vec{b} = 0 \right) = \chi_{\vec{k}_i,l}^B \left(\vec{r}_1, \vec{b} = 0 \right) = \frac{e^{il\varphi_1}}{2\pi} J_l(k_{i\perp}\rho_1)e^{ik_{iz}z_1}$$

(4)

where $J_l(k_{i\perp}\rho_1)$ is the Bessel function with orbital angular momentum l. This expression can conveniently be rewritten as a superposition of tilted plane waves [58], such that

$$\chi^B_{\vec{k}_i,l}\left(\vec{r}_1,\vec{b}=0\right) = \frac{(-i)^l}{(2\pi)^2} \int_0^{2\pi} d\phi_{ki}\, e^{il\phi_{ki}} e^{i\vec{k}_i\cdot\vec{r}_1} \tag{5}$$

For an off-center projectile with $\vec{b}\neq 0$, the Bessel addition theorem [65] can be used to express the Bessel wave function as

$$\chi^B_{\vec{k}_i,l}\left(\vec{r}_1,\vec{b}\right) = \sum_{-\infty}^{\infty} e^{-im\varphi_b} J_m(k_{i\perp}b) \frac{e^{ik_{iz}z}}{\sqrt{2\pi}} J_{l+m}(k_{i\perp}r_1) \frac{e^{il\varphi_1}}{\sqrt{2\pi}} \tag{6}$$

This can, in turn, be expressed in terms of a superposition of plane waves using Equation (5):

$$\chi^B_{\vec{k}_i,l}\left(\vec{r}_1,\vec{b}\right) = \sum_{-\infty}^{\infty} e^{-im\varphi_b} J_m(k_{i\perp}b) \frac{(-i)^{l+m}}{(2\pi)^2} \int_0^{2\pi} d\phi_{ki}\, e^{i(l+m)\phi_{ki}} e^{i\vec{k}_i\cdot\vec{r}_1} \tag{7}$$

The LG beam for $\vec{b}=0$ is given by [58]:

$$\chi^V_{\vec{k}_i}\left(\vec{r}_1,\vec{b}=0\right) = \chi^{LG}_{\vec{k}_i,l}\left(\vec{r}_1,\vec{b}=0\right) = \frac{N}{w_0} e^{il\varphi_1} \left(\frac{\rho_1\sqrt{2}}{w_0}\right)^{|l|} L_n^{|l|}\left(\frac{2\rho_1^2}{w_0^2}\right) e^{-2\rho_1^2/w_0^2} \frac{e^{ik_{iz}z_1}}{\sqrt{2\pi}} \tag{8}$$

where N is a normalization constant[1], w_0 is the beam waist, and $L_n^{|l|}\left(\frac{2\rho_1^2}{w_0^2}\right)$ is an associated Laguerre polynomial with orbital angular momentum l and index n that are related to the number of nodes for a given l. The LG wave function can be written as a convolution of Bessel functions over transverse momentum [58]:

$$\chi^{LG}_{\vec{k}_i,l}\left(\vec{r}_1,\vec{b}=0\right) = \frac{N}{\sqrt{2}} \frac{e^{il\varphi_1}}{n!} \int_0^{\infty} dk_{i\perp}\, e^{-w_0^2 k_{i\perp}^2/8} \left(\frac{w_0 k_{i\perp}}{\sqrt{8}}\right)^{2n+l+1} J_l(k_{i\perp}\rho_1) \frac{e^{ik_{iz}z_1}}{\sqrt{2\pi}} \tag{9}$$

Using Equation (4), the LG wave function can now be expressed as a convolution of Bessel projectile wave functions over transverse momentum:

$$\chi^{LG}_{\vec{k}_i,l}\left(\vec{r}_1,\vec{b}=0\right) = \frac{N\sqrt{\pi}}{n!} \int_0^{\infty} dk_{i\perp} e^{-\frac{k_{i\perp}^2 w_0^2}{8}} \left(\frac{k_{i\perp} w_0}{\sqrt{8}}\right)^{2n+l+1} \chi^B_{\vec{k}_i,l}\left(\vec{r}_1,\vec{b}=0\right) \tag{10}$$

For an off-center LG projectile, Equation (10) becomes

$$\chi^{LG}_{\vec{k}_i,l}\left(\vec{r}_1,\vec{b}\right) = \frac{N\sqrt{\pi}}{n!} \int_0^{\infty} dk_{i\perp} e^{-\frac{k_{i\perp}^2 w_0^2}{8}} \left(\frac{k_{i\perp} w_0}{\sqrt{8}}\right)^{2n+l+1} \chi^B_{\vec{k}_i,l}\left(\vec{r}_1,\vec{b}\right) \tag{11}$$

The transverse coherence $\Delta\rho$ of the incident projectile can be defined using quantum mechanical uncertainty:

$$\Delta\rho = \left[\langle\rho^2\rangle - \langle\rho^2\rangle\right]^{1/2} \tag{12}$$

For Bessel projectiles and plane waves, the transverse uncertainty is infinite, but for LG projectiles, when using Equation (8), it can be shown that the uncertainty is linear with respect to the beam waist:

$$\Delta\rho \sim w_0 \tag{13}$$

Some example values of the uncertainty for LG projectiles used here are listed in Table 1. For comparison, the transverse coherence length for atomic helium is $\Delta \rho = 0.84$ a.u.

Table 1. Transverse coherence length of LG projectiles in atomic units with $n = 0$. Values were calculated using Equation (12).

	$l = 0$	$l = 1$	$l = 2$
$w_0 = 0.5$ a.u.	0.093	0.13	0.16
$w_0 = 2$ a.u.	0.37	0.52	0.64
$w_0 = 4$ a.u.	0.75	1.05	1.28
$w_0 = 8$ a.u.	1.49	2.09	2.55

As is standard for single ionization collisions with fast projectiles [66–74], the initial state target helium atom is represented with a single active electron wave function:

$$\Phi\left(\vec{r}_2\right) = \frac{Z_{eff}^{3/2}}{\sqrt{\pi}} e^{-Z_{eff} r_2} \tag{14}$$

where $Z_{eff} = 1.3443$ [75,76] is the effective nuclear charge of the 1-electron helium atom and is chosen to give the correct ionization potential of helium.

The final state wave function is a product of the scattered projectile wave function $\chi_{\vec{k}_f}\left(\vec{r}_1\right)$, the ionized electron wave function $\chi_{\vec{k}_e}\left(\vec{r}_2\right)$, and the post-collision Coulomb interaction (PCI) M_{ee}:

$$\Psi_f = \chi_{\vec{k}_f}\left(\vec{r}_1\right) \chi_{\vec{k}_e}\left(\vec{r}_2\right) M_{ee} \tag{15}$$

We assume that the scattered projectile leaves the collision as a plane wave given by

$$\chi_{\vec{k}_f}\left(\vec{r}_1\right) = \frac{e^{i\vec{k}_f \cdot \vec{r}_1}}{(2\pi)^{3/2}} \tag{16}$$

The perturbation V_i is the Coulomb interaction between the projectile and target atom, which is given by

$$V_i = \frac{-Z_{eff}}{r_1} + \frac{1}{r_{12}} \tag{17}$$

The ionized electron is modeled as a Coulomb wave:

$$\chi_{\vec{k}_e}\left(\vec{r}_2\right) = \Gamma(1 - i\eta) e^{-\frac{\pi\eta}{2}} \frac{e^{i\vec{k}_e \cdot \vec{r}_2}}{(2\pi)^{\frac{3}{2}}} {}_1F_1\left(i\eta, 1, -ik_e r_2 - i\vec{k}_e \cdot \vec{r}_2\right) \tag{18}$$

where $\Gamma(1 - i\eta)$ is the gamma function and $\eta = Z_{eff} Z_e / k_e$ is the Sommerfeld parameter. We note that the use of $Z_{eff} = 1.3443$ in V_i and $\chi_{\vec{k}_e}\left(\vec{r}_2\right)$ maintains consistency with the treatment of the initial state wave function but does not satisfy asymptotic boundary conditions. This treatment has been used successfully previously for neutral atoms, such as carbon. To be sure that the choice of Z_{eff} does not significantly alter the TDCSs, we performed a few calculations with $Z_{eff} = 1$ for the perturbation and the Coulomb wave and found nearly identical TDCSs to those with $Z_{eff} = 1.3443$.

The post-collision Coulomb repulsion between the two outgoing final state electrons is included through the use of the Ward-Macek factor [77]:

$$M_{ee} = N_{ee} \left| {}_1F_1\left(\frac{i}{2k_{fe}}, 1, -2ik_{fe} r_{ave}\right) \right| \tag{19}$$

where

$$N_{ee} = \sqrt{\frac{\pi}{k_{fe}\left(e^{\frac{\pi}{k_{fe}}} - 1\right)}} \tag{20}$$

The relative momentum is $k_{fe} = \frac{1}{2}\left|\vec{k}_f - \vec{k}_e\right|$ and the average coordinate $r_{ave} = \frac{\pi^2}{16\epsilon}\left(1 + \frac{0.627}{\pi}\sqrt{\epsilon}\ln\epsilon\right)^2$, where $\epsilon = \left(k_f^2 + k_e^2\right)/2$ is the total energy of the two outgoing electrons.

We present the TDCSs for both a fixed impact parameter and an integration of the TDCSs over the impact parameter. The use of a fixed impact parameter allows for the study of projectile-target alignment effects but is not currently experimentally feasible. In a realistic experiment, the projectile's impact parameter cannot be determined or controlled, and theory must integrate over the impact parameter for an accurate comparison with the experiment. For a Bessel projectile, the TDCS integrated over the impact parameter is given by [46,59]:

$$\left.\frac{d^3\sigma_B}{d\Omega_1 d\Omega_2 dE_2}\right|_{int\,b} = \mu_{pa}^2\mu_{ie}\frac{k_f k_e}{k_{iz}(2\pi)}\int\left|T_{fi}^{PW}\right|^2 d\phi_{k_i} \tag{21}$$

where T_{fi}^{PW} is the transition matrix for an incident plane wave.

For an LG projectile, the TDCS integrated over the impact parameter is given by

$$\left.\frac{d^3\sigma_{LG}}{d\Omega_1 d\Omega_2 dE_2}\right|_{int\,b} = \mu_{pa}^2\mu_{ie}\frac{k_f k_e}{k_i}\int\frac{e^{-\frac{k_{i\perp}^2 w_0^2}{4}}}{k_{i\perp}^2}\left(\frac{k_{i\perp}^2 w_0^2}{8}\right)^{2n+l+1}\left|T_{fi}^{PW}\right|^2 k_{i\perp}dk_{i\perp}d\phi_{k_i} \tag{22}$$

3. Results

For plane wave projectiles, the shape of the TDCS can largely be explained by classical momentum conservation. There is a large, dominant forward binary peak that results from a single collision between the projectile and the atomic electron. This binary peak is located along the direction of the momentum transfer vector $\vec{q} = \vec{k}_i - \vec{k}_f$. Directly opposite the binary peak is a smaller recoil peak that results from the atomic electron first undergoing a binary collision with the projectile and then a second deflection by the nucleus that results in backward emission. The top plot in Figure 2 shows the coplanar TDCS as a function of ejected electron angle for a 1 keV plane wave electron colliding with helium. The ionized electron energy was 100 eV, and the scattering angle was 100 mrad (5.7°). These energies and scattering angles ensure that the kinematics are within the applicable range of the FBA. The binary and recoil peaks are clearly visible along and opposite to the momentum transfer vector direction ($\theta_q = 54°$).

The transverse profiles of the projectile beam and target atomic electron density are shown in Figure 3 for different projectile orbital angular momenta l and beam waists w_0. Note that the $l = 1$ projectile profiles are similar to the $l = 2$ profiles and, therefore, are not shown in Figure 3. As the impact parameter $\vec{b}$ increases, the relative distance between the center of the projectile beam and the atomic center increases. An increasing impact parameter is depicted in Figure 3 as the projectile beam shifting to the right relative to the atomic center (blue arrow in Figure 3a). Figure 3 shows that the helium target electron density (red) decays exponentially with a maximum density at the nucleus (origin). Figure 3 also shows the normalized overlap between the transverse beam profile and the atomic electron density as a function of impact parameter calculated using

$$\int |\chi_{\vec{k}_i}^V|^2 |\Phi|^2 d\rho \tag{23}$$

with $\varphi_b = 180°$.

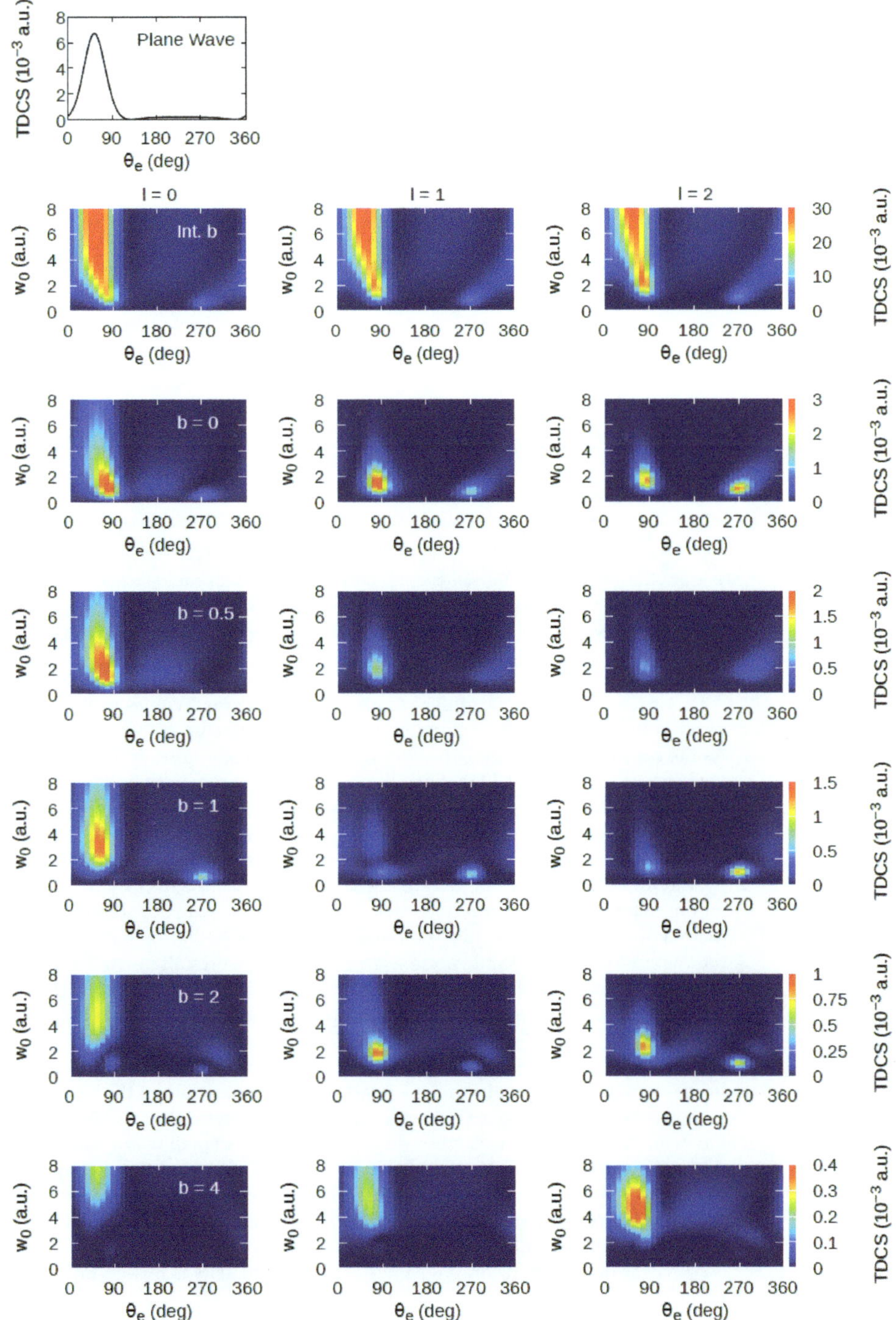

Figure 2. TDCSs for plane wave projectiles (top row) and LG projectiles (rows 2–7). For LG projectiles, the TDCSs are plotted as a function of beam waist w_0 and ejected electron angle θ_e for different impact

parameters and orbital angular momenta (labeled in figure). Row 2 shows the TDCSs integrated over the impact parameter, and rows 3–7 show the TDCSs for fixed values of the impact parameter with $\varphi_b = 180°$. The TDCS is shown in color, with the warmer colors representing larger TDCSs.

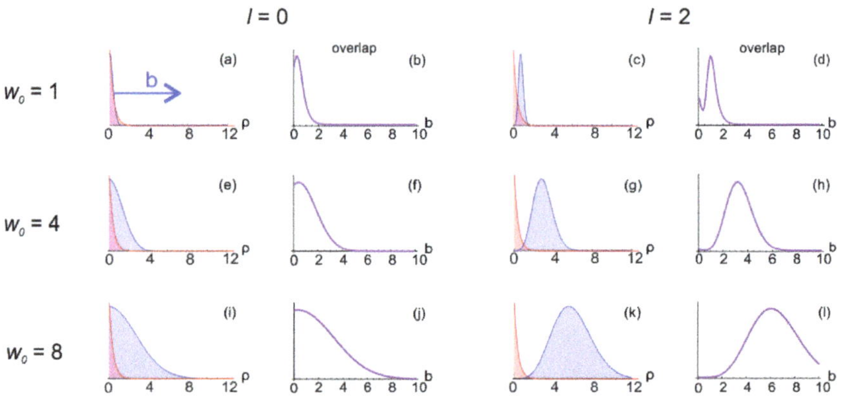

Figure 3. (**a,e,i,c,g,k**) Transverse profiles of the He(1s) electron density (red) and the LG beam (blue) with $n = 0$, orbital angular momentum l, and beam waist w_0 (in a.u.) as a function of transverse distance ρ (in a.u.). All profiles are normalized to 1 to provide a qualitative comparison. The target atomic electron density is the same for all cases. As the impact parameter $\vec{b}$ changes, the projectile beam shifts right, as denoted by the blue arrow in (**a**). (**b,f,j,d,h,l**) Normalized overlap between the transverse LG beam profile and the target atom electron density as a function of the impact parameter b (in a.u.) calculated using Equation (23).

For $l = 0$, the beam profile is Gaussian, and at small values of the beam waist, the overlap between the target electron density and projectile beam is sharply peaked in b. As the beam waist increases, the overlap broadens in b (Figure 3b,f,j). In the case of large w_0, it is expected that the TDCSs will change more slowly with b. It is also expected that for large w_0, the TDCS for an LG projectile will be similar to that of a plane wave since the projectile has become delocalized enough to completely overlap the target out to large radial distances. As the impact parameter increases, the beam profile shifts to the side, and the amount of overlap changes. For a small beam waist, the maximum overlap occurs for an impact parameter of approximately $b = 0.5$ a.u. (Figure 3b). For larger beam waists, the maximum overlap occurs at increasing values of b (Figure 3f,j). Note that for $l = 0$, the maxima of the projectile and target electron densities align for $b = 0$; however, the overlap is maximum at $b \neq 0$. This is due to the contribution to the overlap of the tail of the projectile density (only visible on logscale), causing a maximum in the overlap for a finite value of b.

For $l = 2$, the beam profile has a node at the origin, and the width of this node increases with increasing beam waist. Additionally, the width of the beam peak increases with increasing beam waist (Figure 3c,g,k). For on-center collisions, the largest overlap of the beam and the atomic electron density occurs for a small beam waist (Figure 3c). As the beam waist increases, the central node becomes wider, and the overlap with the target electron density decreases. For off-center collisions, as the impact parameter increases, the overlap increases to a maximum value before decreasing at large b (Figure 3h,l). The value of b for which the overlap is maximum depends on the specific value of the beam waist but occurs near the radial distance where the LG beam density is maximum.

3.1. LG Projectiles with Zero Orbital Angular Momentum

In rows two through seven in Figure 2, we present the TDCSs for $n = 0$ LG projectiles with different beam waists, impact parameters, and orbital angular momenta. We note that the results for $n \neq 0$ are qualitatively similar to those with $n = 0$ and are not shown here. In Figure 2, the TDCS magnitude is presented in color, with the warmer colors representing larger TDCS values. The vertical axis in each panel shows the beam waist, while the horizontal axis depicts the ejected electron angle. The second row shows TDCSs integrated over the impact parameter, while rows three–seven show the TDCSs for a fixed impact parameter (labeled in the left column). For the TDCSs with a fixed impact parameter, we assume that the projectile center is shifted by a distance of b along the negative x-axis (i.e., $\varphi_b = 180°$), and note that the TDCSs calculated for $\varphi_b = 0°$ were identical to those for $\varphi_b = 180°$. For the TDCSs integrated over $\vec{b}$, all radial values and azimuthal angles were included in the integration. Each column shows the TDCSs for a different orbital angular momentum (labeled above row two).

Consider first the on-center $b = 0$ and zero orbital angular momentum $l = 0$ TDCSs (third row, first column in Figure 2). In this case, the beam has a Gaussian transverse profile with no nodes. The largest cross-section occurs with a beam waist of $w_0 = 1$, with the binary peak located at an ejected electron angle of $85°$. As the beam waist increases, the binary peak location shifts to smaller angles until it is located at the plane wave momentum transfer direction of $54°$. This indicates that for a large beam waist, the TDCS more closely resembles that of the plane wave TDCS, as predicted from the complete overlap between the beam and target electron density. The shift of the binary peak from the momentum transfer direction for small beam waists could be caused by two factors—the PCI between the outgoing electrons or the projectile's non-zero transverse momentum. Given the large relative momentum between the two outgoing electrons, it is unlikely that the shift in binary peak location is due to the PCI, and a calculation that does not include PCI (not shown) confirmed this expectation. Therefore, we conclude that the shift in binary peak location is due to the transverse momentum of the projectile. Additional details are provided in Section 3.5.

For off-center collisions and zero angular momentum (rows four–seven, column one in Figure 2), the maximum TDCS occurs with increasing beam waist as the impact parameter increases. For $b \geq 1$, the TDCS only becomes observable in the colormap plots for $w_0 \geq b$ when the overlap between the beam profile and target electron density is non-negligible. Because the TDCS with a large impact parameter is observable only when the beam waist is large, the coherence length of the projectile is also necessarily large. This leads to a more plane wave-like TDCS shape, with the binary peak located at the momentum transfer direction. As the impact parameter increases, the overall magnitude of the TDCS decreases (see changing color scale for different rows), indicating that ionization becomes less likely for larger impact parameters. This correlates with the amount of overlap between the projectile and target electron density. Regardless of beam waist, the overlap decreases with increasing impact parameters (see Figure 3b,f,j).

For most values of w_0 and b, there is almost no recoil peak observable, indicating that rescattering by the nucleus is unlikely. This is primarily due to the energy of the ejected electron, which is fast enough to not experience much Coulomb pull from the nucleus. The notable exception is for $b = 1$ and small beam waists, in which case the recoil peak is larger than the binary peak. This corresponds to a narrow projectile beam impinging on the target. For a small beam waist, there is a maximum in the overlap near $b = 1$ to 1.5 a.u., which corresponds with the kinematical conditions that yield an enhanced recoil peak.

For the TDCSs integrated over $\vec{b}$, a strong binary peak is observed for nearly all w_0. At the smallest beam waist values ($w_0 \lesssim 0.5$ a.u.), the binary peak is very small and virtually invisible on the scale used in Figure 2. As the beam waist increases, a narrow binary peak is visible near $\theta_e = 90°$. This binary peak broadens and shifts to approximately the plane wave binary peak location for $w_0 \geq 4$ a.u. For a beam waist greater than 4 a.u., the transverse coherence length of the projectile is equal to or larger than that of the target atom. Therefore, the incident projectile wave packet fully overlaps the target, and coherent emission of the ionized electron occurs, as in the case of a plane wave or Bessel wave. This leads to the TDCSs for LG projectiles with large transverse coherence resembling the TDCSs of the plane wave and Bessel projectiles.

3.2. LG Projectiles with Non-Zero Orbital Angular Momentum

For on-center collisions with non-zero orbital angular momentum (row three, columns two and three in Figure 2), the binary and recoil peaks are more similar in magnitude, and both peaks are more localized to small beam waists. This is consistent with conditions for maximum overlap between the projectile density and the atomic electron density. For a large beam waist, the node in the center of the projectile density results in very little overlap between the projectile and the target electron density, resulting in very small TDCSs.

For small beam waists, the binary and recoil peaks are located at approximately $90°$ and $270°$, which is shifted from the classical momentum transfer directions. We have previously shown that for delocalized Bessel beam projectiles, the influence of the transverse momentum of the projectile alters the angular distribution of the TDCS [46,54]. Because the TDCS for a Bessel beam was represented as a superposition over tilted plane waves, the TDCS resulting from the smallest momentum transfer dominated the sum. This determined the location of the binary and recoil peaks, which were shifted to approximately $90°$ and $270°$. For LG beams, the transition matrix is a convolution over Bessel transition matrices, and therefore the same effect is present here, but with some averaging of location, as discussed in Section 3.5.

As the beam waist increases, the recoil peak decreases in magnitude more rapidly than the binary peak. This is due to the projectile probing the outer part of the target electron wave function, where the influence of the nucleus is reduced. For $l = 2$, the binary and recoil peak magnitudes are more similar than for $l = 1$, indicating that larger projectile orbital angular momentum results in more secondary scattering from the nucleus and, thus, a larger recoil peak. As was the case for collisions with zero orbital angular momentum, for off-center collisions with non-zero orbital angular momentum, the magnitude of the cross-sections decreases with increasing impact parameter.

The TDCSs averaged over the impact parameter for non-zero orbital angular momentum show similar qualitative features to those with zero orbital angular momentum. The width of the binary peak remains narrow for larger beam waists as orbital angular momentum increases, despite the fact that the transverse coherence length is larger for larger values of orbital angular momentum. This indicates that while transverse coherence length alters the magnitude and binary peak locations of TDCSs, it is not the only factor. Orbital angular momentum also plays a role in the ejected electron distribution.

3.3. Bessel Projectiles with Zero Orbital Angular Momentum

LG projectiles differ from their plane wave counterparts not just in their localization but in their ability to carry quantized orbital angular momentum. For $l \neq 0$, a comparison of TDCSs for LG projectiles with Bessel projectiles can more reasonably isolate localization effects because the LG and Bessel projectiles can carry the same orbital angular momentum. Unlike the localized LG projectiles, Bessel projectiles have infinite transverse extent, and the beam waist parameter does not exist. Comparison of TDCSs for LG and Bessel projectiles allows for the study of coherence effects between projectiles with the same orbital angular momentum.

Bessel projectiles are characterized by their orbital angular momentum l and their opening angle α, which is related to the incident transverse momentum by

$$k_{\perp i} = k_i \sin \alpha. \tag{24}$$

As the opening angle increases, the transverse momentum increases and the peaks in the density become narrower. For $\alpha = 0$ and $l = 0$, the Bessel projectile is identical to a plane wave.

In Figure 4, we plot the normalized transverse projectile density for Bessel projectiles (blue) and the target electron (red). Also shown is the normalized overlap between the projectile and target electron density as a function of impact parameter. As with LG projectiles, the overlap between the Bessel projectile and the target atom varies significantly as orbital angular momentum and opening angle change. In general, the transverse profile of the Bessel projectile has a series of decreasing peaks as the transverse distance increases. For $l = 0$, there is a single peak at the center of the beam, but for $l \neq 0$, there is a node. Unlike the localized LG projectile, the overlap between the Bessel projectile and the target electron density can be significantly non-zero for large values of the impact parameter. This is most notable for small opening angles, where the overlap is appreciable beyond $b = 100$ (Figure 4b,d).

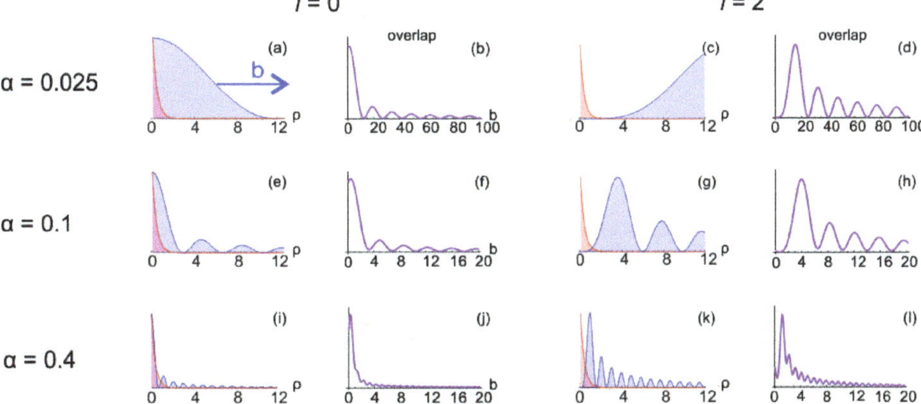

Figure 4. (a,e,i,c,g,k) Transverse profiles of the He(1s) electron density (red) and the Bessel beam (blue) with orbital angular momentum l and opening angle α (in rad) as a function of transverse distance ρ (in a.u.). All profiles are normalized to 1 to provide a qualitative comparison. The target atomic electron density is the same for all cases. As the impact parameter $\vec{b}$ changes, the projectile beam shifts right, as denoted by the blue arrow in (a). (b,f,j,d,h,l) Normalized overlap between the transverse Bessel beam profile and the target atom electron density as a function of impact parameter b (in a.u.) calculated using Equation (23).

For zero orbital angular momentum, Figure 4 shows that for small opening angles, the central peak is quite broad and the overlap between the beam and the target is large. In this case, the TDCSs are expected to resemble those of the plane wave and the LG projectile with large beam waist. As the impact parameter increases, the overlap between the target and the beam decreases until one of the nodes of the Bessel beam overlaps significantly with the target electron density, creating a minimum in the overlap. The overlap then increases again as the next lobe in the Bessel wave function overlaps the target density, creating a peak in the overlap. For each side lobe of the Bessel wave function that overlaps the target density, an additional peak structure is present in the overlap (Figure 4b).

As the opening angle increases, the central peak of the Bessel beam narrows and the overlap with the target electron density decreases. For larger values of α, as the impact parameter increases, the peak structures observed in the overlap are less pronounced (Figure 4j). The overlap function for $\alpha = 0.4$ most closely resembles that of the LG projectile with $w_0 = 1$, and thus the TDCSs for these parameters are expected to be similar.

For non-zero orbital angular momentum, the node at the center of the projectile density is largest for small opening angles (Figure 4c). This results in nearly zero overlap between the target electron density and the projectile for on-center collisions with small opening angles. As the impact parameter increases for small α, a series of peak structures are observed in the overlap, each corresponding to a side lobe of the Bessel projectile overlapping the target (Figure 4d).

As opening angle increases, the width of the central node decreases and the overlap between projectile and target electron density increases for on-center collisions. At the largest opening angle shown in Figure 4k,l, the largest overlap is observed for collisions with a small impact parameter, contrary to what is present for small opening angles. For large opening angles, as the impact parameter increases, the overlap decreases more rapidly than for small opening angles (Figure 4l). In general, for $l \neq 0$, the overlap for Bessel projectiles with large opening angles most closely resembles that of LG projectiles with small beam waists and the TDCSs for these parameters are expected to be similar. As was the case with LG projectiles, the overlap is crucial to interpreting the structures observed in the TDCSs for Bessel projectiles.

Figure 5 shows the TDCSs for the ionization of helium by Bessel projectiles as a function of the opening angle and ejected electron angle. Similar to Figure 2, row one shows the TDCS integrated over impact parameter. For Bessel projectiles, the integration over impact parameter washes out any dependence on orbital angular momentum, and the TDCS is independent of l. Rows two–six show the TDCSs for a fixed impact parameter (labeled in the left column), and each column shows results for a different orbital angular momentum (labeled above row two).

Consider first the on-center $b = 0$ and zero orbital angular momentum $l = 0$ TDCSs (first row, first column in Figure 5). In this case, there is a central lobe in the projectile density (Figure 4a,e,i). The TDCSs are largest for small opening angles and closely resemble those of the plane wave (Figure 2). There is a dominant peak located at the momentum transfer direction and no noticeable recoil peak. As the opening angle increases, the cross-sections drop in magnitude due to the reduced overlap between the projectile and target electron density. For off-center collisions (rows three–six, column one in Figure 5), as the impact parameter increases, there is not much change in the TDCS shape or magnitude. For $\alpha = 0$ and $l = 0$, the Bessel wave function is identical to a plane wave, and, therefore, the TDCSs at a small opening angle with $l = 0$ closely mirror the TDCSs for a plane wave projectile, regardless of impact parameter. This results in the binary peak at the momentum transfer direction that is observed in the TDCSs at small α for fixed impact parameters in Figure 5. As the impact parameter increases, the TDCS drops off more quickly with increasing opening angle, which is again consistent with the decreased overlap between the projectile and target atom densities.

3.4. Bessel Projectiles with Non-Zero Orbital Angular Momentum

For the on-center projectiles with non-zero orbital angular momentum (row two, columns two and three in Figure 5), the peaks in the TDCS occur at larger values of α, and, in some cases, a small recoil peak is present. This is directly related to the overlap between the projectile and target electron density. For the Bessel projectiles with $l \neq 0$, there is a node at the center of the projectile density. For small α values, this node is quite large, and there is almost no overlap between the projectile and target electron density. It is only for $\alpha > 0.1$ rad that some overlap occurs. For even larger values of α, the TDCS again decreases in magnitude as the nodes become narrower and the overlap decreases. For both $l = 1$ and $l = 2$, there is a small binary peak present for $\alpha > 0.1$ rad. This binary peak shifts to larger ejected electron angles and increases in magnitude as the opening angle increases. This effect was observed in some of our previous calculations for the ionization of hydrogen by Bessel projectile and was traced to the transverse momentum component of the incident projectile [54]. At a large opening angle, a recoil peak is present, and its location moves to smaller ejected electron angles as the opening angle increases. The shift in recoil peak location with increasing opening angle can also be traced to the transverse momentum of the incident projectile. In fact, as we showed in [54], at very large values of α, the dominant momentum transfer direction occurs at $\theta_e > 180°$, and what appears to be the recoil peak is, in reality, the binary peak. Thus, the observed TDCS peak at $\theta_e \approx 270°$ for a large α is, in fact, the binary peak.

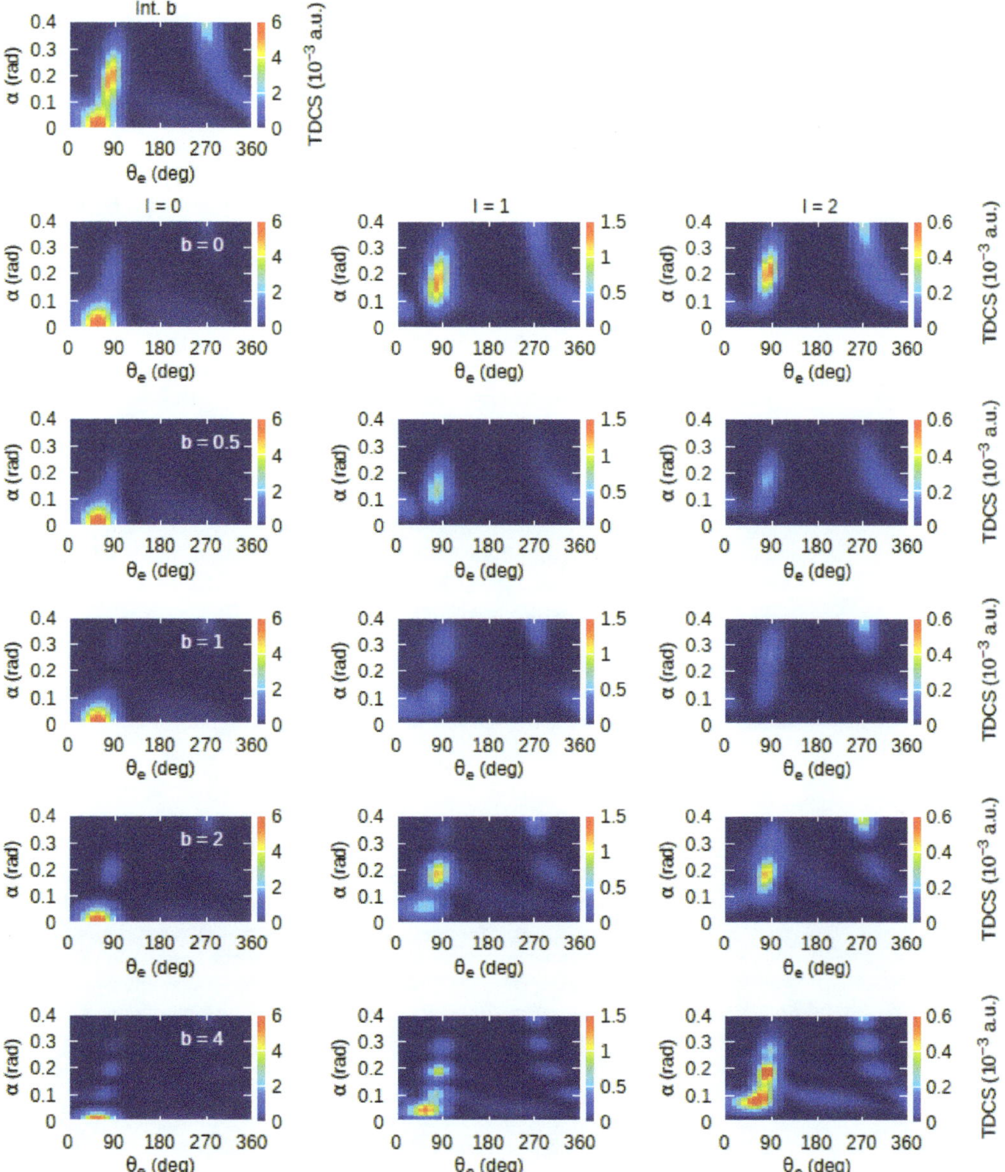

Figure 5. TDCSs for Bessel projectiles plotted as a function of opening angle α and ejected electron angle θ_e for different impact parameters and orbital angular momenta (labeled in figure). The first row contains the TDCSs integrated over the impact parameter. Rows two–six show the TDCSs for fixed impact parameter ($\varphi_b = 180°$) as labeled in the figure. The TDCS is shown in color, with the warmer colors representing larger TDCSs.

As the impact parameter increases, the magnitude of the TDCS decreases, and very few features are observable in the color plots for $b = 0.5$ or 1 a.u. As the impact parameter increases beyond 1 a.u., the side lobes of the Bessel projectile, again, overlap with the target electron density, resulting in clearly observable binary peak structures. At the largest impact parameter of $b = 4$ a.u., oscillations are observed in the binary peak magnitude as the opening angle increases. These are a direct result of the side lobes of the Bessel projectile overlapping the target electron. Each peak in the TDCS corresponds to one of the Bessel lobes overlapping the target. As the opening angle increases, the location of the binary peak shifts to larger ejected electron angles. This can, again, be traced to the projectile's transverse momentum, which increases with the opening angle. As the projectile's transverse momentum increases, the location of the classically predicted momentum transfer direction changes, which shifts the location of the binary peak. Recoil peak structures are observed again at large opening angles for non-zero impact parameters, and at $b = 4$ a.u., oscillations are observed in the recoil peak magnitude as the opening angle increases. As was the case with the binary peak magnitude oscillations, the recoil peak oscillations are due to the Bessel wave function side lobes. Additionally, as was the case for $b = 0$, the recoil peak observed for $b = 4$ a.u. is, in reality, the binary peak.

The TDCSs integrated over the impact parameter show a forward binary peak at the plane wave momentum transfer direction, which then shifts to larger ejected electron angles as the opening angle increases. Because the TDCSs integrated over the impact parameter include contributions from all impact parameters and orbital angular momentum values, some of the features can be traced to TDCS contributions from specific impact parameters or orbital angular momentum values. For example, the binary peak at small α at the plane wave momentum transfer direction is predominantly caused by the $l = 0$ TDCSs, while the binary peak at larger α values is a result of the TDCSs for projectiles with non-zero l. The enhanced recoil peak at the largest opening angles also results from TDCSs of projectiles with non-zero orbital angular momentum.

3.5. Relation of LG to Bessel Projectiles

As Equation (11) shows, the LG projectile can be written as a convolution over the transverse momentum, with each Bessel wave function weighted by the factor

$$
e^{-\frac{k_{i\perp}^2 w_0^2}{8}} \left(\frac{k_{i\perp} w_0}{\sqrt{8}} \right)^{2n+l+1}
\tag{25}
$$

For large values of the beam waist, this weighting factor becomes more localized in $k_{i\perp}$ at smaller values of the transverse momentum. This results in the convolution favoring a few Bessel wave functions with small transverse momenta (i.e., small α). Because these are the Bessel wave functions that most resemble the plane wave function, the resulting LG projectile is delocalized in space with a large coherence length. In other words, the highly localized weighting factor of Equation (25) in transverse momentum space results in a delocalized wave function in position space. This weighting results in the TDCSs for LG projectiles with large beam waists being a sort of average over the Bessel TDCSs for small α. Therefore, the LG TDCSs at large w_0 resemble those of the Bessel projectile at small opening angles with a strong binary peak and negligible recoil peak.

For small values of the beam waist, the weighting factor in Equation (25) is a broader function of $k_{i\perp}$ and centered at larger values of transverse momentum. This results in many Bessel wave functions with large transverse momenta contributing to the convolution. In this case, the broad weighting factor of Equation (25) in transverse momentum space results in a localized wave function in position space. Thus, the LG TDCSs at small w_0 are similar to an average over the Bessel TDCSs at large α with approximately equal magnitude binary and recoil peaks.

4. Summary

We have presented TDCSs for the ionization of helium by LG and Bessel electron projectiles. A comparison of the localized LG TDCSs and the fully coherent, delocalized Bessel TDCSs provides insight into the role of projectile coherence. This allowed for the direct study of coherence effects independent of orbital angular momentum, which is not possible if plane wave projectiles are used. For LG projectiles, we examined the effects of transverse coherence length, orbital angular momentum, and the impact parameter on the TDCSs. The transverse coherence length was altered by changing the projectile beam waist. For localized projectiles with a small coherence length, the location of the binary peak was shifted to larger ejected electron angles from the classical momentum transfer direction. Additionally, a small recoil peak was observed. As the coherence length increased, the recoil peak magnitude decreased, and the location of the binary peak shifted to the classical momentum transfer direction. At a large coherence length, the TDCSs resembled that of the plane wave and Bessel TDCSs for a completely delocalized projectile. A comparison of the TDCSs for the LG and Bessel projectiles with non-zero orbital angular momentum showed that a localized projectile resulted in an enhanced recoil peak, which was most pronounced for larger orbital angular momentum values. These features were traced to the projectile's transverse momentum and the different contributions of the Bessel wave functions that result from writing the LG wave function as a convolution over transverse momentum.

The overlap between the projectile's transverse density and the target electron's density correlated with the magnitude of the TDCSs. For a large overlap, a large TDCS was observed. The impact parameter, beam waist, and orbital angular momentum all affected the overlap and, correspondingly, the conditions that resulted in large cross-sections. In general, a large cross-section was observed when the projectile's maximum density aligned with the target electron's maximum density.

For Bessel projectiles, the location and magnitude of the binary and recoil peaks were dependent upon the opening angle of the incident projectile momentum. As the opening angle increased, the binary peak shifted to larger ejected electron angles, while the recoil peak shifted to smaller angles. This was expected from previous studies on ionization by Bessel projectiles and had been shown to result from the projectile's transverse momentum. By writing the LG wave function as a convolution of the Bessel wave functions over the transverse momentum, we showed that the features of the Bessel TDCSs contributed to the shape of the LG TDCSs. In particular, at a small coherence length, many Bessel wave functions with large transverse momentum contribute, resulting in an enhanced recoil peak for LG projectiles. In contrast, for a large coherence length, only a few Bessel wave functions with small transverse momentum contribute, and the LG TDCSs resembled the plane wave TDCS.

Overall, our results demonstrate that coherence length can be controlled for electron projectiles through the use of sculpted wave packets and that the shape and magnitude of the TDCSs depend on the projectile coherence length. A comparison of TDCSs for Bessel and LG projectiles isolated the effects of coherence from orbital angular momentum and demonstrated that coherence effects persist regardless of l. We anticipate that these results may open the door to future studies on projectile coherence effects using sculpted electrons, in particular for molecular targets where the interference effects are strongly dependent upon projectile coherence.

Funding: This research was funded by the National Science Foundation grant number PHY-1912093. As an invited article, the APC was waived.

Data Availability Statement: The data presented in this study are available on request from the corresponding author.

Acknowledgments: We gratefully acknowledge the support of the National Science Foundation under Grant No. PHY-1912093.

Conflicts of Interest: The authors declare no conflict of interest.

References

1. Egodapitiya, K.N.; Sharma, S.; Hasan, A.; Laforge, A.C.; Madison, D.H.; Moshammer, R.; Schulz, M. Manipulating Atomic Fragmentation Processes by Controlling the Projectile Coherence. *Phys. Rev. Lett.* **2011**, *106*, 153202. [CrossRef]
2. Sharma, S.; Hasan, A.; Egodapitiya, K.N.; Arthanayaka, T.P.; Sakhelashvili, G.; Schulz, M. Projectile Coherence Effects in Electron Capture by Protons Colliding with H_2 and He. *Phys. Rev. A* **2012**, *86*, 022706. [CrossRef]
3. Schneider, K.; Schulz, M.; Wang, X.; Kelkar, A.; Grieser, M.; Krantz, C.; Ullrich, J.; Moshammer, R.; Fischer, D. Role of Projectile Coherence in Close Heavy Ion-Atom Collisions. *Phys. Rev. Lett.* **2013**, *110*, 113201. [CrossRef] [PubMed]
4. Sharma, S.; Arthanayaka, T.P.; Hasan, A.; Lamichhane, B.R.; Remolina, J.; Smith, A.; Schulz, M. Fully Differential Study of Interference Effects in the Ionization of H_2 by Proton Impact. *Phys. Rev. A* **2014**, *90*, 052710. [CrossRef]
5. Járai-Szabó, F.; Nagy, L. Theoretical investigations on the projectile coherence effects in fully differential ionization cross sections. *Eur. Phys. J. D* **2015**, *69*, 4. [CrossRef]
6. Sarkadi, L.; Fabre, I.; Navarrete, F.; Barrachina, R.O. Loss of wave-packet coherence in ion-atom collisions. *Phys. Rev. A* **2016**, *93*, 032702. [CrossRef]
7. Navarrete, F.; Ciappina, M.F.; Sarkadi, L.; Barrachina, R.O. The Role of the Wave Packet Coherence on the Ionization Cross Section of He by P^+ and C^{6+} Projectiles. *Nucl. Instrum. Methods Phys. Res. Sect. B Beam Interact. Mater. At.* **2017**, *408*, 165. [CrossRef]
8. Kouzakov, K.A. Theoretical analysis of the projectile and target coherence in COLTRIMS experiments on atomic ionization by fast ions. *Eur. Phys. J. D* **2017**, *71*, 63. [CrossRef]
9. Karlovets, D.V.; Kotkin, G.L.; Serbo, V.G. Scattering of wave packets on atoms in the Born approximation. *Phys. Rev. A* **2015**, *92*, 052703. [CrossRef]
10. Wang, X.; Schneider, K.; LaForge, A.; Kelkar, A.; Grieser, M.; Moshammer, R.; Ullrich, J.; Schulz, M.; Fischer, D. Projectile coherence effects in single ionization of helium. *J. Phys. B At. Mol. Opt. Phys.* **2012**, *45*, 211001. [CrossRef]
11. Gassert, H.; Chuluunbaatar, O.; Waitz, M.; Trinter, F.; Kim, H.-K.; Bauer, T.; Laucke, A.; Müller, C.; Voigtsberger, J.; Weller, M.; et al. Agreement of Experiment and Theory on the Single Ionization of Helium by Fast Proton Impact. *Phys. Rev. Lett.* **2016**, *116*, 073201. [CrossRef]
12. Moshammer, R.; Fainstein, P.D.; Schulz, M.; Schmitt, W.; Kollmus, H.; Mann, R.; Hagmann, S.; Ullrich, J. Initial State Dependence of Low-Energy Electron Emission in Fast Ion Atom Collisions. *Phys. Rev. Lett.* **1999**, *83*, 4721. [CrossRef]
13. Schulz, M.; Moshammer, R.; Fischer, D.; Kollmus, H.; Madison, D.H.; Jones, S.; Ullrich, J. Three-dimensional imaging of atomic four-body processes. *Nature* **2003**, *422*, 6927. [CrossRef] [PubMed]
14. Maydanyuk, N.V.; Hasan, A.; Foster, M.; Tooke, B.; Nanni, E.; Madison, D.H.; Schulz, M. Projectile–Residual-Target-Ion Scattering after Single Ionization of Helium by Slow Proton Impact. *Phys. Rev. Lett.* **2005**, *94*, 243201. [CrossRef]
15. Dörner, R.; Mergel, V.; Ali, R.; Buck, U.; Cocke, C.L.; Froschauer, K.; Jagutzki, O.; Lencinas, S.; Meyerhof, W.E.; Nüttgens, S.; et al. Electron-electron interaction in projectile ionization investigated by high resolution recoil ion momentum spectroscopy. *Phys. Rev. Lett.* **1994**, *72*, 3166. [CrossRef] [PubMed]
16. Madison, D.H.; Fischer, D.; Foster, M.; Schulz, M.; Moshammer, R.; Jones, S.; Ullrich, J. Probing Scattering Wave Functions Close to the Nucleus. *Phys. Rev. Lett.* **2003**, *91*, 253201. [CrossRef]
17. Schulz, M.; Moshammer, R.; Perumal, A.N.; Ullrich, J. Triply Differential Single-Ionization Cross Sections in Fast Ion-Atom Collisions at Large Perturbation. *J. Phys. B At. Mol. Opt. Phys.* **2002**, *35*, L161. [CrossRef]
18. Fischer, D.; Moshammer, R.; Schulz, M.; Voitkiv, A.; Ullrich, J. Fully differential cross sections for the single ionization of helium by ion impact. *J. Phys. B At. Mol. Opt. Phys.* **2003**, *36*, 3555. [CrossRef]
19. Moshammer, R.; Unverzagt, M.; Schmitt, W.; Ullrich, J.; Schmidt-Böcking, H. A 4π Recoil-Ion Electron Momentum Analyzer: A High-Resolution "Microscope" for the Investigation of the Dynamics of Atomic, Molecular and Nuclear Reactions. *Nucl. Instrum. Methods Phys. Res. Sect. B Beam Interact. Mater. At.* **1996**, *108*, 425. [CrossRef]
20. McGovern, M.; Whelan, C.T.; Walters, H.R.J. C^{6+}-Impact Ionization of Helium in the Perpendicular Plane: Ionization to the Ground State, Excitation-Ionization, and Relativistic Effects. *Phys. Rev. A* **2010**, *82*, 032702. [CrossRef]
21. Pedlow, R.T.; O'Rourke, S.F.C.; Crothers, D.S.F. Fully differential cross sections for 3.6 MeV u^{-1} $Au^{Z}p^+$ + He collisions. *Phys. Rev. A* **2005**, *72*, 062719. [CrossRef]
22. Voitkiv, A.B.; Najjari, B. Projectile-Target Core Interaction in Single Ionization of Helium by 100-MeV/u C^{6+} and 1-GeV/u U^{92+} Ions. *Phys. Rev. A* **2009**, *79*, 022709. [CrossRef]
23. Fiol, J.; Otranto, S.; Olson, R.E. Critical comparison between theory and experiment for C^{6+} + He fully differential ionization cross sections. *J. Phys. B At. Mol. Opt. Phys.* **2006**, *39*, L285. [CrossRef]
24. Harris, A.L.; Madison, D.H.; Peacher, J.L.; Foster, M.; Bartschat, K.; Saha, H.P. Effects of the final-state electron-ion interactions on the fully differential cross sections for heavy-particle-impact ionization of helium. *Phys. Rev. A* **2007**, *75*, 032718. [CrossRef]
25. Schulz, M.; Dürr, M.; Najjari, B.; Moshammer, R.; Ullrich, J. Reconciliation of measured fully differential single ionization data with the first Born approximation convoluted with elastic scattering. *Phys. Rev. A* **2007**, *76*, 032712. [CrossRef]
26. Dürr, M.; Najjari, B.; Schulz, M.; Dorn, A.; Moshammer, R.; Voitkiv, A.B.; Ullrich, J. Analysis of experimental data for ion-impact single ionization of helium with Monte Carlo event generators based on quantum theory. *Phys. Rev. A* **2007**, *75*, 062708. [CrossRef]
27. Sharma, S.; Arthanayaka, T.P.; Hasan, A.; Lamichhane, B.R.; Remolina, J.; Smith, A.; Schulz, M. Complete Momentum Balance in Ionization of H_2 by 75-KeV-Proton Impact for Varying Projectile Coherence. *Phys. Rev. A* **2014**, *89*, 052703. [CrossRef]

28. Feagin, J.M.; Hargreaves, L. Loss of wave-packet coherence in stationary scattering experiments. *Phys. Rev. A* **2013**, *88*, 032705. [CrossRef]

29. Olson, R.E.; Fiol, J. Dynamics underlying fully differential cross sections for fast C6++He collisions. *J. Phys. B At. Mol. Opt. Phys.* **2003**, *36*, L365. [CrossRef]

30. Járai-Szabó, F.; Nagy, L. Semiclassical description of kinematically complete experiments. *J. Phys. B At. Mol. Opt. Phys.* **2007**, *40*, 4259. [CrossRef]

31. Colgan, J.; Pindzola, M.S.; Robicheaux, F.; Ciappina, M.F. Fully differential cross sections for the single ionization of He by C^{6+} ions. *J. Phys. B At. Mol. Opt. Phys.* **2011**, *44*, 175205. [CrossRef]

32. Schulz, M.; Hasan, A.; Lamichhane, B.; Arthanayaka, T.; Dhital, M.; Bastola, S.; Nagy, L.; Borbély, S.; Járai-Szabó, F. Projectile Coherence Effects in Simple Atomic Systems. *J. Phys. Conf. Ser.* **2020**, *1412*, 062007. [CrossRef]

33. Gulyás, L.; Egri, S.; Igarashi, A. Theoretical investigation of the fully differential cross sections for single ionization of He in collisions with 75-keV protons. *Phys. Rev. A* **2019**, *99*, 032704. [CrossRef]

34. Navarrete, F.; Ciappina, M.F.; Barrachina, R.O. Coherence properties of the projectile's beam: The missing piece of the C^{6+} + He ionization puzzle. *J. Physics Conf. Ser.* **2020**, *1412*, 152033. [CrossRef]

35. Barrachina, R.O.; Navarrete, F.; Ciappina, M.F. Atomic Concealment Due to Loss of Coherence of the Incident Beam of Projectiles in Collision Processes. *Atoms* **2021**, *9*, 5. [CrossRef]

36. Voss, T.; Lamichhane, B.R.; Dhital, M.; Lomsadze, R.; Schulz, M. Differential Study of Projectile Coherence Effects on Double Capture Processes in p + Ar Collisions. *Atoms* **2020**, *8*, 10. [CrossRef]

37. Navarrete, F.; Barrachina, R.; Ciappina, M.F. Distortion of the Ionization Cross Section of He by the Coherence Properties of a C^{6+} Beam. *Atoms* **2019**, *7*, 31. [CrossRef]

38. Nagy, L.; Járai-Szabó, F.; Borbély, S.; Tőkési, K. Projectile Coherence Effects Analyzed Using Impact Parameters Determined by Classical Trajectory Monte Carlo Calculations. *J. Phys. Conf. Ser.* **2020**, *1412*, 152032. [CrossRef]

39. Bastola, S.; Dhital, M.; Majumdar, S.; Hasan, A.; Lomsadze, R.; Davis, J.; Lamichhane, B.; Borbély, S.; Nagy, L.; Schulz, M. Interference Effects in Fully Differential Ionization Cross Sections near the Velocity Matching in P + He Collisions. *Atoms* **2022**, *10*, 119. [CrossRef]

40. Keller, C.; Schmiedmayer, J.; Zeilinger, A. Requirements for Coherent Atom Channeling. *Opt. Commun.* **2000**, *179*, 129–135. [CrossRef]

41. Cronin, A.D.; Schmiedmayer, J.; Pritchard, D.E. Optics and interferometry with atoms and molecules. *Rev. Mod. Phys.* **2009**, *81*, 1051. [CrossRef]

42. Uchida, M.; Tonomura, A. Generation of electron beams carrying orbital angular momentum. *Nature* **2010**, *464*, 737. [CrossRef]

43. McMorran, B.J.; Agrawal, A.; Anderson, I.M.; Herzing, A.A.; Lezec, H.J.; McClelland, J.J.; Unguris, J. Electron Vortex Beams with High Quanta of Orbital Angular Momentum. *Science* **2011**, *331*, 192. [CrossRef] [PubMed]

44. Grillo, V.; Gazzadi, G.C.; Mafakheri, E.; Frabboni, S.; Karimi, E.; Boyd, R.W. Holographic Generation of Highly Twisted Electron Beams. *Phys. Rev. Lett.* **2015**, *114*, 034801. [CrossRef] [PubMed]

45. Voloch-Bloch, N.; Lereah, Y.; Lilach, Y.; Gover, A.; Arie, A. Generation of electron Airy beams. *Nature* **2013**, *494*, 331. [CrossRef]

46. Harris, A.; Plumadore, A.; Smozhanyk, Z. Ionization of hydrogen by electron vortex beam. *J. Phys. B At. Mol. Opt. Phys.* **2019**, *52*, 094001. [CrossRef]

47. Van Boxem, R.; Partoens, B.; Verbeeck, J. Inelastic electron-vortex-beam scattering. *Phys. Rev. A* **2015**, *91*, 032703. [CrossRef]

48. Lloyd, S.M.; Babiker, M.; Yuan, J. Interaction of electron vortices and optical vortices with matter and processes of orbital angular momentum exchange. *Phys. Rev. A* **2012**, *86*, 023816. [CrossRef]

49. Lloyd, S.; Babiker, M.; Yuan, J. Quantized Orbital Angular Momentum Transfer and Magnetic Dichroism in the Interaction of Electron Vortices with Matter. *Phys. Rev. Lett.* **2012**, *108*, 074802. [CrossRef]

50. Schattschneider, P.; Schaffer, B.; Ennen, I.; Verbeeck, J. Mapping spin-polarized transitions with atomic resolution. *Phys. Rev. B* **2012**, *85*, 134422. [CrossRef]

51. Asenjo-Garcia, A.; de Abajo, F.J.G. Dichroism in the Interaction between Vortex Electron Beams, Plasmons, and Molecules. *Phys. Rev. Lett.* **2014**, *113*, 066102. [CrossRef] [PubMed]

52. Babiker, M.; Bennett, C.R.; Andrews, D.L.; Dávila Romero, L.C. Orbital Angular Momentum Exchange in the Interaction of Twisted Light with Molecules. *Phys. Rev. Lett.* **2002**, *89*, 143601. [CrossRef] [PubMed]

53. Verbeeck, J.; Tian, H.; Van Tendeloo, G. How to Manipulate Nanoparticles with an Electron Beam? *Adv. Mater.* **2012**, *25*, 1114. [CrossRef] [PubMed]

54. Plumadore, A.; Harris, A.L. Projectile transverse momentum controls emission in electron vortex ionization collisions. *J. Phys. B: At. Mol. Opt. Phys.* **2020**, *53*, 205205. [CrossRef]

55. Harris, A.L. Single and double scattering mechanisms in ionization of helium by electron vortex projectiles. *J. Phys. B: At. Mol. Opt. Phys.* **2021**, *54*, 155203. [CrossRef]

56. Dhankhar, N.; Choubisa, R. Electron impact single ionization of hydrogen molecule by twisted electron beam. *J. Phys. B: At. Mol. Opt. Phys.* **2020**, *54*, 015203. [CrossRef]

57. Dhankhar, N.; Mandal, A.; Choubisa, R. Double ionization of helium by twisted electron beam. *J. Phys. B: At. Mol. Opt. Phys.* **2020**, *53*, 155203. [CrossRef]

58. Van Boxem, R.; Partoens, B.; Verbeeck, J. Rutherford scattering of electron vortices. *Phys. Rev. A* **2014**, *89*, 032715. [CrossRef]

59. Serbo, V.; Ivanov, I.P.; Fritzsche, S.; Seipt, D.; Surzhykov, A. Scattering of twisted relativistic electrons by atoms. *Phys. Rev. A* **2015**, *92*, 012705. [CrossRef]

60. Ivanov, I.P.; Seipt, D.; Surzhykov, A.; Fritzsche, S. Elastic scattering of vortex electrons provides direct access to the Coulomb phase. *Phys. Rev. D* **2016**, *94*, 076001. [CrossRef]

61. Matula, O.; Hayrapetyan, A.G.; Serbo, V.G.; Surzhykov, A.; Fritzsche, S. Radiative capture of twisted electrons by bare ions. *New J. Phys.* **2014**, *16*, 053024. [CrossRef]

62. Kosheleva, V.P.; Zaytsev, V.A.; Surzhykov, A.; Shabaev, V.M.; Stöhlker, T. Elastic scattering of twisted electrons by an atomic target: Going beyond the Born approximation. *Phys. Rev. A* **2018**, *98*, 022706. [CrossRef]

63. Lei, C.; Dong, G. Chirality-dependent scattering of an electron vortex beam by a single atom in a magnetic field. *Phys. Rev. A* **2021**, *103*, 032815. [CrossRef]

64. Plumadore, A.; Harris, A.L. Electron spectra for twisted electron collisions. *J. Phys. B At. Mol. Opt. Phys.* **2022**, *54*, 235204. [CrossRef]

65. Abramowitz, M.; Stegun, I.A. *Handbook of Mathematical Functions with Formulas, Graphs, and Mathematical Tables*; National Bureau of Standards: Dover, UK; New York, NY, USA, 1972.

66. Jones, S.; Madison, D.H.; Franz, A.; Altick, P.L. Three-body distorted-wave Born approximation for electron-atom ionization. *Phys. Rev. A* **1993**, *48*, R22. [CrossRef]

67. Berakdar, J.; Briggs, J.S. Three-body Coulomb continuum problem. *Phys. Rev. Lett.* **1994**, *72*, 3799. [CrossRef]

68. Srivastava, M.K.; Sharma, S. Triple-differential cross sections for the ionization of helium by fast electrons. *Phys. Rev. A* **1988**, *37*, 628. [CrossRef]

69. Bellm, S.; Lower, J.; Bartschat, K.; Guan, X.; Weflen, D.; Foster, M.; Harris, A.L.; Madison, D.H. Ionization and Ionization–Excitation of Helium to the n = 1–4 States of He + by Electron Impact. *Phys. Rev. A* **2007**, *75*, 042704. [CrossRef]

70. Dupre, C.; Lahmam-Bennani, A.; Duguet, A.; Mota-Furtado, F.; O'Mahony, P.F.; Cappello, C.D. (e,2e) triple differential cross sections for the simultaneous ionization and excitation of helium. *J. Phys. B At. Mol. Opt. Phys.* **1992**, *25*, 259. [CrossRef]

71. Balashov, V.V.; Bodrenko, I.V. Triple coincidence (e,2e) measurements as a 'perfect experiment' instrument for ionization-excitation studies. *J. Phys. B At. Mol. Opt. Phys.* **1999**, *32*, L687. [CrossRef]

72. Harris, A.L.; Morrison, K. Comprehensive study of 3-body and 4-body models of single ionization of helium. *J. Phys. B At. Mol. Opt. Phys.* **2013**, *46*, 145202. [CrossRef]

73. Macek, J.H.; Botero, J. Perturbation theory with arbitrary boundary conditions for charged-particle scattering: Application to (e,2e) experiments in helium. *Phys. Rev. A* **1992**, *45*, R8. [CrossRef] [PubMed]

74. Botero, J.; Macek, J.H. Coulomb Born approximation for electron scattering from neutral atoms: Application to electron impact ionization of helium in coplanar symmetric geometry. *J. Phys. B: At. Mol. Opt. Phys.* **1991**, *24*, L405. [CrossRef]

75. DeMars, C.M.; Kent, J.B.; Ward, S.J. Deep minima in the Coulomb-Born triply differential cross sections for ionization of helium by electron and positron impact. *Eur. Phys. J. D* **2020**, *74*, 48. [CrossRef]

76. Botero, J.; Macek, J.H. Coulomb-Born calculation of the triple-differential cross section for inner-shell electron-impact ionization of carbon. *Phys. Rev. A* **1992**, *45*, 154. [CrossRef]

77. Ward, S.J.; Macek, J.H. Wave functions for continuum states of charged fragments. *Phys. Rev. A* **1994**, *49*, 1049. [CrossRef]

Article

Fragmentation Dynamics of CO_2^{q+} (q = 2, 3) in Collisions with 1 MeV Proton

Avijit Duley and Aditya. H. Kelkar *

Department of Physics, Indian Institute of Technology Kanpur, Kanpur 208016, India
* Correspondence: akelkar@iitk.ac.in

Abstract: The fragmentation dynamics of the CO_2^{q+} (q = 2, 3) molecular ions formed under the impact of 1 MeV protons is studied using a recoil ion momentum spectrometer equipped with a multi-hit time- and position-sensitive detector. Both two-body and three-body fragmentation channels arising from the doubly and triply ionized molecular ions of CO_2 are identified and analyzed. Kinetic energy release (KER) distributions have been obtained for various channels. With the help of Dalitz plots and Newton diagrams concerted and sequential processes have been assigned to observed fragmentation channels. In addition, angular correlations are used to determine the molecular geometry of the precursor molecular ion. It is found that the symmetric breakup into $C^+ + O^+ + O^+$ involves asymmetric stretching of the molecular bonds in CO_2^{3+} prior to dissociation via concerted decay implying the fact that collisions with 1 MeV proton induces an asynchronous decay in CO_2.

Keywords: recoil ion momentum spectroscopy; coulomb fragmentation; coincidence imaging

1. Introduction

Over the past few decades, the fragmentation dynamics of multiply charged molecular ions have been studied extensively. These studies are of fundamental interest as they help identify and understand the electronic states of molecular ions. Knowledge of these electronic states works as a verification tool for state-of-the-art theoretical models. These studies are also crucial in plasma and fusion research [1], atmospheric and space physics [2], and radiation therapy [3,4]. When one or more electrons are stripped off from a diatomic or polyatomic molecule, a molecular ion is produced, which might be in a metastable or unstable state depending upon the excitation energy available to the system. A multiply charged (charge state more than 2) molecular ion usually goes to an unstable state and eventually fragments into atomic ions and neutrals due to the Coulomb repulsion between the ionic cores. Investigating the fragmentation dynamics of these molecular ions are important to identify the various electronic states that these molecular ions access after ionization/excitation or both. The fragmentation dynamics of these molecular ions can be studied by detecting the fragments in coincidence and measuring their momenta and KER distributions. The KER distributions of the individual fragments and their angular correlations are crucial to determine the geometry of the molecular ions as well as to detect nuclear motions prior to fragmentation. The dissociation dynamics of a multiply charged polyatomic molecular ion is much more complicated compared to diatomic ions due to the presence of multiple bonds. The carbon dioxide molecule is a prototype system for understanding few-body dissociation dynamics under the impact of particles or photons owing to simple linear geometry of the molecule. Fragmentation dynamics of CO_2 has been studied experimentally using highly charged ions at slow [5], intermediate [6], and swift velocity [7–10], synchrotron radiation [11,12], femtosecond laser pulse [13–15], as well as slow protons [16] and low energy electrons [17–19]. In addition, extensive theoretical studies [17,20–22] complement the experimental results. The CO_2^{2+} and CO_2^{3+} molecular ions are isoelectronic to the isomeric pair NCN and CNN radicals, which also have a linear geometry in the ground state. In photofragmentation studies, it has been

Citation: Duley, A.; Kelkar, A.H. Fragmentation Dynamics of CO_2^{q+} (q = 2, 3) in Collisions with 1 MeV Proton. *Atoms* **2023**, *11*, 75. https://doi.org/10.3390/atoms11050075

Academic Editors: Himadri S. Chakraborty and Hari R. Varma

Received: 9 March 2023
Revised: 16 April 2023
Accepted: 19 April 2023
Published: 23 April 2023

shown that N_2 production from both CNN and NCN radicals is a dominant photodissociation channel [23–25]. This channel is attributed to the bent intermediate states of the free radicals. The NCN and CNN radicals are important in combustion chemistry [24]. Recently, a similar fragmentation channel for CO_2^{2+} molecular ions producing O_2^+ ionic fragments has also been observed in laser-induced ionization and subsequent dissociation study [26]. The C_3^- radical is also part of the same isoelectronic family. It is relevant for plasma physics and hydrocarbon chemistry and is even found in the interstellar space [27,28]. Thus, the study of dissociation dynamics of CO_2^{q+} (q = 2, 3) molecular ions can provide important information about different electronic states of these radicals.

The simplest fragmentation mechanism for a triatomic molecule is the one where the two molecular bonds break in a single step and the charged fragments move away due to mutual Coulomb repulsion. This type of fragmentation is termed as concerted fragmentation [5]. Additionally, the two bonds can break one after another. This decay is termed as sequential fragmentation. In the first step, the parent molecular ion undergoes a two-body breakup. Subsequently, the daughter molecular ions further decay into ionic or neutral fragments. During the dissociation, the unstable molecular ion can also rotate as well as vibrate about its equilibrium geometry. The typical time period for the rotational and vibrational motion of molecular ions is $\approx 10^{-12}$ s and 10^{-14} s, respectively. The fragmentation can happen within or beyond these typical time scales. Hence, we can distinguish between the two extremes of a three body fragmentation, namely concerted and sequential decay, by comparing the two time scales. One is the time difference (Δt) between the cleavage of the two molecular bonds and the other is the mean rotational period τ_{rot} of the primary daughter molecular ion [29]. If $(\Delta t) >> \tau_{rot}$ then the three body decay is called sequential. On the other hand, for a concerted decay, we have $(\Delta t) << \tau_{rot}$. As Δt approaches zero we reach the asymptotic limit of a concerted decay and with $\Delta t = 0$ a three body decay is called a synchronous concerted decay. A situation may also arise where $0 << \Delta t << \tau_{rot}$. This is termed as asynchronous concerted decay. The two concerted decay mechanisms can be illustrated using the symmetric and antisymmetric stretching modes of a linear triatomic molecule [30]. The symmetric stretching causes both the bonds to elongate in phase and eventually break exactly at the same time. This results in Δt being zero, which is the characteristic of a synchronous concerted decay. In the antisymmetric stretching mode, the elongation of one bond happens together with the contraction of the other. If τ_{vib} is the characteristic vibrational period of the parent molecular ion, then during a complete fragmentation process, the second bond will break half a vibrational period later than the first one. As a result Δt would be $\tau_{vib}/2$, characteristic of an asynchronous decay. Hence, in this type of decay, the bonds break in a time span such that the molecular vibration precedes the fragmentation process.

Sequential decay can be further classified into two processes. (a) Initial charge separation *s(i)* , where an atomic ion and a diatomic cation are released in the first step by the break-up of one of the bonds and (b) deferred charge separation *s(d)*, where a neutral atom is released in the first step. The complete kinematics of a concerted decay can only be inferred when all three fragment ions are detected in coincidence. However, if there is a neutral fragment, such as in *s(d)* process, the 3d momenta of this neutral can only be deduced indirectly provided the other two ions are detected in coincidence and their momenta are known completely.

In their pioneering study, Neumann et al. [5] have shown that the amount of energy deposited into a system is the key parameter to determine which pathway will dominate during a molecular fragmentation. In the present experiment, we have used 1 MeV protons ($v_p \approx 6$ a.u.). This projectile charge and energy combination translates into a perturbation strength k (q_p/v_p) of ≈ 0.16 a.u., which falls in the weak perturbative regime. The projectile velocity corresponds to an interaction time (t_{int}) of 37 as which is much shorter than the typical time scale of molecular fragmentation (10 fs) as well as rotational (10^{-12} s) and vibrational (10^{-14} s) time scales. Previous experiments with highly charged ions fall under different values of k as well as t_{int}. For instance, the works by Adoui et al. (8 MeV u^{-1} Ni^{24+}) [7],

Siegmann et al. (5.9 MeV u^{-1} Xe^{18+} and Xe^{43+}) [8], Neumann et al. (3.2 keV u^{-1} Ar^{8+}) [5], Jana et al. (5 MeV u^{-1} Si^{12+}) [9], and Khan et al. (1 MeV Ar^{8+}) [6] correspond to values of 1.34 (13 as), 1.17 (15 as), 2.80 (15 as), 22.3 (657 as), 0.85 (17 as), and 7.94 (233 as), respectively, of k (t_{int}). Recently, Srivastav and Bapat [16] studied the fragmentation of CO$_2^{3+}$ into C$^+$ + O$^+$ + O$^+$ under the impact of protons having velocities of 0.5 a.u. (k = 2.04, t_{int} = 480 as) and 0.83 a.u. (k = 1.21, t_{int} = 285 as).

In the present work, we have studied the fragmentation dynamics of CO$_2^{q+}$ (q = 2, 3) molecular ions produced under the impact of 1 MeV protons using the multiple-hit co-incidence imaging technique. The slopes and shapes of the different islands observed in the ion–ion correlation diagram were used to identify different fragmentation channels. The time-of-flight and position information were used to reconstruct the momenta of each detected fragment ion. The reconstructed momenta were further used to calculate the KER for each fragmentation channel. The momentum distributions, angular correlations and the KER distributions were utilized to identify different fragmentation processes.

2. Experimental Setup

The present experiment has been carried out at the 1.7 MV Tandetron Accelerator Facility at the Indian Institute of Technology Kanpur, India. A newly built recoil ion momentum spectrometer (RIMS) [31,32] equipped with a time and position-sensitive multihit detector was used to obtain the three dimensional momenta of ionic fragments. The details of the experimental setup have been described in detail earlier [33]. Briefly, a beam of 1 MeV proton obtained from the 358 Duoplasmatron source is made to collide with an effusive jet of neutral CO$_2$ gas in a crossed beam geometry. The RIMS is mounted orthogonal to the ion beam and gas jet direction. Acceleration field of 145 V cm^{-1} is used to extract the electrons and recoil ions produced in the interaction zone. The electrons produced in this collision are detected by a channel electron multiplier (CEM). The ions are extracted using an extraction field of 145 V cm^{-1} followed by an accelerating field of 260 V cm^{-1} towards a microchannel plate of 40 mm diameter equipped with a delay line anode. The present spectrometer conditions result in a KER resolution of $\approx$1.2 eV for three-body fragmentation and a 4Π collection efficiency of particles having energy <8 eV/q. The output from the CEM works as the start for the data acquisition system. The background vacuum was better than 5×10^{-8} mBar and the working pressure was kept below 1×10^{-6} mBar. The beam current used in our present experiment was $\approx$200 pA. The time and position data was recorded on an event-by-event basis using a time-to-digital converter. The time-of-flight and the position information of each ion are stored in a list mode file using the CoboldPC software (CoboldPC 2011 R5-2-x64 version 10.1.1412.2, Roentdek Handels GmbH, Frankfurt, Germany) The initial momentum vectors of the fragment ions were reconstructed from the timing and position information.

3. Results and Discussions

In collisions with 1 MeV protons, the CO$_2$ molecule can be ionized to several degrees producing a multiply charged molecular ion, which further dissociates into charged fragments. The correlation diagram or the coincidence plot between the time-of-flights of fragment ions helps to identify different fragmentation channels. Unlike diatomic molecules, the coincidence plot is more complex for triatomic molecules.

3.1. Two-Body Break-Up

In Figure 1, we have shown the coincidence time-of-flight plots between first ion (fragment ion with smallest time-of-flight) vs. second ion (Figure 1a) and second ion vs. third ion (Figure 1b). From the coincidence spectra, one can identify a sharp trace of O$^+$ + CO$^+$ channel arising due to the two body fragmentation of CO$_2^{2+}$. This is the only complete two body break-up channel observed in our present experiment. The slope of the trace is -1.0 ± 0.03, as expected from the momentum conservation for a two-body Coulomb fragmentation. The KER distribution for this channel is shown in Figure 2. This spectrum shows a narrow structure around a peak value of 6 ± 0.3 eV and extends only up to 16 eV. This signifies that this channel arises from a prompt dissociation of the precursor CO_2^{2+} molecular ion and that it only evolves through a few number of PECs. Zhang et al. [22] have obtained the PECs of the 14 low-lying states of CO_2^{2+} using multistate multiconfiguration second-order perturbation theory (MS-CASPT2) and complete active space self-consistent field (CASSCF) methods. A few of the theoretical KER values for the $O^+ + CO^+$ channel are shown as vertical lines on the top axis of the KER spectrum. The most probable KER can be accounted for by considering the decay of CO_2^{2+} from four electronic states: $c^1\Sigma_u^-$, $b^1\Sigma_g^+$, $A^3\Delta_u$, and $A^3\Delta_u$ as shown in Table 1. The decay from $a^1\Delta_g$ can contribute to the KER spectra in the region below the most probable value. Whereas, the range beyond the most probable value can be explained based on the dissociation from the following six electronic states: $a^1\Delta_g$, $E^3\Pi_g$, $b^1\Sigma_g^+$, $D^3\Pi_u$, $E^3\Pi_g$, and $2^3\Pi_g$.

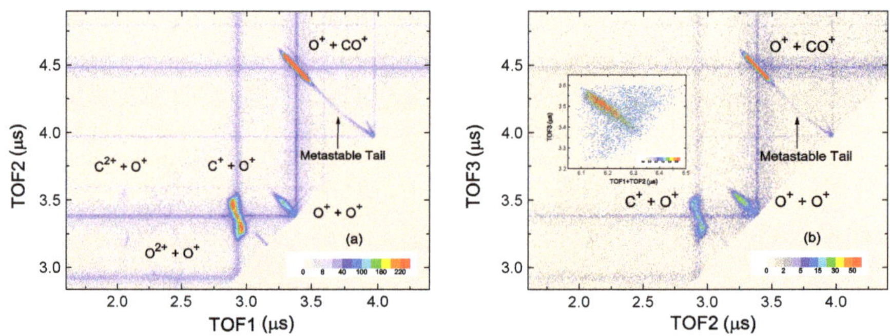

Figure 1. Ion-ion coincidence spectra for the fragmentation of CO_2 under the impact of 1 MeV proton for (**a**) the TOF of the second ion versus the TOF of the first ion and (**b**) the TOF of the third ion versus the TOF of the second ion. The plot for the TOF of third ion versus the sum TOF of the first and the second ion is also shown in the inset.

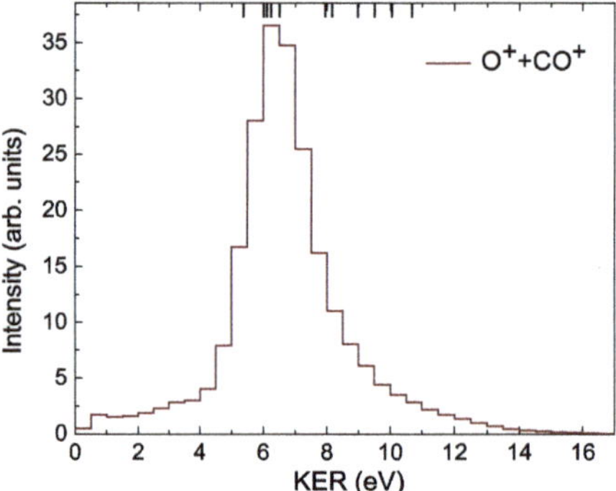

Figure 2. KER distribution for the fragmentation of CO_2^{2+} into $O^+ + CO^+$. The vertical lines on the top axis are the calculated KER values as reported in ref [22].

The KER spectrum can also be compared with previous studies. In their photoion–photoion coincidence (PIPICO) experiment, Dujardin and Winkoun [34] measured three distinct KER values around 4.5 eV, 6.5 eV, and 9.4 eV for this channel. Whereas, under the impact of 5 keV electron, Wang et al. [35] obtained a KER around 6.8 eV. In their experiment with 1.3 keV electron, Sharma et al. [17] obtained a KER around 5.9 eV and from their ab initio calculations assigned this peak to the $^3\Sigma_g^-$ state of the CO_2^{2+} molecular ion dissociating into O^+ (^{4}S) + CO^+ ($X^2\Sigma^+$)channel. Another measurement with 12 keV electron by Bhatt et al. [18] showed a KER value around 4.7 eV. The Coulomb explosion model predicts two possible values (6.2 eV and 12.4 eV) (at an equilibrium distance of 1.16 Å [36] between C and O atoms) as the point charge on the CO^+ molecular ion can be assumed to be either near the C atom or the O atom [18].

Table 1. The possible molecular states of CO_2^{2+} dissociating into $O^+ + CO^+$ along with the theoretically calculated values of KER by Zhang et al. [22] using multistate multiconfiguration second-order perturbation theory (MS-CASPT2) and complete active space self-consistent field (CASSCF) methods.

Molecular States	Dissociation Limit	KER (eV) [22]
$c^1\Sigma_u^-$	$O^+(^4S_u) + CO^+(A^2\Pi)$	6.01
$b^1\Sigma_g^+$	$O^+(^2D_u) + CO^+(X^2\Sigma^+)$	6.11
$A^3\Delta_u$	$O^+(^4S_u) + CO^+(A^2\Pi)$	6.24
$A^3\Delta_u$	$O^+(^2D_u) + CO^+(X^2\Sigma^+)$	6.51
$a^1\Delta_g$	$O^+(^4S_u) + CO^+(A^2\Pi)$	5.37
$a^1\Delta_g$	$O^+(^4S_u) + CO^+(X^2\Sigma^+)$	7.93
$E^3\Pi_g$	$O^+(^4S_u) + CO^+(A^2\Pi)$	8.15
$b^1\Sigma_g^+$	$O^+(^4S_u)CO^+(X^2\Sigma^+)$	8.98
$D^3\Pi_u$	$O^+(^4S_u)CO^+(X^2\Sigma^+)$	9.50
$E^3\Pi_g$	$O^+(^4S_u)CO^+(X^2\Sigma^+)$	10.05
$2^3\Pi_g$	$O^+(^4S_u) + CO^+(A^2\Pi)$	10.68

In the ion–ion coincidence plot (Figure 1) we observe a 'tail' followed by a 'V' shape structure which starts at the end of the sharp trace of $O^+ + CO^+$ channel and extends up to the forward diagonal (the TOF1 = TOF2 line). This particular structure is characteristic of a metastable molecular ion [37]. The time-of-flights of these metastable molecular ions would lie between that of a stable CO_2^{2+} and the fragment ions (CO^+ and O^+). As we go closer to the $O^+ + CO^+$ coincidence along the tail, the time period between the formation and dissociation of the precursor molecular ion gets shorter. In contrast, coincidences closer to the 'V' region signify a longer time period prior to dissociation [38]. Thus, the 'tail' part arises due to the fragmentation of $(CO^{2+})^*$ in the extraction region. Whereas, the 'V' arises when it fragments in the drift tube. This 'V' has two arms, extending from the point on the forward diagonal corresponding to the time-of-flight ($\approx$3972 ns) of a stable CO_2^{2+} molecular ion. The origin of these arms can be explained on the basis of the momentum gained by the O^+ and CO^+ fragment ions. Therefore, in the upper arm the CO^+ was detected first due to its momentum gained towards the detector. On the other hand, in the lower arm, the O^+ gained momentum toward the detector. Lifetime measurements for the metastable $(CO^{2+})^*$ molecular ion have been carried out extensively by various groups [17,37,39–41] and values in the range of 0.9–21 μs have been reported. Field and Eland [37] obtained a lifetime of 0.9 ± 0.2 μs using a set of equations utilizing the charge separation mass spectrometry technique. These set of equations can be modified [6] for our double field system and the intensity in the 'V' and 'tail' region can be used to estimate the lifetime of the $(CO^{2+})^*$ molecular ion. We obtained a metastable lifetime of 1.6 ± 0.2 μs in our experiment.

3.2. Three-Body Break-Up

3.2.1. Fragmentation of CO_2^{2+}

In the last section, we discussed the two-body prompt dissociation of the CO_2^{2+} molecular ion. Here, we will describe its three-body fragmentation. As prescribed by Eland [42], the shape and slope of the coincidence traces can be used to determine the fragmentation dynamics. Thus, for a two-body Coulomb fragmentation, the slope of the island in the coincidence map is $-q_1/q_2$ due to the conservation of momentum. Here, q_1 and q_2 are the charges of the first and the second ion, respectively. Three-body dissociation is much more complex. As already discussed, the dissociation of CO_2^{2+} can happen via either concerted or sequential fragmentation.

1. In the concerted decay, the two C=O bonds break simultaneously:

$$CO_2^{2+} \rightarrow C^+ + O^+ + O \tag{1a}$$

$$\rightarrow C + O^+ + O^+ \tag{1b}$$

As already discussed, the concerted decay can be classified into a synchronous and an asynchronous way depending upon the time period of fragmentation compared to that of molecular vibrational and rotational motion. However, since the CO_2 molecule has a linear geometry, it is expected, in the first case (Equation (1a)), that a C^+ ion would carry much less momentum compared to the O atom and O^+ ion. This implies that the coincidence trace would be predominantly vertical. In the second case (Equation (1b)), the presence of a neutral C atom implies that the two O^+ ions are anticorrelated. Hence, the slope of the coincidence island would be -1.

2. For a sequential or two-step decay, there can be two different situations:

    (a) In the initial charge separation (s(i)) process a charged fragment is released due to the break-up of the C=O bonds. Depending on which ion (C^+ or O^+) is released first, s(i) is further categorized [19] as follows:

    (I) If the lighter ion C^+ is released in the first step:

$$CO_2^{2+} \rightarrow C^+ + O_2^+ \rightarrow C^+ + O^+ + O \qquad : s(i)_1 \tag{2}$$

    In this case, the slope of the coincidence trace should be:

$$-(q_1/q_2)\,\frac{m_2}{m_2 + m_3} \tag{3}$$

    where m_1, m_2, and m_3 are the masses of the lighter ion, the heavier ion, and the neutral atom, respectively.

    (II) Whereas, in the following case

$$CO_2^{2+} \rightarrow CO^+ + O^+ \rightarrow C^+ + O + O^+ \qquad : s(i)_2 \tag{4}$$

    the C^+ ion is released in the second step and the heavier O^+ ion is released in the first step, hence the slope of this coincidence trace should be:

$$-(q_1/q_2)\,\frac{m_1 + m_3}{m_1} \tag{5}$$

    (b) For a deferred charge separation (s(d)) process, a neutral fragment is released due to the break-up of the C=O bonds.

$$CO_2^{2+} \rightarrow CO^{2+} + O \rightarrow C^+ + O^+ + O \tag{6a}$$

$$\rightarrow O_2^{2+} + C \rightarrow O^+ + O^+ + C \tag{6b}$$

In both of these cases, the motion of the two fragment ions produced in the second step, are governed by the mutual Coulomb repulsion and are not affected by the neutral fragment. Thus, similar to a two-body Coulomb fragmentation, the slope of the coincidence trace for a deferred charge separation would be simply $-(q_1/q_2)$. However, the second case (Equation (6)) is special because it demands the isomerization of CO_2^{2+} molecular ion to form an O_2^{2+} intermediate and ejection of a neutral C atom and to the best of our knowledge, this particular channel has never been observed experimentally.

The slopes of the different fragmentation channels from CO_2^{2+} were extracted from the correlation diagram and fitted with the method of least squares. Table 2 shows the comparison between the theoretical predictions and experimental observations.

Table 2. Comparison of the slopes of the best-fit line to the various islands in ion pair coincidence map obtained from CO_2 in collision with 1 MeV proton with theoretical predictions [42,43] and previous experimental results [18,19,44]. Here, $s(i)_1$ ($s(i)_2$) represents the C^+ ion released in the first (second) step of the initial charge separation process, whereas $s(d)$ represents the deferred charge separation process. R is the regression coefficient of the best-fit line.

| Fragmentation Channel | Theoretical Predictions [42,43] | | | | Experimental Results | | | | |
| | $s(i)_1$ | $s(i)_2$ | $s(d)$ | Concerted | Present Experiment | | Electron Impact | | |
					Slope (Fitted)	R	0.2 keV [19]	0.6 keV [44]	12 keV [18]
$O^+ + CO^+$	-	-	-	-1	-1.09 ± 0.03	0.99	-1.01 ± 0.01	-1.00 ± 0.02	-1.00 ± 0.02
$C^+ + O^+ + O$	-0.5	-2.33	-1.0	∞	-2.21 ± 0.01	0.99	-1.75 ± 0.04	-2.75 ± 0.04	-2.75 ± 0.04
$O^+ + O^+ + C$	-0.57	-	-1.0	-1	-1.16 ± 0.02	0.97	-1.03 ± 0.03	-1.00 ± 0.02	-1.00 ± 0.02

$C^+ + O^+ + O$ Channel

For the $C^+ + O^+ + O$ channel, we have obtained a slope of -2.21 ± 0.01. This matches well with the theoretical predictions and earlier measurements. Bhatt et al. [18], in their experiment with 12 keV electron impact, obtained a slope of -2.75 ± 0.04 for the same channel. They attributed the slight departure of the experimental slope from the theoretical value, to the contribution from the concerted decay (Equation (1a)). Wang et al. [19] measured a slope of -1.75 ± 0.04 in their experiment with 200 eV electron. They explained this fragmentation channel to have contribution from both $s(i)_2$ and deferred charge separation.

To better understand the fragmentation process, we take help of the Dalitz plot [45]. The coordinates in a Dalitz plot are defined as $X_{Daliz} = (\epsilon_1 - \epsilon_2)/\sqrt{3}$ and $X_{Daliz} = (\epsilon_3 - 1/3)$, with $\epsilon_i = |P_i|^2 / \sum_i |P_i|^2$. Here, P_i is the momentum of the ith fragment in the center-of-mass frame. However, to obtain these two diagrams, we need all three momenta. Hence, first ultilizing the coincidence technique the momenta of the fragment ions in all three dimensions are obtained. Furthermore, momentum conservation is imposed to deduce the neutral atom momentum. Figure 3a shows the Dalitz plot for the $C^+ + O^+ + O$ channel. It is similar to the result obtained by Laksman et al. [46] with 270 eV photon. We can observe two distinct structures in this diagram.

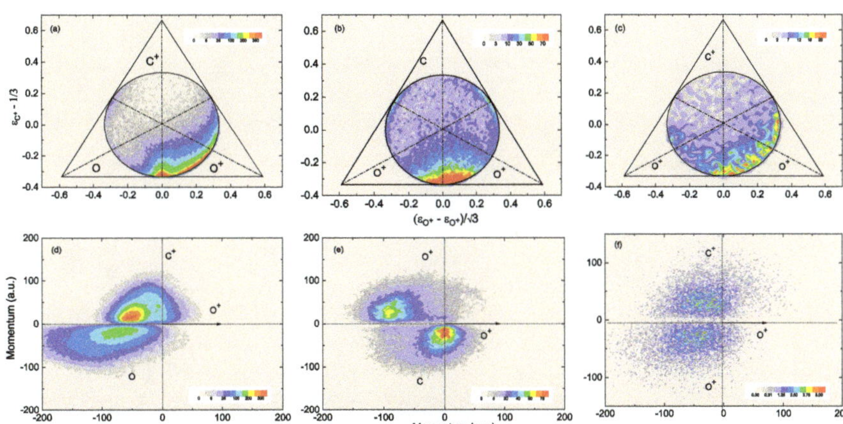

Figure 3. Experimentally observed Dalitz plots of the three-body fragmentation of CO_2^{2+}, (**a**) for $CO_2^{2+} \rightarrow C^+ + O^+ + O$, (**b**) $CO_2^{2+} \rightarrow C + O^+ + O^+$, and of CO_2^{3+}, (**c**) for $CO_2^{3+} \rightarrow C^+ + O^+ + O^+$. The corresponding Newton Diagrams are shown in (**d–f**), respectively.

(i) An intense symmetric structure around the C^+ axis. This symmetry is because of the equal momentum sharing between the O^+ ion and the neutral O atom. The distribution around the minimum C^+ momentum is a clear signature of the linear structure of the CO_2 molecule. The C^+ ion is released with a smaller momentum, while the other two fragments (O^+ and O) are emitted back to back. Thus, this structure corresponds to the concerted process, where the two C=O bonds break simultaneously. (ii) A second structure can also be distinguished in this diagram, distributed in the perpendicular direction to the O^+ axis. This indicates a weak correlation between the O^+ ion and all other fragments [46]. This is a typical signature of a two step $s(i)_2$ process, where the O^+ ion is released in the first step. The CO^+ cation further fragments into C^+ and O after the primary fragments (O^+ and CO^+) have left the Coulomb field region. As a result, the C^+ and O are anticorrelated.

The same data is displayed in a Newton diagram in Figure 3d. The most probable momentum of the O^+ ion is shown by an arrow along the x-coordinate. The relative momentum of the C^+ and O are mapped in the upper and lower half of the diagram, respectively. Although, we could not distinguish between the concerted and the sequential $s(i)_2$ process in this diagram, the anticorrelation between C^+ and O is clearly visible.

To further understand the dynamics we take help of the distributions of the momentum correlation angles (MCAs) α, β, and γ. These angles are shown schematically in Figure 4a and can be obtained from the momentum vectors of the associated ith and jth fragment ions as: $\text{MCA} = \cos^{-1}\left(\frac{\vec{P_i}.\vec{P_j}}{|\vec{P_i}||\vec{P_j}|}\right)$. Figure 4d–f show these angular distributions for the $C^+ + O^+ + O$ channel. Both the O^+ ion and O atom show a double peak structure in the angular distribution with respect to the C^+ ion. The O^+ (neutral O) ion has two peaks around $110°$ ($45°$) and $160°$ ($100°$). These values are in good agreement with measurements reported earlier [18]. The double structure obtained in our present experiment is explained by considering both sequential and concerted decays. In the sequential process, the O atom is released toward C^+ at $45°$, whereas the O^+ ion at $160°$ to balance the $C^+ + O$ center of mass momentum. On the other hand, in the concerted process, both the O^+ and O fragments are released at $110°$ and $100°$ with respect to the C^+ ion. The angle β was found to be around $170°$.

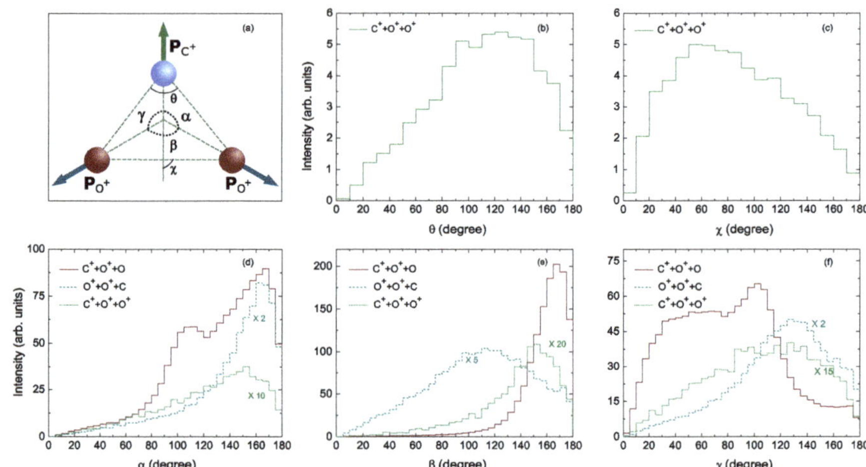

Figure 4. (**a**) Schematic diagram of a CO_2^{q+} ($q = 2, 3$) molecular ion fragmenting into $C^+ + O^+ + O^+$ along with definitions of different angles (α, β, γ, θ, and χ) in the momentum space (discussed in the text). (**b**) The distribution of the momentum space molecular bond angle (θ) and (**c**) the angle χ for the $C^+ + O^+ + O^+$ channel. The distribution of (**d**) α, (**e**) β, and (**f**) γ for all the three-body breakup channels ((1,1,0), (0,1,1), and (1,1,1)). The scaling is performed for visual clarity.

In contrast to the above discussion, the presence of the first structure together with the second one in the Dalitz plot has been attributed to missed three-ion coincidences by Laksman et al. [46]. One factor contributing to this case is the finite dead time ($\sim$35 ns [33]) of the spectrometer, which causes the third one to be missed if the TOFs of the second and the third ion are the same. As a result, a triple coincidence is recorded as a double coincidence. The other contributing factor is the finite detection efficiency of the detector, due to which there is a probability that the third ion could be missed even if the TOFs are very different for all the ions. The contribution from these missed triple coincidences could also be the reason for the deviation of the slopes in the 2D coincidence plots from those reported in the earlier studies.

The KER distribution for the fragmentation of CO_2^{2+} into $C^+ + O^+ + O$ is shown in Figure 5a. The kinetic energies (KEs) of the individual fragments are also shown in the same plot. The KER spectra has a broad distribution around the most probable value 10.8 ± 1 eV, with a small structure around 1.2 ± 0.13 eV, and it extends form 0 eV to around 50 eV. The most probable value of KE are 1.5 ± 0.05 eV, 7.5 ± 0.5 eV, and 1.5 ± 0.12 eV for C^+, O^+, and O, respectively. Additionally, the KE of O^+ ion shows an additional contribution at zero.

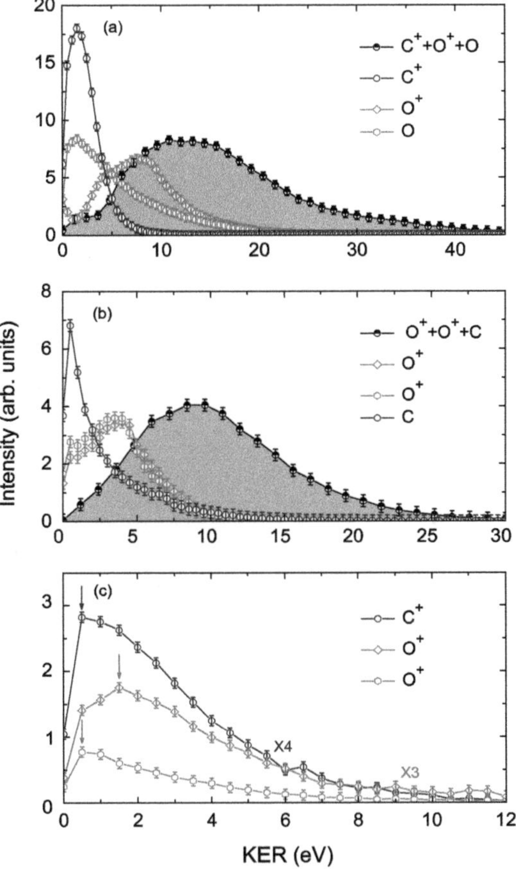

Figure 5. The KER distributions for the three-body fragmentation of CO_2^{q+} (q = 2, 3) into (**a**) (1,1,0) and (**b**) (0,1,1) channels along with the kinetic energies (KEs) of the individual fragments. (**c**) KEs of individual fragments in the (1,1,1) channel. The arrows show the position of the most probable KE of each fragments. The scaling is performed for visual clarity.

$O^+ + O^+ + C$ Channel

For the $O^+ + O^+ + C$ channel, we have obtained a slope of -1.16 ± 0.02. This is in good agreement with the theoretical prediction of -1.0 for a concerted as well as s(d) process. However, as discussed above, the deferred charge separation for this channel demands isomerization of the CO_2 molecule to form an O_2^{2+} cation. In our experiments with 1 MeV proton we have not seen any trace of O_2^{2+} in the TOF spectrum. Therefore, a concerted process (Equation (1b)) seems to be predominantly contributing to this channel [18,19].

Figure 3b shows the Dalitz plot for the $O^+ + O^+ + C$ channel. We can observe two distinct structures in this diagram. (a) An almost symmetric intense structure around the **C** axis. The distribution close to the minimum **C** momentum is again a clear signature of the linear structure of the CO_2 molecule. The C ion is released with a smaller momentum, while most of the momentum is shared between the other two fragments (O^+ and O^+). Hence, this structure corresponds to the concerted process, where the two C=O bonds break simultaneously. This structure is almost identical to that obtained by Wang et el. [19] with 200 eV electron. (b) Two separate structures can also be observed at the two opposite O^+ edges, which are symmetric around each O^+ axis. These correspond to events where one of the O^+ has low momentum, while the C and the other O^+ ion are released one after the other. This momentum sharing is a clear signature of a two step s(i)$_1$ process, where the O^+ ion is released in the first step. While the CO^+ cation further fragments into O^+ and C after the primary fragments (O^+ and CO^+) have left the Coulomb region. As a result, the O^+ and C are anticorrelated.

Figure 3b shows the Newton diagram for this channel where the most probable momentum of the first O^+ ion is plotted along the x-axis. Although, we could not distinguish between the concerted and the sequential s(i)$_1$ process in this diagram, but the anticorrelation between O^+ and C is clearly visible. The low momentum of the C atom can also be seen.

The angular distributions of α, β, and γ for the $O^+ + O^+ + C$ channel is shown in Figure 4d–f. The two O^+ ions show peak structure around $160°$ (α) and $125°$ (γ), with a small contribution around $160°$ in the distribution of the angle γ, whereas the angle β has a broad distribution around $110°$.

The KER distribution for the $O^+ + O^+ + C$ channel is shown in Figure 5b along with the kinetic energies (KEs) of the individual fragments. The KER spectra has a broad distribution around the most probable value of 8.4 ± 0.8 eV. And, it extends form 0 eV to around 30 eV. The most probable values of KE are 4.0 ± 0.1 eV, 3.5 ± 0.2 eV, and 0.5 ± 0.04 eV for O^+, O^+, and C, respectively. The KE of the O^+ ions show an additional contribution at 0.5 eV. The most probable value of KER of the $O^+ + O^+ + C$ channel is smaller than that of $C^+ + O^+ + O$ channel. This difference can be explained by the CE model by noting that due to the linear configuration of the CO_2 molecule the distance between the two oxygen atom (2.32 Å) is larger that that between the carbon and oxygen atoms (1.16 Å).

3.2.2. Fragmentationof CO_2^{3+}

Similar to CO_2^{2+}, we can also observe several sequential and concerted fragmentation channels for the decay of CO_2^{3+} as follows:

(I.) Concerted fragmentation

$$CO_2^{3+} \rightarrow C^+ + O^+ + O^+ \tag{7a}$$

$$\rightarrow C^{2+} + O^+ + O \tag{7b}$$

$$\rightarrow C + O^+ + O^{2+} \tag{7c}$$

(II.) Two-step $s(i)_1$

$$CO_2^{3+} \rightarrow C^+ + O_2^{2+} \rightarrow C^+ + O^+ + O^+ \tag{8a}$$

$$CO_2^{3+} \rightarrow C^{2+} + O_2^+ \rightarrow C^{2+} + O^+ + O \tag{8b}$$

(III.) Two-step $s(i)_2$

$$CO_2^{3+} \rightarrow CO^{2+} + O^+ \rightarrow C^+ + O^+ + O^+ \tag{9a}$$

$$\rightarrow C^{2+} + O + O^+ \tag{9b}$$

$$\rightarrow C + O^{2+} + O^+ \tag{9c}$$

$$CO_2^{3+} \rightarrow CO^+ + O^{2+} \rightarrow C + O^+ + O^{2+} \tag{9d}$$

$$\rightarrow C^+ + O + O^{2+} \tag{9e}$$

(IV.) Two-step $s(d)$

$$CO_2^{3+} \rightarrow CO^{3+} + O \rightarrow C^{2+} + O^+ + O \tag{10a}$$

$$\rightarrow C^+ + O^{2+} + O \tag{10b}$$

$$CO_2^{3+} \rightarrow O_2^{3+} + C \rightarrow C + O^{2+} + O^+ \tag{10c}$$

Among the above fragmentation channels, we could only observe three in Figure 1: (a) $C^+ + O^+ + O^+$, (b) $C^{2+} + O^+ + O$, and (c) $O^{2+} + O^+ + C$. Among these, only the first channel has significant statistics.

$C^+ + O^+ + O^+$ Channel

Figure 3c shows the Dalitz plot for the $C^+ + O^+ + O^+$ channel. The dominant structure is the almost symmetric distribution around the right edge of the circle where the O^+ ion has maximum momentum. As discussed above, it is a signature of two-step $s(i)_2$ process, where an O^+ is released in the first step leaving a metastable CO^{2+}, which decays in the second step producing O^+ and C^+ ions. In addition, there are traces of counts on the left side of the C^+ axis. A sequential process in a three-body break up can be easily identified in a Newton diagram. During the first step, the intermediate CO^{2+} ion acquires some angular momentum. If the lifetime of this CO^{2+} ion is of the order of it's half-rotational time period, then it can rotate while dissociating in the next step [5,15], which shows up as semi-circular structures in the Newton diagram [5,47]. Figure 3f shows the Newton diagram for this channel where the most probable momentum of the first O^+ ion is plotted along the x-axis. It shows two lobes on the upper and the lower half of the diagram, the C^+ ion and the second O^+ ($y < 0$ plane) are anti-correlated, and there is no prominent semicircular structure. This hints towards the fact that either the lifetime of the CO^{2+} ion is less than the half-rotational time period or the vibrational motion precedes the fragmentation process (the asynchronous concerted decay).

To further shed light upon the underlying process, we discuss the MCA distributions. The distributions of α, β, and γ for the $C^+ + O^+ + O^+$ channel is shown in Figure 4d–f. The two O^+ ions show peak structure around $140°$ (α) and $130°$ (γ). While γ has a small contribution around $60°$. The results from Jana et al. using $5 \text{ MeV u}^{-1} \text{ Si}^{12+}$ shows that the angle between the momentum vectors of the two O^+ ions (β) is about $165°$. By comparing with earlier reported studies, they concluded the $C^+ + O^+ + O^+$ fragmentation to be a concerted decay from linear as well as bent structures of CO_2^{3+}. In our data the angle β is around $150°$, which is less compared to the other two fragmentation channels. Therefore, our present data also indicates contribution from bent states. The presence of several bent geometries of CO_2^{3+} molecular ion are also confirmed from the distribution of the momentum-space molecular bond angle θ as shown in Figure 4b, which has a broad distribution around $120°$. Similar results have been also reported by other groups [7,8,18].

In an extreme example of concerted (synchronous) breakup, the molecule dissociates via symmetric stretching around the central C atom. Thus, the C^+ ion would obtain zero momentum and the two O^+ ions are ejected simultaneously with same energy. If the central C^+ ion is released with a finite energy, then the break must have happened from a bent geometry of the precursor molecular ion, whereas in a concerted process, if there is any deviation from the equal sharing of energy between the two terminal ions, then it would correspond to an antisymmetric stretching of the molecule [30]. The kinetic energies (KEs) of the individual fragments for the $C^+ + O^+ + O^+$ channel is shown in Figure 5c. In Figure 6a–c, we have plotted the complete KER distribution for the above channel in three regions: (a) 0–9.6 eV, (b) 9.6–16.8 eV, and (c) 16.8–35.0 eV. The KER spectrum has a most probable value of 7.2 ± 0.4 eV (Figure 6a) with a broad structure around 20 eV (Figure 6c). It extends from 0 eV to about 35 eV. The most probable values of KE (the position of which are depicted as arrows in Figure 5c) are 1.5 ± 0.05 eV, 1.5 ± 0.1eV, and 0.5 ± 0.04 eV for C^+, O^+, and O^+, respectively. The most probable value of KER for this charge symmetric channel is smaller than that of the asymmetric channels. The non-zero kinetic energy of the C^+ ion implies that bent geometries are contributing to the fragmentation. In addition, the unequal energy of the two O^+ ions signifies that vibrational motions precede the fragmentation process, and hence it is an asynchronous concerted decay. To confirm this vibrational stretching we take help of the distribution of the angle χ (Figure 4a). A uniform distribution in χ represents a stepwise sequential process [30], whereas a sharp distribution indicates the involvement of a concerted process. Figure 4c shows the distribution of χ for the $C^+ + O^+ + O^+$ fragmentation channel showing a broad structure around $50°$, which could imply the presence of bending as well as stretching modes during the fragmentation of CO_2^{3+}. These three regions are also shown in the KER spectrum of Figure 5c shaded in yellow, orange, and blue. The corresponding Dalitz plots, Newton diagrams, and the distribution of the angle χ are also shown in Figure 6. For KER range of 0–9.6 eV, most of the counts in the Dalitz plot (Figure 6a) are situated near the bottom of the triangle and right of the C^+ axis indicating the presence of asynchronous decay [48]. As the KER range increases the counts get dispersed away from the central region towards the left and right. In the KER range between 16.8 eV and 35 eV, there are almost no counts around the C^+ axis, while dominant structures are around the two O^+ axes. The unequal energy sharing due to stretching of one of the C=O bonds can easily be identified in both the Dalitz plot and the Newton diagram (Figure 6f). The stretching of the bond is also reflected in the distribution of the angle χ (Figure 6f–h). With increase in the available energy, as the stretching becomes more dominant, an asymmetry takes over the initial isotropic distribution.

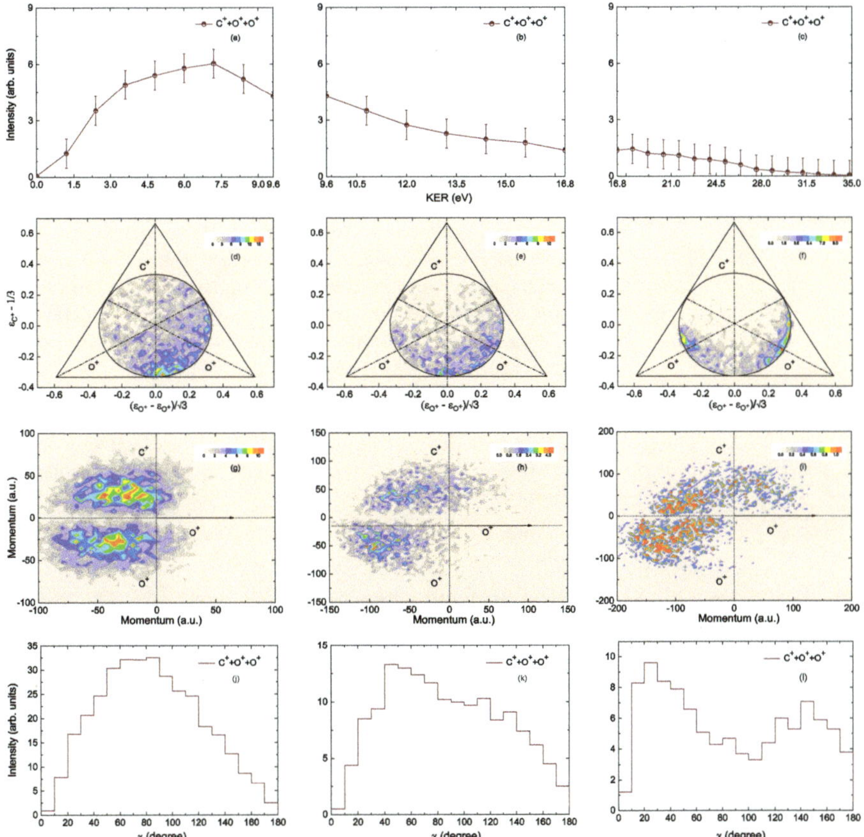

Figure 6. (**a–c**) The KER distributions, (**d–f**) Dalitz plots, (**g–i**) Newton diagrams, and (**j–l**) the distributions of the angle χ for the three-body fragmentation of CO_2^{3+} into (1,1,1) channel. The KER ranges of the three columns are; Left column: 0–9.6 eV, Middle column: 9.6–16.8 eV, Right column: 16.8–35 eV.

4. Conclusions

We have studied the dissociation dynamics of a simple, linear triatomic molecule CO_2 under the impact of 1 MeV protons. We have measured the two- and three-body dissociation of doubly and triply charged molecular ions of CO_2. For the $O^+ + CO^+$ fragmentation channel from the CO_2^{2+} molecular ion, we see a prompt dissociation resulting in narrow KER distribution. This KER distribution can be well explained based on the different electronic states reported by earlier theoretical and experimental studies. The CO_2^{2+} molecular ion also shows a metastable character in the ion–ion correlation diagram as a tail and 'V' structure. Using the intensity of these structures, we have estimated the life time of the metastable $(CO^{2+})^*$ molecular ions. We have also discussed the three-body dissociation of CO_2^{2+}, which produces two ions and a neutral. All three-body dissociation are discussed using Dalitz plots, Newton Diagrams, and angular distributions. For both the $C^+ + O^+ + O$ and $O^+ + O^+ + C$ channels, we have observed contribution from both concerted decay. In addition, for the $C^+ + O^+ + O$ channel, we see signature of an $s(i)_2$ process, whereas in the $O^+ + O^+ + C$ channel contains signature of an $s(i)_1$ process. The contributions from all these processes are also be verified from the angular distributions. We have further discussed the charge symmetric fragmentation of CO_2^{3+} molecular ions producing $C^+ + O^+ + O^+$. The angular distributions for this channel hint toward the fact that the three-body fragmentation is happening from bent molecular geometries of the precursor molecular ions. The Dalitz plots and Newton diagrams further suggest that molecular bond stretching precedes the fragmentation process in this charge symmetric dissociation. The linear triatomic CO_2 molecule has three vibrational modes, namely symmetric stretching, asymmetric stretching, and bending vibration. The typical time scales of these three stretching modes are 25 fs, 14 fs, and 50 fs [49], respectively, which are much larger than the interaction time (t_{int}) of 37 as for the present collision system consisting of 1 MeV protons and CO_2 molecules. The population of these different vibrational modes depends on the available energy of the molecular system. The KER distribution works as a tool to investigate different energy regimes in the fragmentation process. In the lowest KER range, we have observed that the fragmentation is happening due to concerted decay from a linear geometry of the precursor molecular ion (synchronous decay). With the increase in KER values, we observe more contributions from the bending and asymmetric stretching (asynchronous decay) modes.

Author Contributions: Conceptualization, A.D. and A.H.K.; data acquisition, A.D. and A.H.K.; analysis, A.D. and A.H.K.; interpretation, A.D. and A.H.K.; writing—original draft preparation, A.D.; writing—review and editing, A.D. and A.H.K. All authors have read and agreed to the published version of the manuscript.

Funding: This research was funded by Science and Research Engineering Board (SERB), Govt. of India via grant number ECR/2017/002055.

Data Availability Statement: The data presented in this study are available on request from the corresponding author.

Acknowledgments: The authors acknowledge the help from Rohit Tyagi, Sandeep Bari, and Sahan Sykam in accelerator operation during the experiments.

Conflicts of Interest: The authors declare no conflict of interest.

References

1. Janev, R.K. *Atomic and Molecular Processes in Fusion Edge Plasmas*; Springer: Berlin/Heidelberg, Germany, 2013. Available online: https://link.springer.com/book/10.1007/978-1-4757-9319-2 (accessed on 9 April 2023).
2. Tielens, A.G.G.M. The molecular universe. *Rev. Mod. Phys.* **2013**, *85*, 1021. [CrossRef]
3. De Vries, J.; Hoekstra, R.; Morgenstern, R.; Schlathölter, T. Charge driven fragmentation of nucleobases. *Phys. Rev. Lett.* **2003**, *91*, 053401. [CrossRef]
4. López-Tarifa, P.; du Penhoat, M.-A.H.; Vuilleumier, R.; Gaigeot, M.-P.; Tavernelli, I.; Le Padellec, A.; Champeaux, J.-P.; Alcamí, M.; Moretto-Capelle, P.; Martín, F.; et al. Charge driven fragmentation of nucleobases. *Phys. Rev. Lett.* **2011**, *107*, 023202. [CrossRef]
5. Neumann, N.; Hant, D.; Schmidt, L.P.H.; Titze, J.; Jahnke, T.; Czasch, A.; Schöffler, M.S.; Kreidi, K.; Jagutzki, O.; Schmidt-Böcking, H.; et al. Fragmentation Dynamics of $CO_2{}^{3+}$ Investigated by Multiple Electron Capture in Collisions with Slow Highly Charged Ions. *Phys. Rev. Lett.* **2010**, *104*, 103201. [CrossRef] [PubMed]
6. Khan, A.; Tribedi, L.C.; Misra, D. Observation of a sequential process in charge-asymmetric dissociation of $CO_2{}^{q+}$ ($q = 4, 5$) upon the impact of highly charged ions. *Phys. Rev. A* **2015**, *92*, 030701. [CrossRef]
7. Adoui, L.; Tarisien, M.; Rangama, J.; Sobocinsky, P.; Cassimi, A.; Chesnel, J.Y.; Frémont, F.; Gervais, B.; Dubois, A.; Krishnamurthy, M.; et al. HCI-induced molecule fragmentation: Non-coulombic explosion and three-body effects. *Phys. Scr.* **2001**, *2001*, 89. [CrossRef]
8. Siegmann, B.; Werner, U.; Lutz, H.O.; Mann, R. Complete Coulomb fragmentation of CO_2 in collisions with 5.9 MeV u^{-1} Xe^{18+} and Xe^{43+}. *J. Phys. B* **2002**, *35*, 3755. [CrossRef]
9. Jana, M.R.; Ghosh, P.N.; Bapat, B.; Kushawaha, R.K.; Saha, K.; Prajapati, I.A.; Safvan, C.P. Ion-induced triple fragmentation of $CO_2{}^{3+}$ into $C^+ + O^+ + O^+$. *Phys. Rev. A* **2011**, *84*, 062715. [CrossRef]
10. Yan, S.; Zhu, X.L.; Zhang, P.; Ma, X.; Feng, W.T.; Gao, Y.; Xu, S.; Zhao, Q.S.; Zhang, S.F.; Guo, D.L.; et al. Observation of two sequential pathways of $(CO_2)^{3+}$ dissociation by heavy-ion impact. *Phys. Rev. A* **2016**, *94*, 032708. [CrossRef]
11. Singh, R.K.; Lodha, G.S.; Sharma, V.; Prajapati, I.A.; Subramanian, K.P.; Bapat, B. Triply charged carbon dioxide molecular ion: Formation and fragmentation. *Phys. Rev. A* **2006**, *74*, 022708. [CrossRef]
12. Kushawaha, R.K.; Kumar, S.S.; Prajapati, I.A.; Subramanian, K.P.; Bapat, B. Polarization dependence in non-resonant photo-triple-ionization of CO_2. *J. Phys. B* **2009**, *42*, 105201. [CrossRef]
13. Bryan, W.A.; Sanderson, J.H.; El-Zein, A.; Newell, W.R.; Taday, P.F.; Langley, A.J. Laser-induced Coulomb explosion, geometry modification and reorientation of carbon dioxide. *J. Phys. B* **2000**, *33*, 745. [CrossRef]

14. Brichta, J.P.; Walker, S.J.; Helsten, R.; Sanderson, J.H. Ultrafast imaging of multielectronic dissociative ionization of CO_2 in an intense laser field. *J. Phys. B* **2006**, *40*, 117. [CrossRef]
15. Wu, C.; Wu, C.; Song, D.; Su, H.; Yang, Y.; Wu, Z.; Liu, X.; Liu, H.; Li, M.; Deng, Y.; et al. Nonsequential and sequential fragmentation of CO_2^{3+} in intense laser fields. *Phys. Rev. Lett.* **2013**, *110*, 103601. [CrossRef] [PubMed]
16. Srivastav, S.; Bapat, B. Electron-impact-like feature in triple fragmentation of CO_2^{3+} under slow proton impact. *Phys. Rev. A* **2022**, *105*, 012801. [CrossRef]
17. Sharma, V.; Bapat, B.; Mondal, J.; Hochlaf, M.; Giri, K.; Sathyamurthy, N. Dissociative double ionization of CO_2: Dynamics, energy levels, and lifetime. *J. Phys. Chem. A* **2007**, *111*, 10205–10211. [CrossRef]
18. Bhatt, P.; Singh, R.; Yadav, N.; Shanker, R. Formation, structure, and dissociation dynamics of CO_2^{q+} ($q \leq 3$) ions due to impact of 12-keV electrons. *Phys. Rev. A* **2012**, *85*, 042707. [CrossRef]
19. Wang, X.; Zhang, Y.; Lu, D.; Lu, G.C.; Wei, B.; Zhang, B.H.; Tang, Y.J.; Hutton, R.; Zou, Y. Fragmentation of CO_2^{2+} in collisions with low-energy electrons. *Phys. Rev. A* **2014**, *61*, 062705. [CrossRef]
20. Hogreve, H. Stability properties of CO_2^{2+}. *J. Phys. B* **1995**, *28*, L263. [CrossRef]
21. Hochlaf, M.; Bennett, F.R.; Chambaud, G.; Rosmus, P. Theoretical study of the electronic states of CO_2^{++}. *J. Phys. B* **1998**, *31*, 2163–2175. [CrossRef]
22. Zhang, D.; Chen, B.Z.; Huang, M.B.; Meng, Q.; Tian, Z. Photodissociation mechanisms of the CO_2^{2+} dication studied using multi-state multiconfiguration second-order perturbation theory. *J. Chem. Phys.* **2013**, *139*, 174305. [CrossRef]
23. Bise, R.T.; Choi, H.; Neumark, D.M. Photodissociation dynamics of the singlet and triplet states of the NCN radical. *J. Chem. Phys.* **1999**, *111*, 4923–4932. [CrossRef]
24. Bise, R.T.; Hoops, A.A.; Choi, H.; Neumark, D.M. Photodissociation dynamics of the CNN free radical. *J. Chem. Phys.* **2000**, *113*, 4179–4189. [CrossRef]
25. Bise, R.T.; Hoops, A.A.; Choi, H.; Neumark, D.M. Photoisomerization and Photodissociation Dynamics of the NCN, CNN, and HNCN Free Radicals. *ACS Symp. Ser.* **2001**, *770*, 296–311.
26. Larimian, S.; Erattupuzha, S.; Mai, S.; Marquetand, P.; González, L.; Baltuška, A.; Kitzler, M.; Xie, X. Molecular oxygen observed by direct photoproduction from carbon dioxide. *Phys. Rev. A* **2017**, *95*, 011404. [CrossRef]
27. Weltner, W., Jr.; Van Zee, R.J. Carbon molecules, ions, and clusters. *Chem. Rev.* **1989**, *89*, 1713–1747. [CrossRef]
28. Van Orden, A.; Saykally, R.J.H. Small Carbon Clusters: Spectroscopy, Structure, and Energetics. *Phys. Rep.* **1998**, *98*, 2313–2358. [CrossRef] [PubMed]
29. Maul, C.; Gericke, K.H. Photo induced three body decay. *Int. Rev. Phys. Chem.* **1997**, *16*, 1–79. [CrossRef]
30. Strauss, C.E.M.; Houston, P.L. Correlations without coincidence measurements: Deciding between stepwise and concerted dissociation mechanisms for ABC $\rightarrow$ A + B + C. *J. Phys. Chem.* **1990**, *94*, 8751–8762. [CrossRef]
31. Döner, R.; Mergel, V.; Jagutzki, O.; Spielberger, L.; Ullrich, J.; Moshammer, R.; Schmidt-Böcking, H. Cold Target Recoil Ion Momentum Spectroscopy: A 'Momentum Microscope' to View Atomic Collision Dynamic. *Phys. Rep.* **2000**, *330*, 95–192 [CrossRef]
32. Ullrich, J.; Moshammer, R.; Dorn, A.; Dörner, R.; Schmidt, L.P.H.; Schmidt-Böcking, H. Recoil-ion and electron momentum spectroscopy: Reaction-microscopes. *Rep. Prog. Phys.* **2003**, *66*, 1463–1545. [CrossRef]
33. Duley, A.; Tyagi, R.; Bari, S.B.; Kelkar, A.H. Design and characterization of a recoil ion momentum spectrometer for investigating molecular fragmentation dynamics upon MeV energy ion impact ionization. *Rev. Sci. Instrum.* **2022**, *93*, 113308. [CrossRef]
34. Dujardin, G.; Winkoun, D. State to state study of the dissociation of doubly charged carbon dioxide cations. *J. Chem. Phys.* **1985**, *83*, 6222–6228. [CrossRef]
35. Wang, E.; Shan, X.; Shi, Y.; Tang, Y. Chen, X. Momentum imaging spectrometer for molecular fragmentation dynamics induced by pulsed electron beam. *Rev. Sci. Instrum.* **2013**, *84*, 123110. [CrossRef] [PubMed]
36. Itikawa, Y. Cross Sections for Electron Collisions with Carbon Dioxide. *J. Phys. Chem. Ref. Data* **2002**, *31*, 749–767. [CrossRef]
37. Field, T.A.; Eland, J.H.D. Lifetimes of metastable molecular doubly charged ions. *Chem. Phys. Lett.* **1993**, *211*, 436–442. [CrossRef]
38. Slattery, A.E.; Field, T.A.; Ahmad, M.; Hall, R.I.; Lambourne, J.; Penent, F.; Lablanquie, P.; Eland, J.H.D. Spectroscopy and metastability of CO_2^{2+} molecular ions. *J. Chem. Phys.* **2005**, *122*, 084317. [CrossRef]
39. Newton, A.S.; Sciamanna, A.F. Metastable Dissociation of the Doubly Charged Carbon Monoxide Ion. *J. Chem. Phys.* **1970**, *53*, 132–136. [CrossRef]
40. Tsai, B.P.; Eland, J.H.D. Mass spectra and doubly charged ions in photoionization at 30.4 nm and 58.4 nm. *Int. J. Mass Spectrom. Ion Phys.* **1980**, *36*, 143–165. [CrossRef]
41. Alagia, M.; Candori, P.; Falcinelli, S.; Lavollée, M.; Pirani, F.; Richter, R.; Stranges, S.; Vecchiocattivi, F. Double Photoionization of CO_2 Molecules in the 34–50 eV Energy Range. *J. Phys. Chem. A* **2009**, *113*, 14755–14759. [CrossRef]
42. Eland, J.H.D. Dynamics of fragmentation reactions from peak shapes in multiparticle coincidence experiments. *Laser Chem.* **1991**, *11*, 259–263. [CrossRef]
43. Eland, J.H.D. The dynamics of three-body dissociations of dications studied by the triple coincidence technique PEPIPICO. *Mol. Phys.* **1987**, *61*, 725–745. [CrossRef]
44. Tian, C.; Vidal, C.R. Single to quadruple ionization of CO_2 due to electron impact. *Phys. Rev. A* **1998**, *58*, 3783. [CrossRef]
45. Dalitz, R.H. On the analysis of -meson data and the nature of the -meson. Philos. Mag. J. Sci. **1953**, *44*, 1068–1080. [CrossRef]
46. Laksman, J.; Månsson, E.P.; Grunewald, C.; Sankari, A.; Gisselbrecht, M.; Céolin, D.; Sorensen, S.L. Fragmentation of CO_2^{2+} in collisions with low-energy electrons. *Phys. Rev. A* **2012**, *136*, 104303.

47. Guillemin, R.; Decleva, P.; Stener, M.; Bomme, C.; Marin, T.; Journel, L.; Marchenko, T.; Kushawaha, R.K.; Jänkälä, K.; Trcera, N.; et al. Selecting core-hole localization or delocalization in CS_2 by photofragmentation dynamics. *Nat. Comm.* **2015**, *6*, 1–6. [CrossRef] [PubMed]
48. Wang, E.; Shan, X.; Shen, Z.; Gong, M.; Tang, Y.; Pan, Y.; Lau, K.C.; Chen, X. Pathways for nonsequential and sequential fragmentation of CO_2^{3+} investigated by electron collision. *Phys. Rev. A* **2015**, *91*, 052711. [CrossRef]
49. Herzberg, G. Molecular Spectra and Molecular Structure II. Infrared and Raman Spectra of polyatomic molecules. *Van Nostrand* **1945**, *200*, 300–400.

Article

Temporal Response of Atoms Trapped in an Optical Dipole Trap: A Primer on Quantum Computing Speed

S. Baral [1], Raghavan K. Easwaran [1], J. Jose [1,*], Aarthi Ganesan [2] and P. C. Deshmukh [3,4]

[1] Department of Physics, Indian Institute of Technology Patna, Bihta, Bihar 801103, India
[2] Department of Physics, JBAS College for Women, Teynampet, Chennai 600018, India
[3] Department of Physics, Dayananda Sagar University, Bengaluru 560114, India
[4] Department of Physics and CAMOST, Indian Institute of Technology Tirupati, Tirupati 517506, India
* Correspondence: jobin.jose@iitp.ac.in

Abstract: An atom confined in an optical dipole trap is a promising candidate for a qubit. Analyzing the temporal response of such trapped atoms enables us to estimate the speed at which quantum computers operate. The present work models an atom in an optical dipole trap formed using crossed laser beams and further examines the photoionization time delay from such confined atoms. We study noble gas atoms, such as Ne (Z = 10), Ar (Z = 18), Kr (Z = 36), and Xe (Z = 54). The atoms are considered to be confined in an optical dipole trap using X-ray Free Electron Lasers (*XFEL*). The present work shows that the photoionization time delay of the trapped atoms is *different* compared with that of the free atoms. This analysis alerts us that while talking about the speed of quantum computing, the temporal response of the atoms in the trapped environment must also be accounted for.

Keywords: optical dipole trap; RRPA; photoionization; time delay

1. Introduction

With the advent of quantum computing, the second quantum revolution has been ushered in [1]. Quantum memory [2], quantum information processing [3], etc., are rapidly becoming modernized to improvise the performance of computing in this era. The fundamental building block of such quantum tools is isolated atoms or molecules; the isolation is achieved through quantum confinement. Entrapment of atoms in fullerene molecule is one such successful confinement mechanism [4], and the Paul trap is another mechanism by which a cluster of molecules can be isolated [5]. Crossed laser beams create a dipole field, and atoms can also be isolated in such traps [6]. Atoms encapsulated in fullerenes, dipole traps, Paul traps, etc., are potential candidates for qubits, which can be used in quantum computers [7–9]. Experiments to realize quantum computers using isolated atoms are rapidly being developed [10].

Benioff pointed out in an early work that triggering a quantum computer's register involves a physical process; it is, of course, not just mathematical manipulation by matrices that represents the quantum gates [11,12]. A physical process here refers to the interaction of a qubit with any probe, such as a photon, electron, etc. A study of the temporal response of the quantum system under external perturbation, therefore, is in dire need of deciphering the speed of quantum information processing. This is highlighted in Benioff's *second* design of a quantum computer [11,12]. The present work attempts to study the temporal response of an atom trapped in a crossed laser beam, which can be considered a qubit.

During the last three decades, the developments in the field of laser cooling and trapping have been steadfast [13–19]. In 1962, Askar'yan envisaged that the optical dipole force can trap neutral atoms [20]. The probability of trapping atoms with the dipole force was considered by Letokhov [21], who recommended that atoms might be confined one-dimensionally at the nodes or antinodes of standing waves far detuned with the atomic transition frequency. Further, a neutral atom trapped by dipole force was demonstrated

Citation: Baral, S.; Easwaran, R.K.; Jose, J.; Ganesan, A.; Deshmukh, P.C. Temporal Response of Atoms Trapped in an Optical Dipole Trap: A Primer on Quantum Computing Speed. *Atoms* **2023**, *11*, 72. https://doi.org/10.3390/atoms11040072

Academic Editor: Himadri S. Chakraborty

Received: 17 February 2023
Revised: 5 April 2023
Accepted: 7 April 2023
Published: 10 April 2023

by Bjorkholm employing a focused laser beam [22]. In an outstanding breakthrough in 1986, Chu et al. utilized this force to realize the first optical trap for neutral atoms [23]. A very small optical dipole trap of microscopic size has been designed to store, analyse, and manipulate individual atoms [24–26]. For instance, the axial oscillation frequency of the atom and the atomic energy distribution in the dipole trap have been measured by isolating a single cesium atom in a standing wave optical dipole trap [27]. There have been investigations into the applicability of a single atom trapped in laser for quantum memories [28–30].

As mentioned above, a trapped atom in an optical dipole trap is identified to be a potential candidate for qubits in quantum computers [31]. To retrieve information from such a system, one needs to consider the interaction of a qubit or a trapped atom with external stimuli. One can intuitively see that the time scale of such an interaction defines the quantum information processing time. The atom–field interaction due to the optical dipole trap modifies the intrinsic nuclear field, leading to changes in the electron transition time. This effect is particularly interesting for quantum memory applications [32], as the storage time of quantum information depends on the electronic transition time [33]. In other words, the speed of quantum computers using a qubit would depend on its interaction time with a stimulus, say a photon. The present work pivots to investigating the interaction time of a trapped atom in a crossed laser beam keeping electromagnetic radiation as the probe. Most of the studies on coherent light–atom interaction consider a natural atomic system that has a set of intrinsic energy levels. Finding a suitable transition for a particular application in an experimental setting is very difficult. Engineering the atomic level energy and its transition and de-coherence rates [34] can be accomplished using an optical dipole trap.

In the present work, noble gas atoms are modeled to be trapped in the field of X-ray free electron lasers (XFEL), and the temporal response of such trapped atoms to an external stimulus (Photon) is investigated. Hereafter, the quantum system of interest in the present study is denoted as A@XFEL, where A is the trapped atom. The external electromagnetic field would photoionize the trapped atom in the XFEL field, and the photoionization time delay is studied in the present work. Due to the short wavelength range, XFEL [35] can be focused to a few nano-meters, or even below, employing various experimental techniques by which atoms can be trapped and isolated. In one of the earlier studies, an X-ray beam of photon energy 8.2 keV having a wavelength of approximately 0.151 nm has been focused to 50 nm [36]. In another study, an X-ray beam having photon energy 9.1 keV ($\lambda = 0.136$ nm) has been focused to 10 nm [37]. The present work employs the XFEL having wavelength 0.785 nm ($E = 1.58$ keV), which is focused to 1 nm, to trap atoms. This laser is far detuned with the atomic transition frequency of all atoms considered here so that the trapping field does not ionize the atoms. Further, the power of the laser field is also chosen low so that strong field ionization does not occur. We study photoionization parameters, such as cross section, angular distribution asymmetry parameter, and photoionization time delay, employing the relativistic random phase approximation (RRPA) [38]. Although alkali metal atoms, such as Rb, Na, etc., are commonly used in the dipole trap experiments, the open shell nature of the alkali atoms makes them unsuitable for the application of RRPA. Therefore, as a pilot study, the noble gas atoms, such as Ne ($Z = 10$), Ar ($Z = 18$), Kr ($Z = 36$), and Xe ($Z = 54$), are considered in the present work. Furthermore, a study of the bound-to-bound transition's temporal response is desirable to indicate the lifetime of the qubit. However, the response time of the bound-to-continuum transition (photoionization) investigated in this preliminary work is also an indicator of the bound-to-bound transitions.

Section 2 contains the theoretical details regarding modeling the dipole trap; Section 3 discusses the results; and Section 4 summarizes the results.

2. Theory

The mechanism of optical dipole trapping of neutral atoms using a laser field is well described using a semi-classical picture where the atom is treated as a simple dipole oscillator [39]. Atoms do not have a permanent electric dipole moment in the ground state.

However, a dipole moment can be induced in the atom when it is subjected to an external electric field. In the classical picture of an atom in a laser field, the oscillating electric field of the laser having frequency ω induces an oscillating dipole moment, $\vec{d}$, at the driving frequency, ω, itself. The oscillating electric field of the laser can be written as

$$\vec{E}(\vec{r},t) = \hat{e}E(\vec{r})e^{-i\omega t} + c.c.,\tag{1}$$

where $\hat{e}$ gives the direction of polarization.

The induced oscillating dipole moment of the atom can be written as

$$\vec{d}(\vec{r},t) = \hat{e}d(\vec{r})e^{-i\omega t} + c.c.\tag{2}$$

The relation between the amplitude of the induced oscillating dipole moment of the atom and the driving electric field is given by

$$\vec{d} = \alpha(\omega)\vec{E},\tag{3}$$

where $\alpha(\omega)$ is the (driving) frequency-dependent complex polarizability of the neutral atom.

The interaction between the induced dipole moment, $\vec{d}$, of the atom and the oscillating electric field, $\vec{E}$, gives rise to an interaction potential given by the relation [39]

$$U_{dip} = -\frac{1}{2}\left\langle \vec{d}.\vec{E} \right\rangle\tag{4}$$

The intensity profile of the focused XFEL laser beam in one direction (say in the z-direction) is expressed in cylindrical polar coordinates as [39]

$$I(\rho,z) = \frac{2P}{\pi w^2(z)}e^{-2\frac{\rho^2}{w^2(z)}},\tag{5}$$

where P is the power of the laser beam, ρ denotes the radial coordinate, and $w(z) = w_0\sqrt{1+\left(\frac{z}{z_0}\right)^2}$ is the beam waist radius. The z_0 is popularly known as Rayleigh length: $z_0 = \pi w_0^2/\lambda$, where w_0 is the waist radius of the trapping beam at the focal point.

For dipole trapping, crossed laser beams from all six directions are used, and they are focused in a narrow trapping region [40]. To a good approximation, the intensity profile of the crossed laser beam is considered spherically symmetric within the trap and, therefore, has a spherical Gaussian profile indicated as

$$I(r) = \frac{2P'}{\pi w^2(r)}e^{-2\frac{r^2}{w^2(r)}}\tag{6}$$

In Equation (6), $w(r)$ controls the intensity profile, which is given as

$$w(r) = w_0\sqrt{1+\left(\frac{r}{r_0}\right)^2},\tag{7}$$

where $r_0 = \pi w_0^2/\lambda$. In Equation (6), P' indicates the cumulative power due to all the focused laser beams.

Note that, because of the symmetry considerations, the intensity profile of the crossed laser beam (Equation (6)) is presented in spherical polar coordinates. The graph of I(r) varying with radial distance r is shown in Figure 1, where a crossed laser beam of wavelength $\lambda = 0.785$ nm and power $P' = 3$ Watt is focused to 1 nm radius to form a dipole trap. At $r = 0$, the intensity is a maximum, and with increasing r, the intensity is reduced.

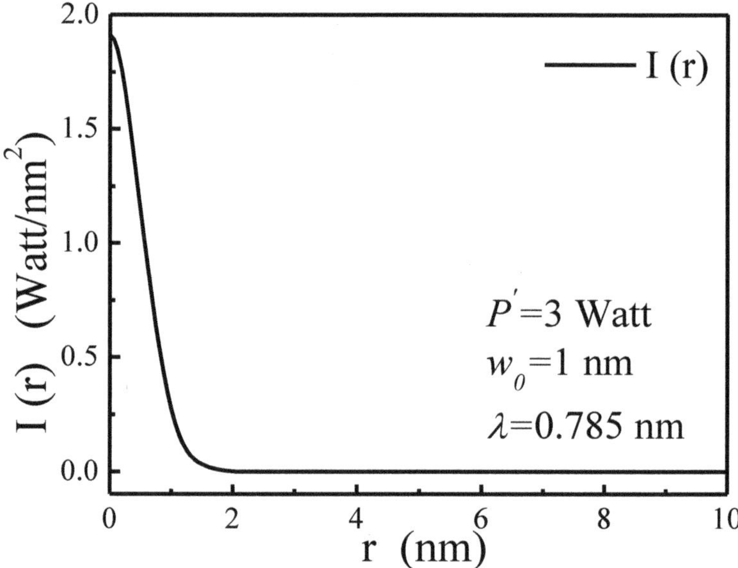

Figure 1. Intensity I(r) of the crossed beam varying with radial distance within the dipole trap.

Concerning the trapping of an alkali atom using a far-detuned laser beam having wavelength λ = 0.785 nm, the following points have to be taken into account. As mentioned in the ref. [39], if we prepare the atoms in the excited state, for instance, F = 2 in Rb, a blue detuned laser beam will create a trappable potential. However, the intensity has to be tuned in such a way that the trap depth is appreciable to hold the atom. Hence, selectively one has to choose the atoms and laser intensity appropriately and prepare them in a specific excited state to achieve optical dipole trapping.

Note that Equation (4) depicts the dipole trap the atom experiences. However, the potential felt by an atomic electron is calculated as

$$U(r) = \int_{\infty}^{r} -qEr^2 dr, \tag{8}$$

where q depicts the charge of the electron. Since the local intensity of the optical field is $I = 2\varepsilon_0 c|E|^2$ [39], the average electric field used in Equation (8) is expressed as

$$E = \sqrt{\frac{P}{\pi\varepsilon_0 c}}\,\frac{1}{w(r)}\exp\left(-r^2/w^2(r)\right) \tag{9}$$

As a model case, the potential experienced by the $1s$ electron of the hydrogen atom trapped in the crossed XFEL field (H@XFEL) having wavelength λ = 0.785 nm ($E = 1.58\,keV$) is computed when it is focused to $w_0 = 1$ nm. Figure 2 shows the effective potential (in a.u.): $V_{eff} = -\frac{1}{r} + U(r)$, where $U(r)$ is the potential felt by the atomic electron due to the laser trapping given in Equation (8). Here, the power of the laser beam is taken to be 3 Watts. The plot also compares the pure Coulombic potential of a free H atom: $U_{coulomb} = \frac{-1}{r}$. One can see that the laser field does alter the depth of the potential. The modification due to the crossed laser beam tends to change the binding energy of the trapped atom.

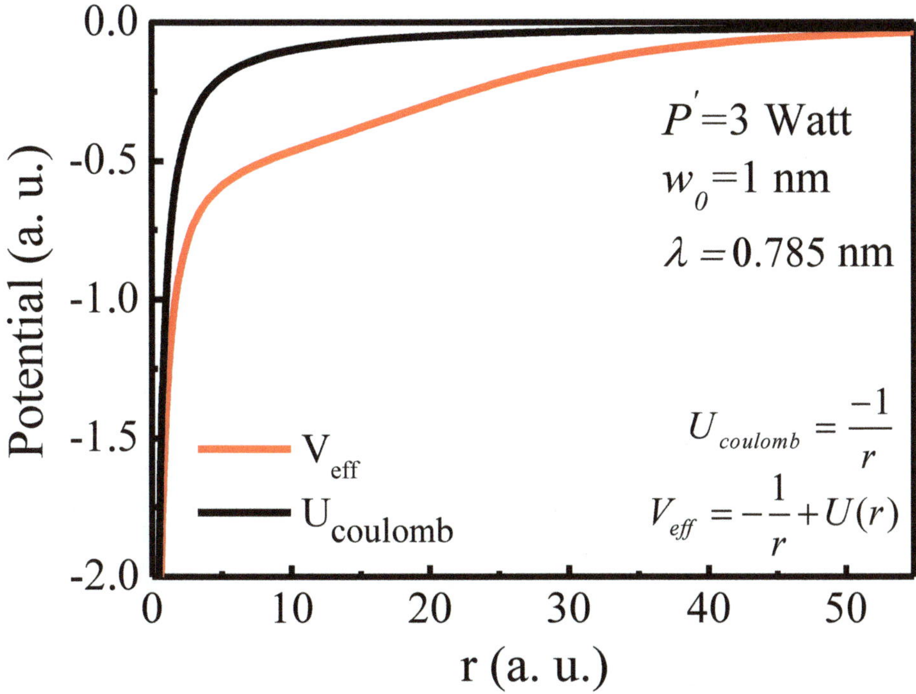

$$U_{coulomb} = \frac{-1}{r}$$

$$V_{eff} = -\frac{1}{r} + U(r)$$

Figure 2. The effective potential of H (black) and H@XFEL (red) atom.

For a multi-electron system, the potential given in Equation (6) is added to the original Dirac–Hartree–Fock (DHF) equation [41] and then solved by the equations using the self-consistent method. The modified Dirac–Hartree–Fock (DHF) orbital's wavefunction $u_i(\vec{r})$ of an N-electron atomic system in the dipole trap satisfies [41]

$$\left(c\,\vec{\alpha}.\vec{p} + \beta mc^2 - \frac{Z}{r} + V + U(r)\right)u_i(\vec{r}) = \varepsilon_i u_i(\vec{r}), i = 1, 2, \ldots, N, \tag{10}$$

where ε_i is the DHF energy eigenvalue of the *i-th* orbital, and V represents the inter-electron interaction term composed of direct and exchange terms defined as

$$Vu_i(\vec{r}) = \sum_{j=1}^{N} \int \frac{d^3r'}{\left|\vec{r} - \vec{r'}\right|}\left[\left(u_j^\dagger u_j\right)' u_i - \left(u_j^\dagger u_i\right)' u_j\right]. \tag{11}$$

Thus, the confined atom in the laser field is simulated, and the structural properties can be evaluated. For the present work, the noble gas atoms, such as Ne (Z = 10), Ar (Z = 18), Kr (Z = 36), and Xe (Z = 54), are considered. The corresponding ionization potentials of valence orbitals of Ne, Ar, Kr, and Xe for both free and confined cases are given in Table 1. One can notice that the laser confinement increases the threshold by roughly a constant amount ~0.39 a. u. This shift in the energy of the ionization threshold is due to the alterations in the depth of the potential.

Table 1. Binding energy of valence subshells of neutral atom and A@XFEL.

Atom	Subshell	Binding Energy (a. u.)	
		Neutral Atom	A@XFEL
Ne	2s	1.935	2.328
	$2p_{1/2}$	0.852	1.245
	$2p_{3/2}$	0.848	1.240
Ar	3s	1.286	1.678
	$3p_{1/2}$	0.595	0.987
	$3p_{3/2}$	0.587	0.980
Kr	4s	1.187	1.579
	$4p_{1/2}$	0.541	0.933
	$4p_{3/2}$	0.514	0.906
Xe	5s	1.010	1.401
	$5p_{1/2}$	0.492	0.884
	$5p_{3/2}$	0.439	0.831

The DHF wavefunction is considered as the initial state of the target atom, which is photoionized. The dipole-trapped atom is subjected to an external time-dependent external field: $v_+ e^{-i\omega t} + v_- e^{i\omega t}$. The modified RRPA equations with the inclusion of laser potential, $U(r)$, can be obtained from the time-dependent DHF method given as [38,42,43]

$$
\left(c\vec{\alpha}.\vec{p} + \beta mc^2 - \frac{Z}{r} + V + U(r) - \varepsilon_i \mp \omega \right) w_{i\pm}\left(\vec{r} \right) = \\
\left(v_\pm - V_\pm^{(1)} \right) u_i\left(\vec{r} \right) + \sum_j \lambda_{ij\pm} u_j\left(\vec{r} \right), \ i = 1, 2, \ldots, N,
\tag{12}
$$

where the Lagrangian multipliers $\lambda_{ij\pm}$ are incorporated to guarantee that the perturbed orbitals $w_{ij\pm}$ are orthogonal to the occupied orbitals u_i. The RRPA includes many-electron correlation effects in both the initial and the final states through the terms $V_\pm^{(1)}$ in the above equation; all possible two-electron two-hole excitations in the initial state and the interchannel coupling of the final-state channels are accounted for. In the present work, relevant interchannel coupling effects are included in the RRPA for the photoionization of the laser-cooled noble gas atoms. The number of dipole channels coupled in the RRPA is 7 for Ne (channels from the 2p and 2s subshell), 14 for Ar (3p, 3s, 2p, and 2s subshell), 20 for Kr (4p, 4s, 3d, 3p, and 3s subshell), and 20 for Xe (5p, 5s, 4d, 4p, and 4s subshell).

In photoionization, for a particular transition from an initial state $|n, \kappa\rangle$ to a final state $|\varepsilon, \overline{\kappa}\rangle$, the radial dipole matrix element is given by [44]

$$
\langle \varepsilon, \overline{\kappa} | \hat{d} | n, \kappa \rangle = i^{1-\bar{l}} e^{i\delta_{\overline{\kappa}}} \langle \varepsilon, \overline{\kappa} | Q_1^{(1)} | n, \kappa \rangle
\tag{13}
$$

Here, $\langle \varepsilon, \overline{\kappa} | Q_1^{(1)} | n, \kappa \rangle$ is the reduced dipole matrix element, and $\delta_{\overline{\kappa}}$ is the phase shift of the final continuum wavefunction. Since matrix element is generally complex in nature, the phase shift of the photoelectron is defined by

$$
\delta_{\overline{\kappa}}(\varepsilon) = \tan^{-1}\left\{ \frac{\mathrm{Im}\langle \varepsilon, \overline{\kappa} | \hat{d} | n, \kappa \rangle}{\mathrm{Re}\langle \varepsilon, \overline{\kappa} | \hat{d} | n, \kappa \rangle} \right\}.
\tag{14}
$$

For a dipole transition, indicated by $\kappa \to \overline{\kappa}$, the total subshell cross section $\sigma_{n\kappa}$ is given as [40,44]

$$
\sigma_{n\kappa} = \frac{4\pi^2\alpha}{3}\, \omega \left(|D_{\kappa \to \kappa-1}|^2 + |D_{\kappa \to \kappa}|^2 + |D_{\kappa \to \kappa+1}|^2 \right),
\tag{15}
$$

where $D_{\kappa \to \overline{\kappa}}$ is the dipole transition matrix element present in Equation (13).

The dipole angular distribution asymmetry parameter $\beta_{n\kappa}(\omega)$ is given by [38,42]

$$\beta_{n\kappa}(\omega) = \left\{ \frac{1}{2}\frac{(2\kappa-3)}{2\kappa}|D_{\kappa\to\kappa-1}|^2 - \frac{2}{2\kappa}\sqrt{\left(\frac{2\kappa-1}{2(2\kappa+2)}\right)}\left[D_{\kappa\to\kappa-1}D^*_{\kappa\to\kappa+1} + c.c.\right] - \right.$$
$$\frac{(2\kappa-1)(2\kappa+3)}{2\kappa(2\kappa+2)}|D_{\kappa\to\kappa}|^2 - \frac{3}{2}\sqrt{\left(\frac{(2\kappa-1)(2\kappa+3)}{2\kappa(2\kappa+2)}\right)}\left[D_{\kappa\to\kappa-1}D^*_{\kappa\to\kappa+1} + c.c.\right] +$$
$$\left. \frac{1}{2}\frac{(2\kappa+5)}{(2\kappa+2)}|D_{\kappa\to\kappa+1}|^2 + \frac{3}{2\kappa+2}\sqrt{\left(\frac{2\kappa+3}{2(2\kappa)}\right)}\left[D_{\kappa\to\kappa}D^*_{\kappa\to\kappa+1} + c.c.\right] \right\} * \left\{|D_{\kappa\to\kappa-1}|^2 + \right.$$
$$\left. |D_{\kappa\to\kappa}|^2 + |D_{\kappa\to\kappa+1}|^2 \right\}^{-1} \tag{16}$$

The photoionization time delay of a particular transition is obtained as the energy derivative of the phase of the photoionization complex transition matrix element [45,46]. This quantity represents the temporal response of the atomic electron while photoionizing. The average time delay in photoionization of a particular subshell is presented in the current work. It is defined as the sum of the individual channel time delays weighted by the ratio of the respective individual channel cross sections to the total of the cross sections. Hence, the present study computes and analyses the photoionization cross section, angular distribution asymmetry parameters, and the photoionization time delay for both the laser-trapped atom as well as for the free atom. This work focuses on the valance ns and np subshells of the noble gas atoms considered.

3. Results and Discussion

In this section, the results for the photoionization cross section, angular distribution asymmetry parameter, and time delay of the valence shells, ns and np, of noble gas atoms (Ne, Ar, Kr, and Xe) trapped by XFEL dipole trap are presented. A comparison of the results for the neutral and that of the A@XFEL is facilitated. As the speed of qubit used in quantum techniques application depends on its interaction time with photon, the photoionization time delay provides a benchmark estimate of the temporal response.

3.1. Neon

Figure 3 shows the photoionization cross section of the $2p$ and the $2s$ subshells of the free Ne (solid black) and Ne@XFEL (solid red). For the RRPA, seven dipole channels from the $2s$ and $2p$ subshells are coupled. The photoionization cross section of $2s$ and $2p$ subshells of Ne exhibit a shape resonance. Since the Ne atom is less relativistic, results for the spin–orbit split $2p_{1/2}$ and the $2p_{3/2}$ subshells are similar, except for their magnitudes; the ratio of both cross sections indicates the ratio of the number of electrons in the subshells, known as the branching ratio [44]. For Ne@XFEL, the photoionization thresholds are offset by 0.39 a.u., and, therefore, the onset of photoionization occurs at higher energy. Nevertheless, the Ne@XFEL cross section also exhibits the delayed maximum.

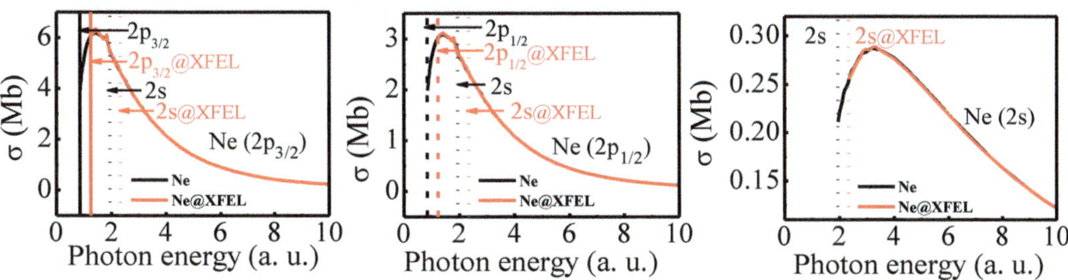

Figure 3. Photoionization cross section (σ) of $2p_{3/2}$ (**Left**), $2p_{1/2}$ (**Middle**), and $2s$ (**Right**) subshells of Ne (black) and Ne@XFEL (red). Solid, dashed, and dotted vertical lines represent the threshold for $2p_{3/2}$, $2p_{1/2}$, and $2s$ thresholds of free (black) and confined atom (red), respectively.

Figure 4 shows the comparison of angular distribution asymmetry parameter β of the $2p_{3/2}$, $2p_{1/2}$, and $2s$ subshells of the free (solid black) and confined Ne (solid red). The figure indicates that apart from the delayed onset of the β parameter, there is no change induced by the optical dipole trap. This is understandable from the analysis of the cross section. Nevertheless, one may also note that the angular distribution asymmetry parameter has an additional dependence on the relative phase shift of different pair channels $(\delta_k - \delta_{\bar{k}})$, as is evident from Equation (16). Figure 4 indicates that the relative phase shift difference is also unaffected by the laser trapping. One may ask at this juncture whether the relative phase shift of two dipole channels is unaffected by the optical trapping, i.e., will the individual time delay of the dipole channels be altered? A naive answer is 'possibly not', as the relative phases are not affected. However, a detailed scrutiny of the individual channel's time delay requires an affirmative answer. Figure 5 shows the average photoionization time delay from the $2p_{3/2}$, $2p_{1/2}$, and $2s$ subshells of Ne. The time delay for the free and confined Ne are quantitatively as well as qualitatively different; the former is larger compared with the latter. For instance, while the time delay in $2s$ photoionization is negative and attains a minimum at ~2.5 a.u., the same in Ne@XFEL shows a higher positive value and it does not showcase any symptom of a minimum. A similar quantitative difference is seen in the case of time delay in the $2p_{3/2}$ and $2p_{1/2}$ cases. The additional time delay due to the laser coupling varies from tens of attoseconds to hundreds. Note here that although the relative phase difference is unaltered due to the laser trapping, the phases of the complex transition matrix elements are affected, and it leads to a significantly altered time delay of individual subshell photoionization time delay.

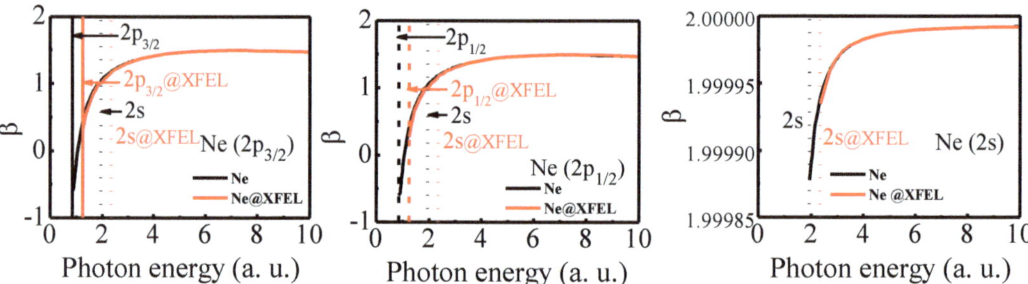

Figure 4. Angular distribution asymmetry parameter (β) of $2p_{3/2}$ (**Left**), $2p_{1/2}$ (**Middle**), and $2s$ (**Right**) subshells of Ne (black) and Ne@XFEL (red). Solid, dashed, and dotted vertical lines represent the threshold for $2p_{3/2}$, $2p_{1/2}$, and $2s$ thresholds of free (black) and confined atom (red), respectively.

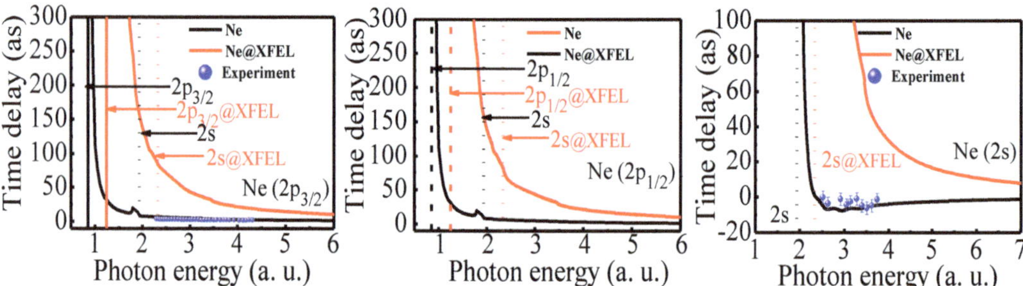

Figure 5. Time delay (τ) of $2p_{3/2}$ (**Left**), $2p_{1/2}$ (**Middle**), and $2s$ (**Right**) subshells of Ne (black) and Ne@XFEL (red). Solid, dashed, and dotted vertical lines represent the threshold for $2p_{3/2}$, $2p_{1/2}$, and $2s$ thresholds of free (black) and confined atom (red), respectively. Blue scattered points in left and right panel of the figure show the experimental results [47].

The present results are enriching in two ways. Firstly, the electrons in the atoms in the optical dipole trap suffer more time delay while responding to the external impulse. This observation will have consequences on the performance of such atoms when used in the quantum techniques application. Secondly, the phase shift difference appearing in the angular distribution parameter is unaffected due to the laser coupling, but the individual phase shift on the other hand is altered. This observation asserts that individual channel time delay is more sensitive to the external perturbation compared with the other dynamical variables. A similar remark is made in earlier work on photoionization from Xe [48].

From an experimental perspective, individual subshell time delay and the relative time delay between 2p and 2s photoionization of Ne is measured by several groups [47,49]. In the latest experimental attempt, the relative time delay ($\tau(2s)$-$\tau(2p)$) is directly measured [47]. In addition, the earlier work obtained Wigner time delay for 2p subshell to obtain more details about the relative delay difference measurements. Furthermore, the earlier work has obtained 2s Wigner time delay data by subtracting the $\tau(2p)$ from $\tau(2s)$-$\tau(2p)$. The left panel of Figure 5 shows the comparison of time delay for 2p from the experiment with the present calculation of average time delay for 2$p_{3/2}$. Since the Ne is less relativistic, the comparison of the average time delay of the 2p subshell with that of 2$p_{3/2}$ is justified. The comparison shows good agreement between theory and experiment. Similarly, the 2s time delay from the earlier experiment and the present work is also compared, which is shown in the right panel of Figure 5. The comparison of the 2s time delay also renders an encouraging comparison, especially in the region of minimum in the time delay. Comparison of theory and experiment encourages us to anticipate that the time delay would be enhanced when probing an atom isolated in a dipole trap.

Figure 6 compares the relative time delay difference ($\tau(2s)$-$\tau(2p)$) of the Ne and Ne@XFEL obtained in the RRPA. The theoretical result is also compared with the available experimental result [47]. Although the RRPA overestimates the relative time delay compared with the experiment at the minimum, an overall qualitative agreement is found. Of course, the Ne@XFEL has enhanced delay difference, as is evident from the figure. As photon energy increases, the time delay difference between free and confined Ne vanishes. This is true even in the case of individual channel time delays. This is understandable, as the highly energetic photoelectron does not see the details of the confinement potential, and, therefore, the Ne and @Ne time delay is more or less the same.

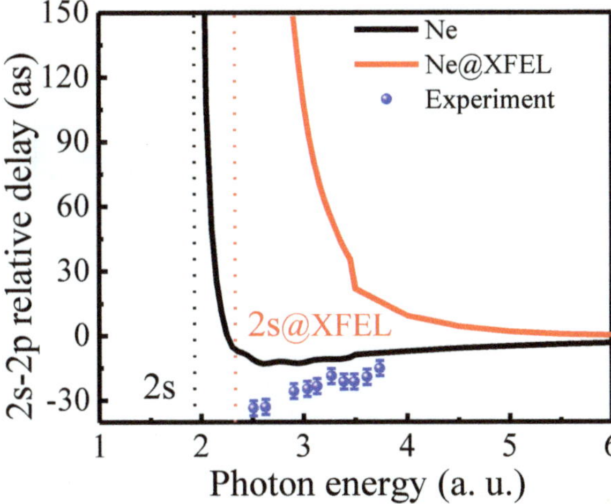

Figure 6. The time delay difference of 2s and 2p subshell ($\tau(2s)$-$\tau(2p)$) of Ne (black) and Ne@XFEL (red) in the RRPA calculation is compared with that from the experiment (blue) [47]. Vertical lines show the 2s subshell threshold for Ne (black) and Ne@XFEl (red).

3.2. Argon

Figures 7 and 8 show the photoioinization cross section and angular distribution asymmetry parameters of spin–orbit split valence subshells of Ar and Ar@XFEL. Since Ar is less relativistic, the $3p_{3/2}$ and the $3p_{1/2}$ subshells exhibit similar photoionization features. The $3p$ cross section drops from a high value, as there is a shape resonance. At ~1.85 a.u., the $3p \rightarrow \varepsilon d$ dipole channels undergo a Cooper minimum. Likewise, the $3s$ subshell cross section also exhibits a Cooper minimum at 1.55 a.u. Note that the Cooper minimum in the $3s$ photoionization channel in the 1.5 a.u. region arises solely due to interchannel coupling with the 3p photoionization channels [50]. The angular distribution asymmetry parameter has additional dependence on the relative phase shift of different photoionizing channels. The β_{3p} rises to a maximum value at 1.35 a.u. and displays a minimum at 1.85 a.u. Note that the minimum in the β_{3p} occurs at the location of the Cooper minimum. In the $3s$ case, because of the Cooper minimum, there is a dip in the β; the deviation of β from 2.00 in the ns case shows the impact of the relativistic effect on the CM in the spin–orbit split subshell channels. The reliability of the RRPA results are well established through a comparison with experimental results.

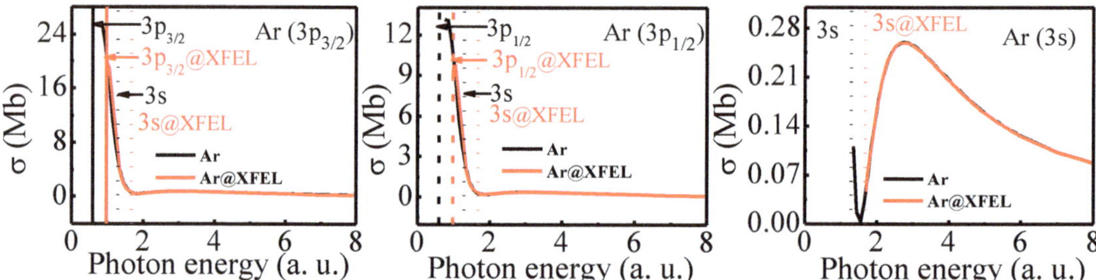

Figure 7. Photoionization cross section (σ) of 3p$_{3/2}$ (**Left**), 3p$_{1/2}$ (**Middle**), and 3s (**Right**) subshells of Ar (black) and Ar@XFEL (red). Solid, dashed, and dotted vertical lines represent the threshold for 3p$_{3/2}$, 3p$_{1/2}$, and 3s thresholds of free (black) and confined atom (red), respectively.

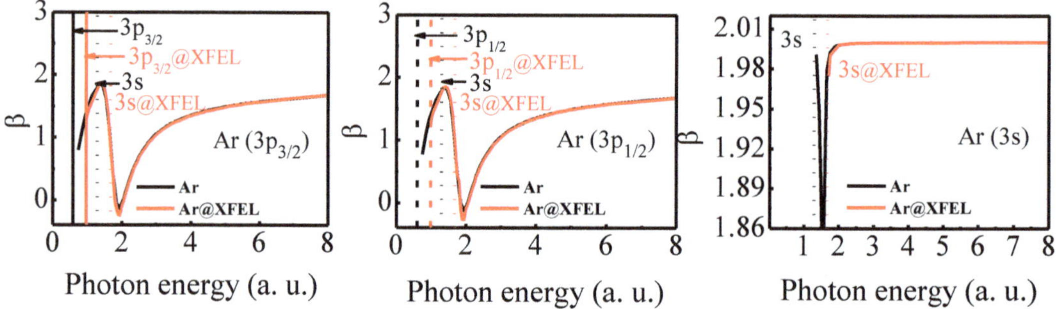

Figure 8. Angular distribution asymmetry parameter (β) of p$_{3/2}$ (**Left**), 3p$_{1/2}$ (**Middle**), and 3s (**Right**) subshells of Ar (black) and Ar@XFEL (red). Solid, dashed, and dotted vertical lines represent the threshold for 3p$_{3/2}$, 3p$_{1/2}$, and 3s thresholds of free (black) and confined atom (red), respectively.

In the case of Ar@XFEL, as already discussed, the subshell thresholds are offset by 0.39 a.u. One can see from Figures 7 and 8 that the photoionization cross section and angular distribution asymmetry parameters follow the same profile of the free Ar atom case, except for the shift in the threshold. The $3p$ cross section has the Cooper minimum at the same location as in the free Ar case. However, in the case of the $3s$ subshell, the Cooper minimum is in the discrete region below the threshold. Therefore, the Cooper minimum is not present in the 3*s* cross section. Accordingly, the dip in the β is also not present in the 3*s* angular distribution asymmetry parameter.

The average time delay (τ) of $3p_{3/2}$, $3p_{1/2}$, and 3*s* subshell photoionization of Ar and Ar@XFEL is shown in Figure 9. Cooper minima in the $p \rightarrow \varepsilon d$ transition matrix element are exhibited as a $-\pi$ jump in the phase shift, which results in a sharp and deeper negative time delay at approximately 1.805 a.u. in the individual spin–orbit split $p \rightarrow \varepsilon d$ channels of free Ar. The $p \rightarrow \varepsilon s$ channels do not have the Cooper minimum; therefore, the time delay of these channels decreases from a positive value smoothly for neutral Ar. The trend of the average time delay of 3*p* subshells of the free Ar follows that of the $3p \rightarrow \varepsilon d$ channels as its matrix element is dominant except in the CM region; at the CM, the $3p \rightarrow \varepsilon s$ channel time delay dominates. As a result, the average time delay of $3p_{1/2}$ and $3p_{3/2}$ subshells exhibit a competition between that of the $p \rightarrow \varepsilon d$ and $p \rightarrow \varepsilon s$ channels. Hence, the average $3p_{3/2}$ and $3p_{1/2}$ time delay shown, respectively, in Figure 9 left and middle panel is wider and less deep compared with the individual time delays in the region of Cooper minima due to the contribution from the $3p \rightarrow \varepsilon s$ channels. The average time delay of $3p_{1/2}$ and $3p_{3/2}$ subshell exhibit, respectively, a minimum at 1.805 a.u. and 1.82 a.u. of photon energies, which corresponds to the Cooper minimum. The Cooper minimum in the $3s \rightarrow \varepsilon p$ channels induces a $+\pi$ jump in the phase shift, which is translated as a positive peak in the individual channel time delay of the free Ar atom. Upon averaging, the peak widens and results in a maximum at 1.51 a.u. The results for free Ar atoms have been discussed in great detail using RRPA and other theories earlier [51,52].

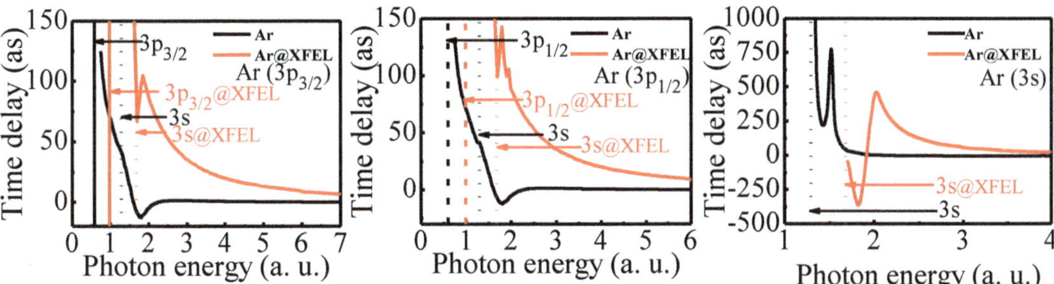

Figure 9. Time delay (τ) of $3p_{3/2}$ (**Left**), $3p_{1/2}$ (**Middle**), and 3*s* (**Right**) subshells of Ar (black) and Ar@XFEL (red). Solid, dashed, and dotted vertical lines represent the threshold for $3p_{3/2}$, $3p_{1/2}$, and 3*s* thresholds of free (black) and confined atom (red), respectively.

In quite a contrast to the free Ar case, the photoionization time delay in the Ar@XFEL is widely different. Firstly, the Ar atom in the optical dipole trap experiences a larger time delay compared with the free case. Although the Cooper minimum is present in the 3*p* cross section of Ar@XFEL, the features of the Cooper minimum are overshadowed by the confinement effects. As discussed in the free Ar case, the average time delay of $3p_{1/2}$ and $3p_{3/2}$ also follows the competition between individual $p \rightarrow \varepsilon d$ and $p \rightarrow \varepsilon s$ channels in the Ar@XFEL case. However, the $p \rightarrow \varepsilon s$ channel time delay is considerably larger in the Ar@XFEL case at the CM compared with the free Ar case. Therefore, the CM features are visible in the 3*p* time delay as a kink near CM. For instance, the time delay of $3p_{3/2}$ (left panel of Figure 9) and $3p_{1/2}$ (middle panel of Figure 9) subshell of Ar@XFEL shows a bump at 1.7 a.u.. The variation of the time delay in $3p_{1/2}$ and $3p_{3/2}$ subshell photoionization illustrates the importance of relativistic effects.

Concerning the 3*s* case, as the CM in the $3s \rightarrow \varepsilon p$ channels is moved to the discrete and, therefore, is absent in the continuum energy range in the Ar@XFEL case, the peak in the time delay is missing. Rather, it decreases to a negative minimum and then becomes positive. Further, the 3*s* time delay decreases from a high positive time delay. Note that the time delay in photoionization from the Ar@XFEL is, in general, large compared with that of free Argon.

It is to be asserted that the cross section and angular distribution are unaltered due to the dipole trapping, except for the shifting of the threshold. However, the individual phase shift and the time delay are modified due to the laser confinement. At this point, it is important to check and verify whether the time delay differences are altered due to trapping the atom. Because the time delay differences can be measured [52], Figure 10 presents the time delay difference between 3s and 3p photoelectrons in free and confined Ar. Experimental results of relative time delay ($\tau(3s)$-$\tau(3p)$) are included for comparison [53]. From the experimental work, the time delay for the single-photon ionization channel is plotted. There is a good qualitative agreement between RRPA and experimental data; the minimum in the relative time delay is in good agreement, although the magnitude is different. While the 3p electron time delay dominates over 3s electrons at higher energy value, near the 3s threshold, the 3s electrons escape more slowly compared with the 3p electrons in the case of free Ar case. In the Ar@XFEL case, the time delay difference is modified due to the laser trapping. The peak in the time delay difference is missing due to the absence of the Cooper minimum in the 3s subshell channels.

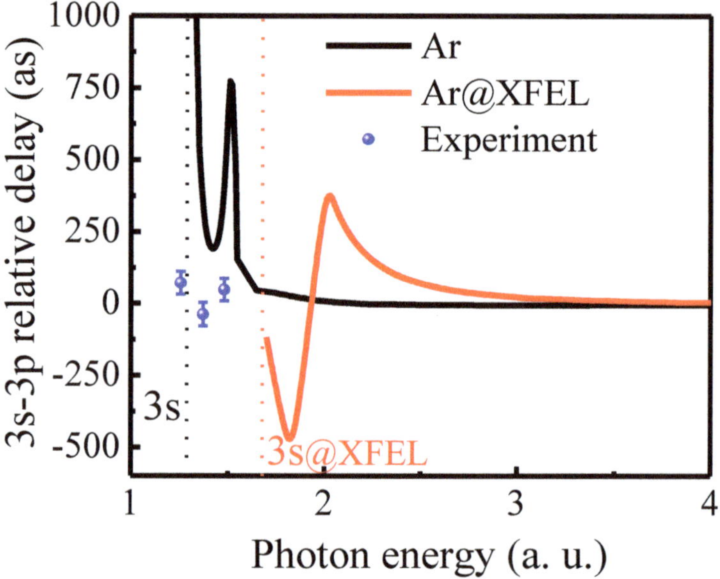

Figure 10. The time delay difference of 3s and 3p subshell (τ(3s)-τ(3p)) of Ar (black) and Ar@XFEL (red) in the RRPA calculation is compared with that from the experiment (blue) [53]. Vertical lines show the 3s subshell threshold for Ar (black) and Ar@XFEl (red).

3.3. Krypton

We display the partial photoionization cross section of the $4p_{3/2}$, $4p_{1/2}$, and 4s subshells of Kr calculated in RRPA in Figure 11. A comparison between free Kr and Kr@XFEL is also rendered in Figure 11. A similar comparison for the angular distribution asymmetry parameter β of the $4p_{3/2}$, $4p_{1/2}$, and 4s subshells is given in Figure 12. The 4p subshell cross section of the free Kr displays a shape resonance at the threshold, and the σ drops from the threshold. The 4s subshell cross section exhibits a Cooper minimum at 1.65 a.u., which is due to the correlation effects [54]. The inclusion of correlation effects using RRPA has demonstrated excellent agreement with experimental results [52]. Considering the Kr@XFEL case, since the thresholds for the 4p and 4s are shifted in the confined case, the cross section has a different onset in the RRPA case, as shown in Figure 11. Because of this, the Cooper minimum in the 4*s* cross section is absent in the Kr@XFEL case, rather the cross section appears to recover from the CM at the threshold. This will have implications in the angular distribution asymmetry parameter as well. From Figure 12, one can see that the $4p_{3/2}$ and $4p_{1/2}$ angular distribution asymmetry parameter of Kr@XFEL agrees with that of free Kr, except for the shift in the threshold. In the 4*s* case, the free Kr exhibits a dip in the β at the CM location. However, in the Kr@XFEL case, the β_{4s} increases from the threshold and reaches the non-relativistic value of 2. Since the CM in the 4*s* subshell is below the threshold in the confined case, the dip in the angular distribution parameter is missing.

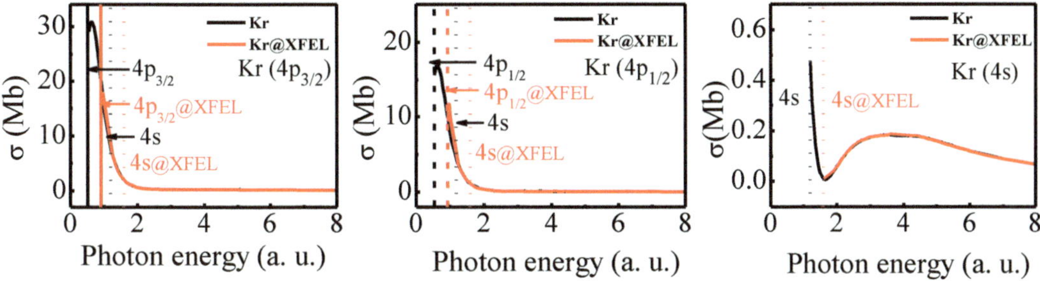

Figure 11. Photoionization cross section (σ) of $4p_{3/2}$ (**Left**), $4p_{1/2}$ (**Middle**), and 4s (**Right**) subshells of Kr (black) and Kr@XFEL (red). Solid, dashed, and dotted vertical lines represent the threshold for $4p_{3/2}$, $4p_{1/2}$, and 4s thresholds of free (black) and confined atom (red), respectively.

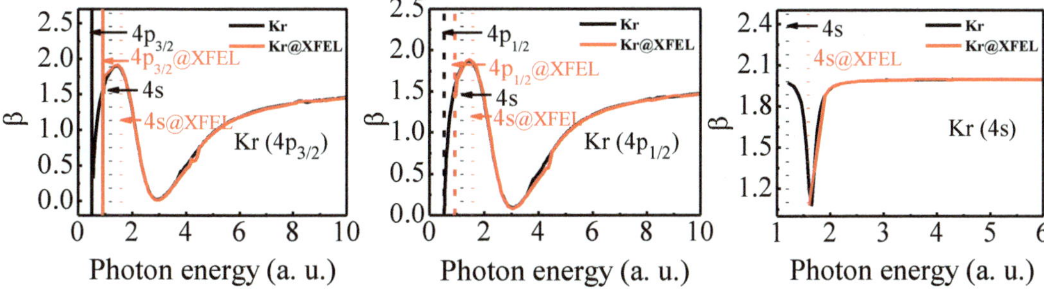

Figure 12. Angular distribution asymmetry parameter (β) of $4p_{3/2}$ (**Left**), $4p_{1/2}$ (**Middle**), and 4s (**Right**) subshells of Kr (black) and Kr@XFEL (red). Solid, dashed, and dotted vertical lines represent the threshold for $4p_{3/2}$, $4p_{1/2}$, and 4s thresholds of free (black) and confined atom (red), respectively.

The average time delay of the $4p_{3/2}$, $4p_{1/2}$, and the 4*s* subshells of the Kr and Kr@XFEL are shown in Figure 13. The individual channel time delays are qualitatively and quantitatively different in the two cases; the Kr@XFEL shows a larger time delay consistently in all cases. At the CM, the 4*p* subshells exhibit a minimum time delay in the free Kr case. The CM features are shadowed in the Kr@XFEL case; the atom in the dipole trapping shows enhanced time delay in the entire region. The alterations in the time delay due to the laser trapping follow a trend similar to that of the Ar case.

The induced Cooper minimum in the 4*s* subshell cross section manifests as a peak in the corresponding subshell's time delay. As in the 4*p* cases, the confinement vanishes the features of the Cooper minimum in the time delay, as the CM is absent in the Kr@XFEL case.

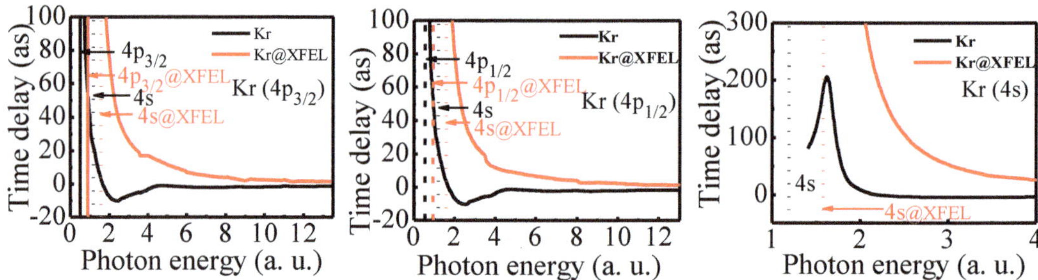

Figure 13. Time delay (τ) of $4p_{3/2}$ (**Left**), $4p_{1/2}$ (**Middle**), and 4s (**Right**) subshells of Kr (black) and Kr@XFEL (red). Solid, dashed, and dotted vertical lines represent the threshold for $4p_{3/2}$, $4p_{1/2}$, and 4s thresholds of free (black) and confined atom (red) respectively.

Figure 14 shows the difference between $4s$ and $4p$ subshells' time delay in the free and confined cases. Although experimental data on photoionization time delay in Kr are not available for comparison, we provide theoretical results for completion and as a reference for experimentalists. Figure 14 shows that the $4s$ electron takes a longer time than $4p$ electrons near the $4s$ threshold of the free Kr. Due to the effect of $4s$ CM, the relative time delay is a maximum. The confinement modifies the time delay difference; the relative time delay attains a maximum near the $4s$ threshold, and it decreases monotonically.

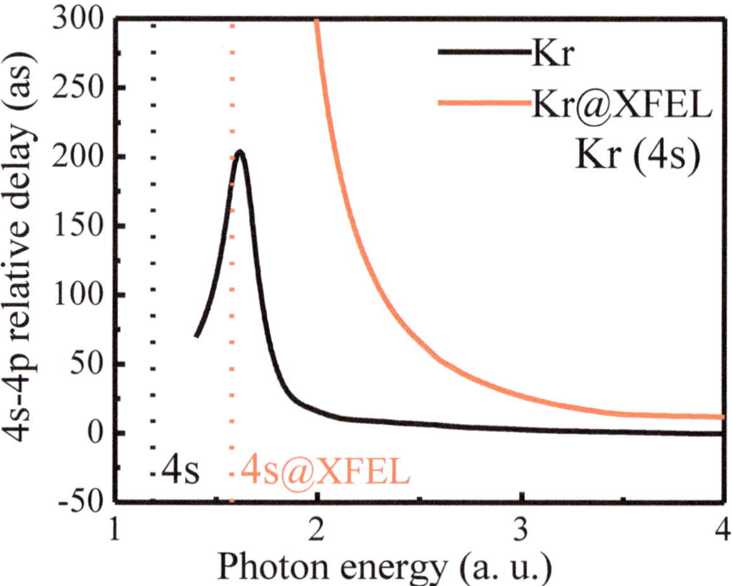

Figure 14. The time delay difference of 4s and 4p subshell ($\tau(4s)-\tau(4p)$) of Kr (black) and Kr@XFEL (red) in the RRPA calculation. Vertical lines show the 4s subshell threshold for Kr (black) and Kr@XFEl (red).

Similar to the other noble gas atoms, Kr also exhibits differences in the time delay due to the dipole trapping. At the same time, the dipole cross section and angular distribution asymmetry parameters are more or less the same except for the shift in the threshold. The current observations reassert the observation in the ref. [55] that time delay is more susceptible to external perturbations compared with the angular distribution of the photoelectrons.

3.4. Xenon

Figures 15 and 16 show, respectively, the photoionization cross section and angular distribution asymmetry parameter of the $5p_{3/2}$, $5p_{1/2}$, and the $5s$ subshells of Xenon. The $5p_{3/2}$ and the $5p_{1/2}$ cross section of the free Xe exhibit a Cooper minimum at approximately 2.3 a.u. Further, there is a second Cooper minimum, which is at relatively higher energies, at 5.7 a.u. In the $5s$ case of free Xe also, two Cooper minima are, respectively, observed at photon energy 1.3 a.u. and 5.65 a.u. It has been observed that the effects of coupling with the $4d$ photoionization channels are quite important in the region of both Cooper minima. It is also established that the second Cooper minima in Xe valance subshells are due to the interchannel coupling correlation effects. Considering the Xe@XFEL case, the $5p$ subshell exhibits delayed onset of cross section due to the shifting of the threshold. Likewise, in the $5s$ case, the first Cooper minimum is absent in the confined Xe case. Apart from the shift in the threshold, the cross section profiles of the Xe@XFEL are qualitatively and quantitatively the same as that of the free one.

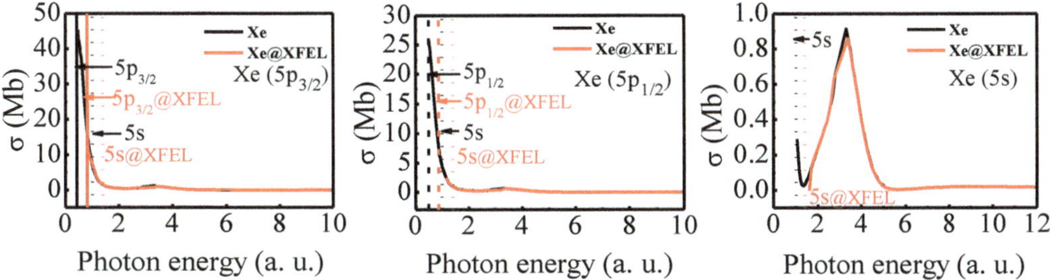

Figure 15. Photoionization cross section (σ) of $5p_{3/2}$ (**Left**), $5p_{1/2}$ (**Middle**), and $5s$ (**Right**) subshells of Xe (black) and Xe@XFEL (red). Solid, dashed, and dotted vertical lines represent the threshold for $5p_{3/2}$, $5p_{1/2}$, and $5s$ thresholds of free (black) and confined atom (red), respectively.

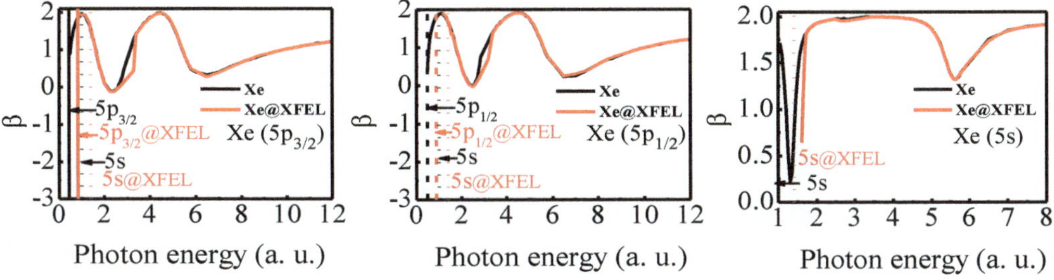

Figure 16. Angular distribution asymmetry parameter (β) of $5p_{3/2}$ (**Left**), $5p_{1/2}$ (**Middle**), and $5s$ (**Right**) subshells of Xe (black) and Xe@XFEL (red). Solid, dashed, and dotted vertical lines represent the threshold for $5p_{3/2}$, $5p_{1/2}$, and $5s$ thresholds of free (black) and confined atom (red), respectively.

The $5p$ and $5s$ angular distribution asymmetry parameters for free and confined Xe are shown in Figure 16. The β is dependent on the ratio of the magnitudes of the matrix elements of the relativistic dipole channels along with their relative phases. From Figure 16 left and middle panel, the comparison of $5p$ β of free Xe and Xe@XFEL suggests no dramatic modification due to laser trapping, except for the shift in the threshold. Corresponding to the Cooper minima in the cross section, a dip in the β is obtained. Since the CM is present in both cases in $5p$, the features of the angular distribution are alike. However, the $5s$ case is remarkably different. Since the first Cooper minimum is missing in the Xe@XFEL case, the β_{5s} appears to rise from the minimum at the threshold. The additional dip at 5.65 a.u. is due to the CM at that location.

A set of corresponding figures of the time delay (τ) is shown in Figure 17. Two dips in the $5p$ photoionization time delay of free Xe correspond to a $-\pi$ jump in the phase shift at the Cooper minimum. Note that the inclusion of the dipole trap reserves the qualitative nature of the $5p$ time delay, although a quantitative shift has occurred. In the $5s$ case also, there is a qualitative similarity between the time delay of the photoionization from the free and confined Xe. Since the CM occurs below the threshold for the $5s$ subshell of Xe @XFEL, the first dip in the time delay is present only as a kink. Apart from the shift in the threshold, there is a dramatic change in the individual channel phase shift and, therefore, in the time delay also.

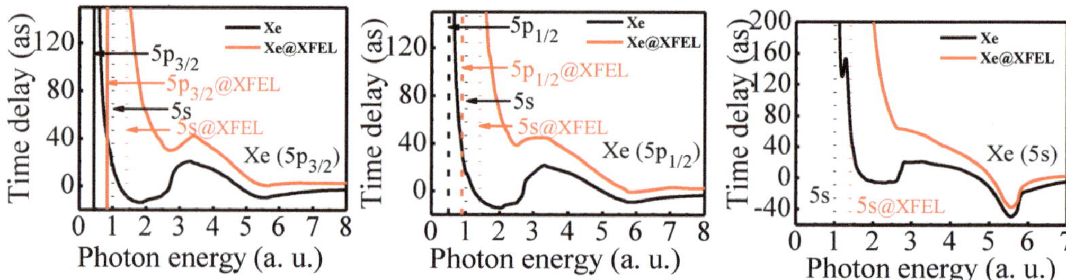

Figure 17. Time delay (τ) of $5p_{3/2}$ (**Left**), $5p_{1/2}$ (**Middle**), and $5s$ (**Right**) subshells of Xe (black) and Xe@XFEL (red). Solid, dashed, and dotted vertical lines represent the threshold for $5p_{3/2}$, $5p_{1/2}$, and $5s$ thresholds of free (black) and confined atom (red), respectively.

Figure 18 shows the time delay difference between the $5s$ and $5p$ subshell photoionization. The relative time delay difference is enhanced upon crossed laser beam confinement, which is evident from Figure 18. Due to the presence of CM, there is a dip in the time delay difference at the location of the second Cooper minimum of the 5s and $5p$ subshells. As the photon energy increases, the difference between both the free and confined time delay difference is reduced.

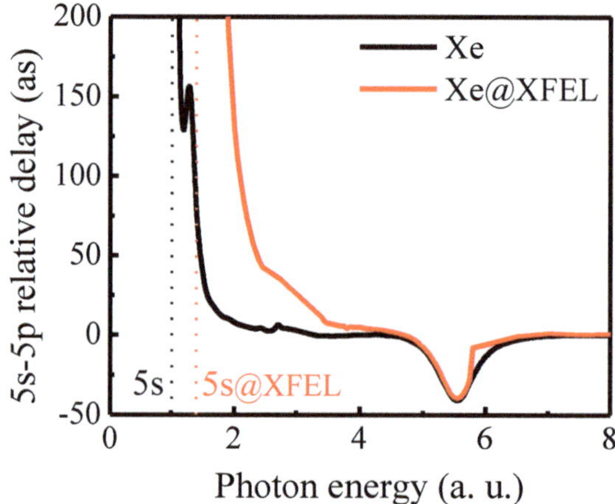

Figure 18. The time delay difference of 5s and 5p subshell (τ(5s)-τ(5p)) of Xe (black) and Xe@XFEL (red) in the RRPA calculation. Vertical lines show the 5s subshell threshold for Xe (black) and Xe@XFEl (red).

The current set of results on the noble gas atoms hints that the interaction time of the dipole-trapped atom is noticeably modified compared with that of the free atoms. For applications in the quantum information side, a pump laser is made to interact with a trapped atom. In the context of a two-level system, Rabi frequency is defined as $\Omega = \frac{\vec{d}.\vec{E}}{\hbar}$, where $\vec{d}$ is the atomic dipole moment, and $\vec{E}$ is the electric field of the pump laser. The speed of such a quantum device (quantum readout time) depends on the Rabi frequency of the pump laser. In other words, Rabi frequency gives the electronic transition rate between the two levels considered. Since we interpret the interaction time scale of the atom (qubit) as the readout time, the speed of the quantum device depends on the Rabi frequency. Hence, the present investigation serves as a primer roadmap to assess the speed of quantum devices.

4. Conclusions

The current work concludes that, except for the shift in the threshold, the photoionization parameters (cross section and angular distribution asymmetry) are very similar in the case of free atoms and those trapped in cross laser beams. However, due to the shift in the ionization threshold, a few signature features of the Cooper minimum and the shape resonances are missing in σ and β. It is already believed that photoionization time delay is very sensitive to external perturbations [56]. The present work shows that the time delay in photoionization is quantitatively and qualitatively different in the case of free and laser-trapped atoms. In a generic sense, the time delay in photoelectron ejection is increased due to laser trapping. This observation is consistent in all the cases we have studied. In addition, qualitative features are also altered; Cooper minimum features are mostly masked by the changes due to the confinement. Further, the difference between the *ns* and *np* subshells' time delay ($\tau(ns)$-$\tau(np)$) is also obtained, which is a measurable quantity. Our results show that the ($\tau(ns)$-$\tau(np)$) is also modified due to the spatial confinement. The present work underscores the importance of also considering interaction delays when estimating the speed of quantum information processing.

The present study is a seminal analysis of the speed of quantum devices in the presence of external perturbations. The work is limited in two ways. Firstly, in mimicking the quantum devices and their responses, investigating a bound-to-bound hyperfine split transition would be ideal. Secondly, the alkali metal atoms are the ideal test bed for such applications, and, hence, dealing with such atoms would be desirable. Nevertheless, the present bound-to-continuum studies on the noble gas atoms are indicators of the fact that the trap environment is capable of altering the temporal response. The results from this work allow us to anticipate changes in the response of the atoms in dipole traps. We are not aware of any other work of this kind that addresses the speed of a quantum computing device using the temporal response of a prospective quantum register to probes. The speed of quantum computing is determined by considerations such as those investigated in the present work; we hope that the results will be of consequence in the general field of quantum information science.

Author Contributions: Conceptualization, S.B., P.C.D., J.J. and R.K.E.; methodology, J.J., P.C.D. and R.K.E.; software, S.B.; validation, S.B., J.J. and R.K.E.; formal analysis, S.B., J.J., P.C.D. and R.K.E.; investigation, S.B.; resources, S.B. and A.G.; data curation, S.B. and A.G.; writing—original draft preparation, S.B.; writing—review and editing, P.C.D., R.K.E. and J.J.; visualization, J.J. and P.C.D.; supervision, J.J.; project administration, J.J.; funding acquisition, J.J. All authors have read and agreed to the published version of the manuscript.

Funding: J.J. acknowledges the funding from SERB through project No: CRG/2022/000191.

Data Availability Statement: The data generated part of this study are all used and are available in this manuscript.

Conflicts of Interest: The authors declare no conflict of interest.

References

1. Yanofsky, N.S. An introduction to quantum computing. In *Proof, Computation, and Agency*; Springer: Berlin/Heidelberg, Germany, 2011; pp. 145–180.
2. Lvovsky, A.I.; Sanders, B.C.; Tittel, W. Optical quantum memory. *Nat. Photonics* **2009**, *3*, 706–714. [CrossRef]
3. Zoller, P.; Beth, T.; Binosi, D.; Blatt, R.; Briegel, H.; Bruss, D.; Calarco, T.; Cirac, J.I.; Deutsch, D.; Eisert, J.; et al. Quantum information processing and communication. *Eur. Phys. J. D-At. Mol. Opt. Plasma Phys.* **2005**, *36*, 203–228. [CrossRef]
4. Benjamin, S.C.; Ardavan, A.; Briggs, G.A.D.; Britz, D.A.; Gunlycke, D.; Jefferson, J.; Jones, M.A.G.; Leigh, D.F.; Lovett, B.W.; Khlobystov, A.N.; et al. Towards a fullerene-based quantum computer. *J. Phys. Condens. Matter* **2006**, *18*, S867. [CrossRef]
5. Olmschenk, S.; Younge, K.C.; Moehring, D.L.; Matsukevich, D.N.; Maunz, P.; Monroe, C. Manipulation and detection of a trapped Yb+ hyperfine qubit. *Phys. Rev. A* **2007**, *76*, 052314. [CrossRef]
6. DeMille, D. Quantum computation with trapped polar molecules. *Phys. Rev. Lett.* **2002**, *88*, 067901. [CrossRef] [PubMed]
7. Wakabayashi, T. Fullerene C60: A possible molecular quantum computer. In *Molecular Realizations of Quantum Computing 2007*; World Scientific: Singapore, 2009; pp. 163–192.
8. Debnath, S. A Programmable Five Qubit Quantum Computer Using Trapped Atomic Ions. Ph.D. Thesis, University of Maryland, College Park, MD, USA, 2016.
9. Linke, N.M.; Maslov, D.; Roetteler, M.; Debnath, S.; Figgatt, C.; Landsman, K.A.; Wright, K.; Monroe, C. Experimental comparison of two quantum computing architectures. *Proc. Natl. Acad. Sci. USA* **2017**, *114*, 3305–3310. [CrossRef] [PubMed]
10. Ladd, T.D.; Jelezko, F.; Laflamme, R.; Nakamura, Y.; Monroe, C.; O'Brien, J.L. Quantum computers. *Nature* **2010**, *464*, 45–53. [CrossRef]
11. Benioff, P. The computer as a physical system: A microscopic quantum mechanical Hamiltonian model of computers as represented by Turing machines. *J. Stat. Phys.* **1980**, *22*, 563–591. [CrossRef]
12. Benioff, P. Quantum mechanical models of Turing machines that dissipate no energy. *Phys. Rev. Lett.* **1982**, *48*, 1581. [CrossRef]
13. Stenholm, S. The semiclassical theory of laser cooling. *Rev. Mod. Phys.* **1986**, *58*, 699. [CrossRef]
14. Minogin, V.G.; Letokhov, V.S. *Laser Light Pressure on Atoms*; CRC Press: Boca Raton, FL, USA, 1987.
15. Phillips, W. Laser Manipulation of Atoms & Ions. In *Proceedings of the International School of Physics "Enrico Fermi"—Course CXVIII*; IOS Press: Amsterdam, The Netherlands, 1992.
16. Metcalf, H.; van der Straten, P. Cooling and trapping of neutral atoms. *Phys. Rep.* **1994**, *244*, 203–286. [CrossRef]
17. Chu, S. Nobel Lecture: The manipulation of neutral particles. *Rev. Mod. Phys.* **1998**, *70*, 685. [CrossRef]
18. Cohen-Tannoudji, C.N. Nobel Lecture: Manipulating atoms with photons. *Rev. Mod. Phys.* **1998**, *70*, 707. [CrossRef]
19. Phillips, W.D. Nobel Lecture: Laser cooling and trapping of neutral atoms. *Rev. Mod. Phys.* **1998**, *70*, 721. [CrossRef]
20. Askar'yan, G.A. Effects of the Gradient of a Strong Electromagnetic Beam on Electrons and Atoms. *Sov. Phys. JETP* **1962**, *15*, 1088.
21. Letokhov, V.S. Narrowing of the Doppler width in a standing light wave. *JETP Lett.* **1968**, *7*, 272.
22. Bjorkholm, J.E.; Freeman, R.R.; Ashkin, A.; Pearson, D.B. Observation of Focusing of Neutral Atoms by the Dipole Forces of Resonance-Radiation Pressure. *Phys. Rev. Lett.* **1978**, *41*, 1361. [CrossRef]
23. Chu, S.; Bjorkholm, J.E.; Ashkin, A.; Cable, A. Experimental Observation of Optically Trapped Atoms. *Phys. Rev. Lett.* **1986**, *57*, 314. [CrossRef]
24. Schlosser, N.; Reymond, G.; Protsenko, I.; Grangier, P. Sub-poissonian loading of single atoms in a microscopic dipole trap. *Nature* **2001**, *411*, 1024–1027. [CrossRef]
25. Schlosser, N.; Reymond, G.; Grangier, P. Collisional blockade in microscopic optical dipole traps. *Phys. Rev. Lett.* **2002**, *89*, 023005. [CrossRef]
26. Reymond, G.; Schlosser, N.; Protsenko, I.; Grangier, P. Single-atom manipulations in a microscopic dipole trap. *Philos. Trans. R. Soc. Lond. Ser. A Math. Phys. Eng. Sci.* **2003**, *361*, 1527–1536. [CrossRef]
27. Alt, W.; Schrader, D.; Kuhr, S.; Müller, M.; Gomer, V.; Meschede, D. Single atoms in a standing-wave dipole trap. *Phys. Rev. A* **2003**, *67*, 033403. [CrossRef]
28. Weber, M.; Volz, J.; Saucke, K.; Kurtsiefer, C.; Weinfurter, H. Analysis of a single-atom dipole trap. *Phys. Rev. A* **2006**, *73*, 043406. [CrossRef]
29. Specht, H.P.; Nölleke, C.; Reiserer, A.; Uphoff, M.; Figueroa, E.; Ritter, S.; Rempe, G. A single-atom quantum memory. *Nature* **2011**, *473*, 190–193. [CrossRef] [PubMed]
30. Nölleke, C.; Neuzner, A.; Reiserer, A.; Hahn, C.; Rempe, G.; Ritter, S. Efficient teleportation between remote single-atom quantum memories. *Phys. Rev. Lett.* **2013**, *110*, 140403. [CrossRef] [PubMed]
31. García-Ripoll, J.J.; Zoller, P.; Cirac, J.I. Quantum information processing with cold atoms and trapped ions. *J. Phys. B At. Mol. Opt. Phys.* **2005**, *38*, S567. [CrossRef]
32. Ma, L.; Slattery, O.; Tang, X. Optical quantum memory based on electromagnetically induced transparency. *J. Opt.* **2017**, *19*, 043001. [CrossRef] [PubMed]
33. Pachos, J.; Walther, H. Quantum Computation with Trapped Ions in an Optical Cavity. *Phys. Rev. Lett.* **2002**, *89*, 187903. [CrossRef]
34. Verstraete, F.; Wolf, M.M.; Ignacio Cirac, J. Quantum computation and quantum-state engineering driven by dissipation. *Nat. Phys.* **2009**, *5*, 633–636. [CrossRef]
35. Altarelli, M. The European X-ray free-electron laser facility in Hamburg. *Nucl. Instrum. Methods Phys. Res. Sect. B Beam Interact. Mater. At.* **2011**, *269*, 2845–2849. [CrossRef]

36. Seiboth, F.; Schropp, A.; Scholz, M.; Wittwer, F.; Rödel, C.; Wünsche, M.; Ullsperger, T.; Nolte, S.; Rahomäki, J.; Parfeniukas, K.; et al. Perfect X-ray focusing via fitting corrective glasses to aberrated optics. *Nat. Commun.* **2017**, *8*, 14623. [CrossRef] [PubMed]

37. Matsuyama, S.; Inoue, T.; Yamada, J.; Kim, J.; Yumoto, H.; Inubushi, Y.; Osaka, T.; Inoue, I.; Koyama, T.; Tono, K.; et al. Nanofocusing of X-ray free-electron laser using wavefront-corrected multilayer focusing mirrors. *Sci. Rep.* **2018**, *8*, 17440. [CrossRef] [PubMed]

38. Johnson, W.; Lin, C.; Cheng, K.; Lee, C. Relativistic random-phase approximation. *Phys. Scr.* **1980**, *21*, 409. [CrossRef]

39. Grimm, R.; Weidemüller, M.; Ovchinnikov, Y.B. Optical dipole traps for neutral atoms. In *Advances in Atomic, Molecular, and Optical Physics*; Elsevier: Amsterdam, The Netherlands, 2000; Volume 42, pp. 95–170.

40. Adams, C.S.; Lee, H.J.; Davidson, N.; Kasevich, M.; Chu, S. Evaporative Cooling in a Crossed Dipole Trap. *Phys. Rev. Lett.* **1995**, *74*, 3577–3580. [CrossRef] [PubMed]

41. Grant, I.P. *Relativistic Quantum Theory of Atoms and Molecules: Theory and Computation*; Springer: Berlin/Heidelberg, Germany, 2010.

42. Johnson, W.; Lin, C. Multichannel relativistic random-phase approximation for the photoionization of atoms. *Phys. Rev. A* **1979**, *20*, 964. [CrossRef]

43. Dalgarno, A.; Victor, G.A. The time-dependent coupled Hartree-Fock approximation. *Proc. R. Soc. London. Ser. A Math. Phys. Sci.* **1966**, *291*, 291–295.

44. Munasinghe, C.R.; Deshmukh, P.C.; Manson, S.T. Photoionization branching ratios of spin-orbit doublets far above thresholds: Interchannel and relativistic effects in the noble gases. *Phys. Rev. A* **2022**, *106*, 013102. [CrossRef]

45. Deshmukh, P.C.; Banerjee, S. Time delay in atomic and molecular collisions and photoionisation/photodetachment. *Int. Rev. Phys. Chem.* **2021**, *40*, 127–153. [CrossRef]

46. Deshmukh, P.C.; Banerjee, S.; Mandal, A.; Manson, S.T. Eisenbud–Wigner–Smith time delay in atom–laser interactions. *Eur. Phys. J. Spec. Top* **2021**, *230*, 4151–4164. [CrossRef]

47. Isinger, M.; Squibb, R.; Busto, D.; Zhong, S.; Harth, A.; Kroon, D.; Nandi, S.; Arnold, C.L.; Miranda, M.; Dahlström, J.M.; et al. Photoionization in the time and frequency domain. *Science* **2017**, *358*, 893–896. [CrossRef]

48. Bray, A.W.; Naseem, F.; Kheifets, A.S. Photoionization of Xe and Xe@C60 from the 4d shell in RABBITT fields. *Phys. Rev. A* **2018**, *98*, 043427. [CrossRef]

49. Schultze, M.; Fiess, M.; Karpowicz, N.; Gagnon, J.; Korbman, M.; Hofstetter, M.; Neppl, S.; Cavalieri, A.L.; Komninos, Y.; Mercouris, T.; et al. Delay in photoemission. *Science* **2010**, *328*, 1658–1662. [CrossRef] [PubMed]

50. Amusia, M.Y.; Ivanov, V.; Cherepkov, N.; Chernysheva, L. Interference effects in photoionization of noble gas atoms outer s-subshells. *Phys. Lett. A* **1972**, *40*, 361–362. [CrossRef]

51. Saha, S.; Banerjee, S.; Jose, J. Impact of Charge Migration and the Angle-Resolved Photoionization Time Delays of the Free and Confined Atom X@C60. *Atoms* **2022**, *10*, 44. [CrossRef]

52. Saha, S.; Mandal, A.; Jose, J.; Varma, H.R.; Deshmukh, P.; Kheifets, A.; Dolmatov, V.K.; Manson, S.T. Relativistic effects in photoionization time delay near the Cooper minimum of noble-gas atoms. *Phys. Rev. A* **2014**, *90*, 053406. [CrossRef]

53. Guénot, D.; Klünder, K.; Arnold, C.L.; Kroon, D.; Dahlström, J.M.; Miranda, M.; Fordell, T.; Gisselbrecht, M.; Johnsson, P.; Mauritsson, J.; et al. Photoemission-time-delay measurements and calculations close to the 3s-ionization-cross-section minimum in Ar. *Phys. Rev. A* **2012**, *85*, 053424. [CrossRef]

54. Johnson, W.; Cheng, K. Photoionization of the outer shells of neon, argon, krypton, and xenon using the relativistic random-phase approximation. *Phys. Rev. A* **1979**, *20*, 978. [CrossRef]

55. Deshmukh, P.; Mandal, A.; Saha, S.; Kheifets, A.; Dolmatov, V.; Manson, S. Attosecond time delay in the photoionization of endohedral atoms A@C 60: A probe of confinement resonances. *Phys. Rev. A* **2014**, *89*, 053424. [CrossRef]

56. Keating, D.; Manson, S.; Dolmatov, V.; Mandal, A.; Deshmukh, P.; Naseem, F.; Kheifets, A.S. Intershell-correlation-induced time delay in atomic photoionization. *Phys. Rev. A* **2018**, *98*, 013420. [CrossRef]

Communication

Does Carrier Envelope Phase Affect the Ionization Site in a Neutral Diatomic Molecule?

Alex Schimmoller [†], Harrison Pasquinilli [†] and Alexandra S. Landsman *

Department of Physics, The Ohio State University, Columbus, OH 43210, USA; schimmoller.11@osu.edu (A.S.)
* Correspondence: landsman.7@osu.edu
† These authors contributed equally to this work.

Abstract: A recent work shows how to extract the ionization site of a neutral diatomic molecule by comparing Quantum Trajectory Monte Carlo (QTMC) simulations with experimental measurements of the final electron momenta distribution. This method was applied to an experiment using a 40-femtosecond infrared pulse, finding that a downfield atom is roughly twice as likely to be ionized as an upfield atom in a neutral nitrogen molecule. However, an open question remains as to whether an assumption of the zero carrier envelope phase (CEP) used in the above work is still valid for short, few-cycle pulses where the CEP can play a large role. Given experimentalists' limited control over the CEP and its dramatic effect on electron momenta after ionization, it is desirable to see what influence the CEP may have in determining the ionization site. In this paper, we employ QTMC techniques to simulate strong-field ionization and electron propagation from neutral N_2 using an intense 6-cycle laser pulse with various CEP values. Comparing simulated electron momenta to experimental data indicates that the ratio of down-to-upfield ions remains roughly 2:1 regardless of the CEP. This confirms that the ionization site of a neutral molecule is determined predominantly by the laser frequency and intensity, as well as the ground-state molecular wavefunction, and is largely independent of the CEP.

Keywords: strong field ionization; molecular ionization; Quantum Trajectory Monte Carlo

Citation: Schimmoller, A.; Pasquinilli, H.; Landsman, A.S. Does Carrier Envelope Phase Affect the Ionization Site in a Neutral Diatomic Molecule? *Atoms* **2023**, *11*, 67. https://doi.org/10.3390/atoms11040067

Academic Editors: Himadri S. Chakraborty and Hari R. Varma

Received: 28 February 2023
Revised: 30 March 2023
Accepted: 1 April 2023
Published: 4 April 2023

1. Introduction

Tunneling occurs when a laser's strong electric field distorts the Coulombic barrier of an atom enough to allow for the electron to escape [1]. For the case of a diatomic molecule, this picture is complicated by the presence of a double-well potential. This leads to two possible ionization sites, as shown in Figure 1: the upfield (higher energy) atom and the downfield (lower energy) atom [2]. Commonly used theories of molecular ionization, such as molecular ADK [3], molecular SFA, and the partial Fourier transform approach [4], assume implicitly that all ionization is downfield, corresponding to the bound electron wavepacket adiabatically responding to the relatively low-frequency laser field. However, it is known that ionization in charged molecules can occur from either atom depending on the internuclear separation, alignment, and other conditions [2].

When a positively charged diatomic molecule begins to dissociate, the resulting bond softening traps the electron in the upper well and leads to upfield ionization. This process is known as ionization enhancement [5–10] and has been repeatedly confirmed in experiments that examine molecular fragments following a Coulomb explosion [7,8,11,12]. However, until recently, there was no technique for determining the ionization site in neutral atoms. A recent work suggests that the longitudinal photoelectron momentum distributions for charged ions could be looked at to identify the ionization location [13]. The principle behind this technique is that the electron experiences different forces due to the Coulomb potential depending on which atom it is ionized from. If it is ionized from the downfield atom, it will propagate directly into the continuum, but if it is from the upfield atom, it will first have to pass the downfield atom, distorting its trajectory compared to the case of ionization from an atom with the same binding potential. Additionally, if the electron tunnels from the upfield atom to the downfield atom, there will be a delay in ionization, causing a shift in the photoelectron momentum distribution (PMD) for circularly or elliptically polarized light. However, the approach presented in [13] views the ionization process as either all upfield or all downfield, and therefore does not provide a method for quantifying upfield to downfield ionization when both contributions are significant. A more recent work by Ortmann and colleagues [14] presents a method for quantifying the ratio of upfield to downfield ionization events, finding a significant contribution from both under typical experimental conditions that employ infrared light for strong field ionization.

In this work, we focus on the approach presented in [14], which establishes a quantitative procedure for finding the ionization site in a neutral diatomic molecule. This procedure relies on simulating a variety of upfield:downfield ionization site ratios and determining which ratio matches the experimental momentum distribution. We expand upon this technique by examining the effect that the carrier envelope phase (CEP) has upon the results to see if the approach requires stabalizing the CEP or if it can work over a random CEP distribution.

It is important to check the robustness of this model with respect to changing the carrier envelope phase for two reasons. First, it expands the range of applicability of the approach in [14] to few-cycle pulses without requiring CEP averaging or pulse stabilization, which would introduce additional sources of uncertainty. Changing the CEP can change the final PMD, which may affect the technique as it depends on comparing the final transverse momenta of the electrons in order to determine the ionization site. This is not an issue for longer pulses where PMDs are independent of CEP, but it can play an important role for few-cycle pulses where the CEP is not stabilized. Stabilizing the CEP is a non-trivial task experimentally, let alone setting it to a specific value [15]. Second, the robustness of the ionization site calculation to CEP changes supports the view that upfield ionization is a non-adiabatic effect determined by the Keldysh parameter, $\gamma = \omega \sqrt{2I_p}/E_0$, which is independent of the CEP. Here, ω, E_0, and I_p are the laser frequency, peak field strength, and ionization potential, respectively (atomic units are assumed throughout this text).

The remainder of our work is organized as follows. Section 2 describes the techniques and simulation used to calculate the PMDs. Section 3 analyzes the results, finding that CEP does not have a significant impact on the relative contribution of upfield to downfield ionization. Section 4 concludes and summarizes.

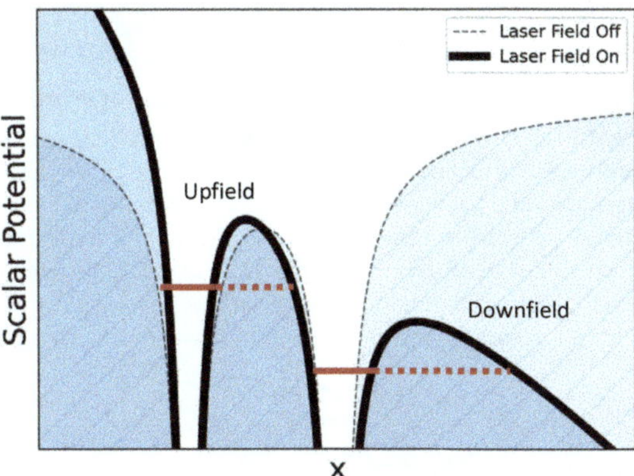

Figure 1. Schematic of the electric potential created by a diatomic molecule with and without a strong laser field present. Turning on the laser electric field allows electrons originating from both atomic sites to tunnel into the continuum. Upfield and downfield electrons experience different potentials due to the molecule's asymmetric Coulomb forces, altering their trajectories.

2. Simulating N_2 Strong Field Ionization

To highlight the role that the carrier envelope phase can play in the final momentum distribution, consider a laser pulse linearly polarized along the x-axis with electric field profile

$$\mathbf{E}(t) = E_0 \cos(\omega t + \phi)\text{env}(t)\hat{\mathbf{x}}, \tag{1}$$

where $\omega = 2\pi c/\lambda$ is the angular frequency of the wave, $\text{env}(t) = \cos^2(\frac{\omega t}{2N})$ is an envelope centered at $t = 0$ containing N laser cycles, and ϕ is the carrier envelope phase. According to the strong field approximation [16], ionization is most likely near the absolute maxima of the laser pulse and the electron's final momentum is largely determined by the vector potential at the time of ionization $\mathbf{A}(t_0) = - \int^{t_0} \mathbf{E}(t')dt'$. Therefore, the CEP influences both the ionization time and final electron momenta. When ionizing diatomic molecules [14], it also controls when parent atoms are considered either upfield or downfield. For long pulses, this poses no problem since the field and vector potential can be approximated as plane waves and ionization takes place over many optical cycles. However, for short pulses when N is on the order of a few optical cycles and the envelope function $\text{env}(t)$ decays quickly away from $t = 0$, only the centermost field peaks contribute to ionization, amplifying the CEP's influence on the photoelectron distribution.

The setup for the simulation closely follows that of reference [14]: a six-cycle, linearly polarized laser pulse of the form (1) with wavelength $\lambda = 800$ nm and peak intensity $I_0 = 1.3 \cdot 10^{14}$ W/cm^2 is incident upon a neutral N_2 molecule with ionization potential $I_p = 15.6$ eV. Focal averaging is applied to the intensity profile by assigning to each intensity I a relative weight $\sim \frac{2I+I_0}{I^{5/2}}\sqrt{I_0 - I}$ [17–19]. In the simulation, intensities are sampled according to these weights and used to determine the peak electric field $E_0 = \sqrt{I}$ for subsets of the simulated electrons. The molecule is tilted $\theta = 45$ degrees against the polarization direction with nitrogen atoms located at positions $\mathbf{r}_A = \frac{R_0}{2\sqrt{2}}(-1,0,-1)$ and $\mathbf{r}_B = \frac{R_0}{2\sqrt{2}}(1,0,1)$ a.u., respectively, where $R_0 = 2$ a.u. is the internuclear distance. This tilt creates an asymmetric Coulomb force acting on electrons originating from each of the parent nuclei. When the laser pulse is incident upon the molecule, electrons may be ionized from either the up- or downfield atom (this designation alternates depending on whether the electric field is positive or negative). They then propagate semiclassically until the end of the laser pulse, when their positions and momenta are recorded and asymptotic momenta are calculated.

Initial conditions for ionized electrons are achieved via Monte Carlo reject sampling [20,21]. With the choice of the up- or downfield parent atom fixed, ionization times t_0 and initial transverse velocities $v_\perp = \sqrt{v_{0,y}^2 + v_{0,z}^2}$ are fed into a reject-sampling algorithm that compares the (normalized) ionization rate to randomly generated values. This ionization rate accounts for the molecular orbital by importing the electronic wavefunction in N_2 from GAMESS [22] and performing a partial Fourier transform to obtain the electron's initial transverse velocity distribution. Electrons tunnel nonadiabatically to the continuum according to the ionization theory presented in reference [23], though with the more general field profile (1) containing both the enveloping function and carrier envelope phase. The atomic ionization rate $W(t_0, v_\perp)$ for an electron ionized at time t_0 and with transverse velocity $v_\perp$ is

$$W(t_0, v_\perp) = \frac{\omega^2(2I_p)^{5/2}}{2[E_0\text{env}(t_0)]^4\gamma^2(t_0,v_\perp)[\gamma^2(t_0,v_\perp)+\cos^2(\omega t_0+\phi)]\cos^2(\omega t_0+\phi)}$$

$$\times \exp\left(-\frac{[E_0\text{env}(t_0)]^2}{\omega^3}\left\{\left[\sin^2(\omega t_0 + \phi) + \gamma^2(t_0,v_\perp) + \tfrac{1}{2}\right] \times \sinh^{-1}\gamma(t_0,v_\perp)\right.\right. \tag{2}$$

$$\left.\left. -\tfrac{1}{2}\gamma(t_0,v_\perp)\sqrt{1+\gamma^2(t_0,v_\perp)}\left(1 + 2\sin^2(\omega t_0 + \phi)\right)\right\}\right),$$

where

$$\gamma(t_0, v_\perp) = \omega \frac{\sqrt{2I_p + v_\perp^2}}{|\mathbf{E}(t_0)|} \tag{3}$$

is the effective Keldysh parameter [23]. Electrons with ionization times t_0 and transverse velocities $v_\perp$ that pass reject sampling are then assigned tunnel exit positions $\mathbf{r}_0 = \mathbf{r}_{A/B} + \mathrm{Re}(x_0, 0, 0)$ and longitudinal velocities $v_{0,x}$ where

$$\mathrm{Re}\{x_0(t_0, v_\perp)\} = \frac{E_0 \mathrm{env}(t_0)}{\omega^2} \cos(\omega t_0 + \phi)\left[1 - \sqrt{1 + \gamma^2(t_0, v_\perp)}\right], \tag{4}$$

$$v_{0,x} = \frac{E_0 \mathrm{env}(t_0) \sin(\omega t_0 + \phi)}{\omega}\left[\sqrt{1 + \gamma^2(t_0, v_\perp)} - 1\right], \tag{5}$$

$\mathbf{r}_{A/B}$ is the up-/downfield atomic site (depending on the field sign) and $\mathrm{Re}\{x_0(t_0, v_\perp)\}$ corresponds to the real part of the tunnel exit along the x-axis.

After ionization, electrons propagate semiclassically. Their dynamical positions and momenta are calculated numerically by solving Newton's equation of motion for an electron interacting with the driving laser electric field (1) and two softcore Coulomb forces from the N_2 ion, each with $1/2$ fundamental charge at their respective centers:

$$\ddot{\mathbf{r}}(t) = -\mathbf{E}(t) - \nabla V(\mathbf{r}), \tag{6}$$

where the potential $V(\mathbf{r})$ is given by

$$V(\mathbf{r}) = \sum_{j=A,B} -\frac{(1/2)}{\sqrt{[\mathbf{r}(t) - \mathbf{r}_j]^2 + SC}}. \tag{7}$$

In the simulation, $SC = 0.01$ to avoid numerical problems created by the singularities at the atomic centers. Electrons are propagated until the end of the laser pulse t_1. During propagation, each electron accumulates a complex phase Φ derivable from its classical action S [20,24]:

$$\Phi = \int_{t_0}^{t_1} \left(\frac{\mathbf{v}^2}{2} + V(\mathbf{r}) - \mathbf{r} \cdot \nabla V(\mathbf{r})\right) dt - I_p t_0 + \mathbf{v}_0 \cdot (\mathbf{r}_0 - \mathbf{r}_{A/B}) + \Phi_0, \tag{8}$$

where the initial phase Φ_0 accounts for the molecular tilt and is given by [17]

$$\tan \Phi_0 = \tan\left(\frac{v_{z,0} R_0 \sin \theta}{2}\right) \tanh\left(\mathrm{sign}[E_x(t_0)]\frac{R_0 \cos \theta}{2}\sqrt{2I_p + v_{z,0}^2}\right). \tag{9}$$

Once the trajectory calculation is complete, Rydberg electrons are filtered out and final momenta and phases are recorded.

Electron momenta at the detector are determined from the continuum electrons' positions $\mathbf{r}_1$ and momenta $\mathbf{v}_1$ at the end of the laser pulse [25]. Assuming that the electron–molecule interaction can now be approximated as a two-body problem, the asymptotic momentum $\mathbf{v} = (v_x, v_y, v_z)$ is given by

$$\mathbf{v} = v\frac{v(\mathbf{L} \times \mathcal{A}) - \mathcal{A}}{1 + v^2 L^2}, \tag{10}$$

where $\mathbf{L} = \mathbf{r}_1 \times \mathbf{v}_1$ is the angular momentum and $\mathcal{A} = \mathbf{v}_1 \times \mathbf{L} - \mathbf{r}_1/r_1$ is the Runge–Lenz vector, both of which are conserved quantities. The asymptotic momentum magnitude

$v = \sqrt{v_x^2 + v_y^2 + v_z^2}$ comes from solving for the electron's kinetic energy far away from the charged molecule:

$$\frac{v^2}{2} = \frac{v_1^2}{2} - \frac{1}{r_1}. \tag{11}$$

Sample results of the simulation are shown in Figure 2, which plots the 2D asymptotic momentum distribution in the v_x-v_z plane. Note that quantum interference is included by attaching a phase to each trajectory, resulting in a complex factor $e^{i\Phi}$ multiplying each trajectory, and accounting for interference between different trajectories that end up with the same final momentum. Additional details about the QTMC simualtions can be found in [14].

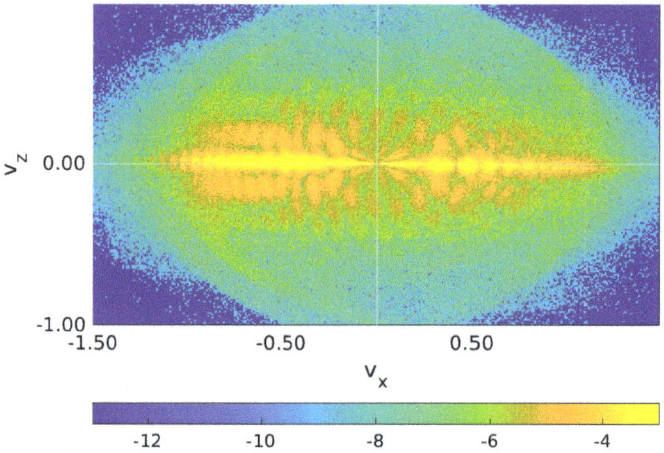

Figure 2. Simulated 2D photoelectron momentum distribution (PMD) when the ionization ratio $q = 0.6$ (Equation (12)) and carrier envelope phase $\phi = 0$. In this simulation, a six-cycle, 800-nm laser pulse with peak intensity $I = 1.3 \cdot 10^{14}$ W/cm^2 is incident upon a neutral N$_2$ molecule tilted 45 degrees with respect to the polarization direction. When $q = 0.6$, electrons are ionized at the upfield atom four times more often than the downfield atom. The color bar is on a logarithmic scale with arbitrary units.

3. Analyzing Momentum Data

To determine the relative number of electrons ionized at either the upfield or downfield locations, we again closely follow the analysis used in reference [14]. First, electron trajectories originating from both atomic sites are calculated. The relative number of up- and downfield electrons used in the analysis is determined by the ionization ratio

$$q = \frac{\#\,\mathrm{up} - \#\,\mathrm{down}}{\#\,\mathrm{up} + \#\,\mathrm{down}}, \tag{12}$$

which is sampled within the range -1 (all downfield) to $+1$ (all upfield). For each set of trajectories, a 2D photoelectron momentum distribution $w(i, j)$ is generated, where i and j index over bins of v_x and v_z, respectively. These momentum distributions are compared to that of experiment [26] by calculating the average offset momentum a for each distribution, where

$$a = \frac{\sum_{i=1}^{m} \mathrm{sign}[v_x(i)] v_{z,\mathrm{mean}}[v_x(i)]}{m}, \tag{13}$$

and

$$v_{z,\text{mean}}[v_x(i)] = \frac{\sum_{j=1}^{n} w(i,j)v_z(j)}{\sum_{j=1}^{n} w(i,j)}. \tag{14}$$

In Figure 3, offset momentum a is plotted versus ionization ratio q for various CEP values and compared to the offset momentum calculated from the experimental data. It appears that changing the CEP creates a slight variation in the offset momentum for different q values. However, these ionization ratios for the different CEP all correspond physically to ionizing roughly two downfield electrons for every one upfield.

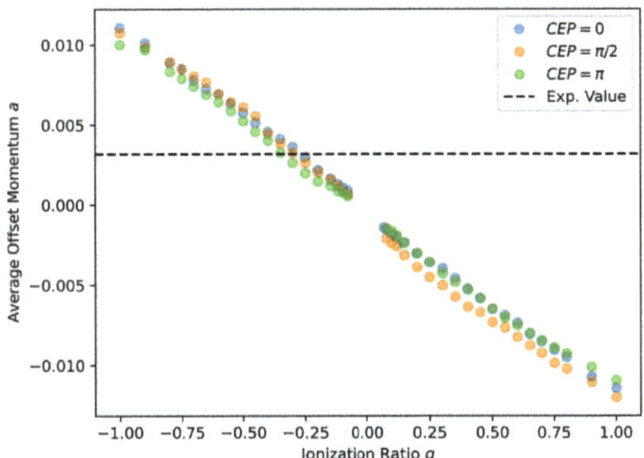

Figure 3. Ionization ratio q (Equation (12)) vs. average offset momentum a (Equation (13)) for various values of carrier envelope phase. The experimental offset momentum determined in reference [14] is indicated by the dashed line. It appears that regardless of the CEP, the ratio of downfield to upfield ionization remains roughly 2:1.

4. Conclusions

We have simulated the ionization of N_2 in a strong electric field through the use of QTMC techniques. By comparing experimental results [26] to simulated electron momentum distributions with various upfield and downfield contributions, we confirm the 2:1 downfield-to-upfield ionization ratio found in [14] regardless of the laser field's carrier envelope phase. Thus, determining the ionization site through this technique does not require experimental CEP stabilization or simulated averaging over CEP values, limiting possible sources of uncertainty.

Importantly, our results support the paradigm of non-adiabatic strong field molecular ionization depending mostly on the Keldysh parameter, γ, which itself depends only on the laser intensity, frequency, and ionization potential. This view is supported by prior analytical calculations in a static electric field, corresponding to $\gamma \ll 1$, which find that all tunneling is downfield in this fully adiabatic limit [27]. Experimental studies of the strong field ionization of neutral diatomic molecules using longer-wavelength mid-IR pulses, combined with the ionization site extraction procedure proposed in [14], could further test the robustness of this paradigm.

Author Contributions: Conceptualization, A.S.L.; simulation and data analysis, A.S. and H.P.; writing—original draft preparation, A.S., H.P. and A.S.L.; writing—review and editing, N/A; supervision, A.S.L.; project administration, A.S., H.P. and A.S.L.; funding acquisition, A.S.L. All authors have read and agreed to the published version of the manuscript.

Funding: This research was funded by the U.S. Department of Energy, Office of Basic Energy Sciences, Atomic, Molecular and Optical Sciences Program, under Award No. DE-SC0022093.

Data Availability Statement: Data can be provided upon reasonable request.

Acknowledgments: We acknowledge prior code development by Lisa Ortmann, which made this work possible.

Conflicts of Interest: The authors declare no conflict of interest.

Abbreviations

The following abbreviations are used in this manuscript:

MDPI	Multidisciplinary Digital Publishing Institute
DOAJ	Directory of Open Access Journals
CEP	Carrier envelope phase
PMD	Photoelectron momentum distribution
QTMC	Quantum Trajectory Monte Carlo

References

1. Keldysh, L. Ionization in the field of a strong electromagnetic wave. *Sov. Phys. JETP* **1965**, *20*, 1307–1314.
2. Zuo, T.; Bandrauk, A.D. Charge-resonance-enhanced ionization of diatomic molecular ions by intense lasers. *Phys. Rev. A* **1995**, *52*, R2511–R2514. [CrossRef] [PubMed]
3. Tong, X.M.; Zhao, Z.; Lin, C.D. Theory of molecular tunneling ionization. *Phys. Rev. A* **2002**, *66*, 033402. [CrossRef]
4. Liu, M.; Liu, Y. Application of the partial-Fourier-transform approach for tunnel ionization of molecules. *Phys. Rev. A* **2016**, *93*, 043426. [CrossRef]
5. Seideman, T.; Ivanov, M.Y.; Corkum, P.B. Role of electron localization in intense-field molecular ionization. *Phys. Rev. Lett.* **1995**, *75*, 2819. [CrossRef] [PubMed]
6. Staudte, A.; Pavičić, D.; Chelkowski, S.; Zeidler, D.; Meckel, M.; Niikura, H.; Schöffler, M.; Schössler, S.; Ulrich, B.; Rajeev, P.P.; et al. Attosecond Strobing of Two-Surface Population Dynamics in Dissociating H_2^+. *Phys. Rev. Lett.* **2007**, *98*, 073003. [CrossRef]
7. Constant, E.; Stapelfeldt, H.; Corkum, P.B. Observation of Enhanced Ionization of Molecular Ions in Intense Laser Fields. *Phys. Rev. Lett.* **1996**, *76*, 4140–4143. [CrossRef]
8. Xin, L.; Qin, H.C.; Wu, W.Y.; He, F. Fraunhofer-like diffracted lateral photoelectron momentum distributions of H_2^+ in charge-resonance-enhanced ionization in strong laser fields. *Phys. Rev. A* **2015**, *92*, 063803. [CrossRef]
9. Posthumus, J.H. The dynamics of small molecules in intense laser fields. *Rep. Prog. Phys.* **2004**, *67*, 623. [CrossRef]
10. Bocharova, I.; Karimi, R.; Penka, E.F.; Brichta, J.P.; Lassonde, P.; Fu, X.; Kieffer, J.C.; Bandrauk, A.D.; Litvinyuk, I.; Sanderson, J.; et al. Charge Resonance Enhanced Ionization of CO_2 Probed by Laser Coulomb Explosion Imaging. *Phys. Rev. Lett.* **2011**, *107*, 063201. [CrossRef]
11. Betsch, K.J.; Pinkham, D.W.; Jones, R.R. Directional Emission of Multiply Charged Ions During Dissociative Ionization in Asymmetric Two-Color Laser Fields. *Phys. Rev. Lett.* **2010**, *105*, 223002. [CrossRef] [PubMed]
12. Wu, J.; Meckel, M.; Voss, S.; Sann, H.; Kunitski, M.; Schmidt, L.P.H.; Czasch, A.; Kim, H.; Jahnke, T.; Dörner, R. Coulomb Asymmetry in Strong Field Multielectron Ionization of Diatomic Molecules. *Phys. Rev. Lett.* **2012**, *108*, 043002. [CrossRef] [PubMed]
13. Liu, K.; Barth, I. Identifying the Tunneling Site in Strong-Field Ionization of H_2^+. *Phys. Rev. Lett.* **2017**, *119*, 243204. [CrossRef] [PubMed]
14. Ortmann, L.; AlShafey, A.; Staudte, A.; Landsman, A.S. Tracking the Ionization Site in Neutral Molecules. *Phys. Rev. Lett.* **2021**, *127*, 213201. [CrossRef] [PubMed]
15. Baltuška, A.; Udem, T.; Uiberacker, M.; Hentschel, M.; Goulielmakis, E.; Gohle, C.; Holzwarth, R.; Yakovlev, V.S.; Scrinzi, A.; Hänsch, T.W.; et al. Attosecond control of electronic processes by intense light fields. *Nature* **2003**, *421*, 611–615. [CrossRef]
16. Ivanov, M.Y.; Spanner, M.; Smirnova, O. Anatomy of strong field ionization. *J. Mod. Opt.* **2005**, *52*, 165–184. [CrossRef]
17. Liu, M.M.; Li, M.; Wu, C.; Gong, Q.; Staudte, A.; Liu, Y. Phase Structure of Strong-Field Tunneling Wave Packets from Molecules. *Phys. Rev. Lett.* **2016**, *116*, 163004. [CrossRef]
18. Kopold, R.; Becker, W.; Kleber, M.; Paulus, G. Channel-closing effects in high-order above-threshold ionization and high-order harmonic generation. *J. Phys. B At. Mol. Opt. Phys.* **2002**, *35*, 217. [CrossRef]
19. Augst, S.; Meyerhofer, D.D.; Strickland, D.; Chin, S.L. Laser ionization of noble gases by Coulomb-barrier suppression. *JOSA B* **1991**, *8*, 858–867. [CrossRef]
20. Shvetsov-Shilovski, N.I.; Lein, M.; Madsen, L.B.; Räsänen, E.; Lemell, C.; Burgdörfer, J.; Arbó, D.G.; Tőkési, K. Semiclassical two-step model for strong-field ionization. *Phys. Rev. A* **2016**, *94*, 013415. [CrossRef]
21. Ortmann, L.; Pérez-Hernández, J.A.; Ciappina, M.F.; Schötz, J.; Chacón, A.; Zeraouli, G.; Kling, M.F.; Roso, L.; Lewenstein, M.; Landsman, A.S. Emergence of a Higher Energy Structure in Strong Field Ionization with Inhomogeneous Electric Fields. *Phys. Rev. Lett.* **2017**, *119*, 053204. [CrossRef] [PubMed]
22. Schmidt, M.W.; Baldridge, K.K.; Boatz, J.A.; Elbert, S.T.; Gordon, M.S.; Jensen, J.H.; Koseki, S.; Matsunaga, N.; Nguyen, K.A.; Su, S.; et al. General atomic and molecular electronic structure system. *J. Comput. Chem.* **1993**, *14*, 1347–1363. [CrossRef]

23. Li, M.; Geng, J.W.; Han, M.; Liu, M.M.; Peng, L.Y.; Gong, Q.; Liu, Y. Subcycle nonadiabatic strong-field tunneling ionization. *Phys. Rev. A* **2016**, *93*, 013402. [CrossRef]
24. Li, M.; Geng, J.W.; Liu, H.; Deng, Y.; Wu, C.; Peng, L.Y.; Gong, Q.; Liu, Y. Classical-Quantum Correspondence for Above-Threshold Ionization. *Phys. Rev. Lett.* **2014**, *112*, 113002. [CrossRef]
25. Shvetsov-Shilovski, N.I.; Dimitrovski, D.; Madsen, L.B. Ionization in elliptically polarized pulses: Multielectron polarization effects and asymmetry of photoelectron momentum distributions. *Phys. Rev. A* **2012**, *85*, 023428. [CrossRef]
26. Meckel, M.; Staudte, A.; Patchkovskii, S.; Villeneuve, D.; Corkum, P.; Dörner, R.; Spanner, M. Signatures of the continuum electron phase in molecular strong-field photoelectron holography. *Nat. Phys.* **2014**, *10*, 594–600. [CrossRef]
27. Bondar, D.I.; Liu, W.K. Shapes of leading tunnelling trajectories for single-electron molecular ionization. *J. Phys. Math. Theor.* **2011**, *44*, 275301. [CrossRef]

Article

Rigorous Negative Ion Binding Energies in Low-Energy Electron Elastic Collisions with Heavy Multi-Electron Atoms and Fullerene Molecules: Validation of Electron Affinities

Alfred Z. Msezane *and Zineb Felfli

Department of Physics and Center for Theoretical Studies of Physical Systems, Clark Atlanta University, Atlanta, GA 30314, USA
* Correspondence: amsezane@cau.edu

Abstract: Dramatically sharp resonances manifesting stable negative ion formation characterize Regge pole-calculated low-energy electron elastic total cross sections (TCSs) of heavy multi-electron systems. The novelty of the Regge pole analysis is in the extraction of rigorous and unambiguous negative ion binding energies (BEs), corresponding to the measured electron affinities (EAs) of the investigated multi-electron systems. The measured EAs have engendered the crucial question: is the EA of multi-electron atoms and fullerene molecules identified with the BE of the attached electron in the ground, metastable or excited state of the formed negative ion during a collision? Inconsistencies in the meaning of the measured EAs are elucidated and new EA values for Bk, Cf, Fm, and Lr are presented.

Keywords: Regge poles; generalized bound states; multi-electron atoms; elastic cross sections; anionic binding energies; electron affinities; electron correlation; core–polarization interaction

Citation: Msezane, A.Z.; Felfli, Z. Rigorous Negative Ion Binding Energies in Low-Energy Electron Elastic Collisions with Heavy Multi-Electron Atoms and Fullerene Molecules: Validation of Electron Affinities. *Atoms* **2023**, *11*, 47. https://doi.org/10.3390/atoms11030047

Academic Editors: Himadri S. Chakraborty and Hari R. Varma

Received: 4 January 2023
Revised: 23 February 2023
Accepted: 27 February 2023
Published: 3 March 2023

1. Introduction

In the electron impact energy range $0.0 \leq E \leq 10.0$ eV, dramatically sharp resonances manifesting stable ground, metastable and excited negative ion formation, shape resonances (SRs), and Ramsauer–Townsend (R-T) minima characterize the Regge pole calculated low-energy electron elastic total cross sections (TCSs) of heavy multi-electron atoms and fullerene molecules [1]. The energy positions of the sharp resonances correspond to the measured electron affinities (EAs) of the considered multi-electron atoms and fullerene molecules. Indeed, the extraction from the TCSs of rigorous and unambiguous negative ion binding energies (BEs), the SRs, and the R-T minima, without any experimental or other theoretical assistance demonstrates the novelty and strength of the Regge pole analysis and its vital importance in the understanding of low-energy electron collisions with complex multi-electron systems through negative ion formation.

The recent theoretical investigation of low-energy electron elastic collisions with heavy multi-electron atoms and fullerene molecules, using Regge pole analysis, also discussed the meaning of the measured EA within two prevailing contexts [1]. The first viewpoint considers the EA to correspond to the electron binding energy when it is attached to the ground state of the formed negative ion during the collision. The second view interprets the EA as corresponding to the BE of the attached electron in an excited state of the formed negative ion. Examples of the first case are the measured EAs of the Au, Pt, and the highly radioactive At atoms [2–7] as well as of the C_{60} and C_{70} fullerene molecules [8–12]. The measured EAs of Nb [13,14], Hf [15], the lanthanide atoms Eu [16,17] and Tm [18], and the actinide atoms Th [19] and U [20,21] as well as the theoretical EAs of Bk, Cf, Fm, and Lr [22–26] represent the second interpretation of the EA. Clearly, whether the measured/calculated EA of heavy multi-electron atoms and fullerene molecules is interpreted as corresponding to the binding energy (BE) of the attached electron in the ground state, the metastable state, or the excited state of the formed negative ion during the collision, the Regge pole-calculated binding energies provide rigorous and unambiguous energy values.

We demonstrate the second viewpoint of the meaning of the EA by using the Nb atom as an example. The EA of atomic Nb is both interesting and revealing because there are two measured EA values, namely 0.917 eV [13] and 0.894 eV [14] as well as two theoretical EAs 0.82 eV [27] and 0.99 eV [28]. Their interpretation notwithstanding, the agreement among these values is quite good and with the Regge pole-metastable BE value of 0.902 eV. Experimental studies of the lanthanide atoms are challenging due to the difficulty of producing sufficient negative ions for use in photodetachment experiments [16]. Problems concerning the interpretation of what is meant by the measured EAs of the lanthanide atoms have already been discussed [29,30]. For the actinide atoms, the experimental breakthrough using a nanogram of Bk and Cf [31] and the recent first ever EA measurements of the highly radioactive elements At [7], Th [19], and U [20,21] represent significant advances in the measurements of the challenging to handle atoms. In addition, more such measurements in other radioactive atoms can be expected in the near future. Of great concern and puzzle, however, is that the measured [7] and the calculated [32–35] EAs of At correspond to the ground state BE of the formed At$^-$ anion during the collision, while the measured EAs of Th and U are identified with the BEs of the attached electron in the excited states of the formed anions during a collision. Consequently, reliable theoretical predictions and guidance are essential for a fundamental understanding and interpretation of what is actually being measured.

The Regge pole method has been benchmarked on the measured EAs of atomic Au, At, and Eu as well as the C_{60} fullerene molecule through the negative ion BEs extracted from the Regge pole-calculated electron elastic TCSs. For clarity, Table 1 compares the Regge pole-calculated negative ion BEs with measured and calculated EAs of various atoms and fullerene molecules to assess the reliability of the existing measured/calculated EAs. Indeed, from the Eu atom through the end of Table 1, the meaning of the EA is ambiguous and riddled with uncertainty. This is particularly the case with the actinide atoms Bk, Cf, Fm, and Lr, the focus of this paper, and, for these atoms there are no measured EAs available. This explains our focus on them. We determine their ground, metastable, and excited state negative ion BEs from the Regge pole-calculated electron elastic scattering TCSs to understand the existing calculated EAs and assess their reliability. We then present unambiguous and reliable BEs to guide the measurements of their EAs.

Table 1. Negative ion binding energies (BEs) and ground state Ramsauer–Townsend (R-T) minima, all in eV extracted from TCSs of the atoms and the fullerene molecules C_{60} and C_{70}. They are compared with the measured electron affinities (EAs) in eV. GRS, MS-*n*, and EXT-*n* (*n* = 1, 2) refer, respectively, to ground, metastable, and excited states. Experimental EAs, EXPT, and theoretical EAs, theory is also included. The numbers in the square brackets are the references.

System Z	BEs GRS	BEs MS-1	BEs MS-2	EAs EXPT	BEs EXT-1	BEs EXT-2	R-T GRS	BEs/EAs Theory	EAs RCI [23]	EAs GW [24]
Au 79	2.26	0.832	-	2.309 [2] 2.301 [3] 2.306 [4]	0.326	-	2.24	2.50 [27] 2.19 [36] 2.313 [37] 2.263 [38]	-	-
Pt 78	2.16	1.197	-	2.128 [2] 2.125 [5] 2.123 [6]	0.136	-	2.15	2.163 [38]	-	-
At 85	2.42	0.918	0.412	2.416 [7]	0.115	0.292	2.43	2.38 [32] 2.42 [33] 2.412 [34] 2.45 [35]	-	-
C_{60}	2.66	1.86	1.23	2.684 [8] 2.666 [9] 2.689 [10]	0.203	0.378	2.67	2.57 [39] 2.63 [40] 2.663 [41]	-	-

Table 1. *Cont.*

System Z	BEs GRS	BEs MS-1	BEs MS-2	EAs EXPT	BEs EXT-1	BEs EXT-2	R-T GRS	BEs/EAs Theory	EAs RCI [23]	EAs GW [24]
C_{70}	2.70	1.77	1.27	2.676 [9] 2.72 [11] 2.74 [12]	0.230	0.384	2.72	3.35 [42] 2.83 [42]	-	-
Nb 41	2.48	0.902	-	0.917 [13] 0.894 [14]	0.356	-	2.47	0.82 [27] 0.99 [28]	-	-
Eu 63	2.63	1.08	-	0.116 [16] 1.053 [17]	0.116	-	2.62	0.117 [22] 0.116 [43]	-	-
Tm 69	3.36	1.02	-	1.029 [18]	0.016	0.274	3.35	-	-	-
Hf 72	1.68	0.525	-	0.178 [15]	0.017	0.113	1.67	0.114 [44] 0.113 [45]	-	-
Th 90	3.09	1.36	0.905	0.608 [19]	0.149	0.549	3.08	0.599 [19] 0.549 [46]	0.368	1.17
U 92	3.03	1.44	-	0.315 [20] 0.309 [21]	0.220	0.507	3.04	0.175 [47] 0.232 [21]	0.373	0.53
Bk 97	3.55	1.73	0.997	N/A	0.267	0.505	3.56	-	0.031	−0.276 −0.503
Cf 98	3.32	1.70	0.955	N/A	0.272	0.577	3.34	-	0.010 0.018	−0.777 −1.013
Fm 100	3.47	1.79	1.02	N/A	0.268	0.623	3.49	-	-	0.354 0.597
Lr 103	3.88	1.92	1.10	N/A	0.321	0.649	3.90	0.160 [25] 0.310 [25] 0.476 [26]	0.295 0.465	−0.212 −0.313

2. Method of Calculation

Understanding the structure and dynamics of low-energy electron elastic collisions with multi-electron atoms and fullerene molecules, resulting in the formation of stable negative ions, is quite challenging for conventional quantum mechanical methods. Most of the sophisticated methods developed in atomic physics were tasked with reproducing experimental results with high accuracies but did very little to unravel and elucidate the intricate details and the precise description of the nature of the different physical effects important to a particular process; they also lacked the predictive power. Expressing the desired solutions to scattering problems, as a partial wave (PW) series where the summation is over the orbital (or total) angular momentum quantum number, presents a major problem to the conventional quantum mechanical approaches. The PW series is notoriously very slowly convergent, particularly when the wavelength of the incoming particle is much smaller than the range of the scattering potential. The PW expansion may contain hundreds, if not thousands, of terms, with the result that the calculated EAs are generally riddled with uncertainties and therefore difficult to interpret.

However, if the angular momentum is allowed to become complex-valued, this slowly convergent PW series is replaced with a more rapidly converging series. This leads us to the concept of Regge poles. Simply put, Regge poles are generalized bound states, i.e., the solution of the Schrödinger equation where the energy E is real, positive and the angular momentum λ is complex. Here, the rigorous Regge pole method has been used to calculate the electron elastic TCSs. Regge poles, singularities of the S-matrix, rigorously define resonances [48,49] and in the physical sheets of the complex plane, they correspond to bound states [50]. The Regge poles formed during low-energy electron elastic scattering become stable bound states [51]. The near-threshold electron–atom/fullerene collision TCS resulting in negative ion formation as resonances are calculated independently of measurements using the Mulholland formula [52]. This formula converts the infinite discrete sum into a background integral plus the contribution from a few poles to the process under consideration. Indeed, the method requires no a priori knowledge of the experimental or any other theoretical data as inputs; hence, its predictive nature.

Electron–electron correlations and core–polarization interactions are both crucial for the existence and stability of most negative ions. The former effects are embedded in the Mulholland formula [53,54] for the TCS, while the latter interactions are incorporated through the well-investigated Thomas–Fermi type model potential. Within the CAM, λ description of scattering, Im λ is used to differentiate between the shape resonances (short-lived resonances) and the stable bound states of the negative ions (long-lived resonances) formed as Regge resonances in the electron–atom (molecule) collision. For the latter, the Im λ is several orders of magnitude smaller than that for the former. The Mulholland formula [52] used here is of the form [53,54] (atomic units are used throughout):

$$\begin{aligned}
\sigma_{tot}(\mathrm{E}) &= 4\pi k^{-2} \int_0^\infty Re[1 - S(\lambda)]\lambda d\lambda \\
&\quad -8\pi^2 k^{-2}\sum_n Im \frac{\lambda_n \rho_n}{1+\exp(-2\pi i \lambda_n)} + I(E)
\end{aligned} \tag{1}$$

In Equation (1) $S(\lambda)$ is the S-matrix, $k = \sqrt{2mE}$, with $m = 1$ being the mass and E the impact energy, ρ_n is the residue of the S-matrix at the nth pole, λ_n and $I(E)$ contains the contributions from the integrals along the imaginary λ-axis (λ is the complex angular momentum); its contribution has been demonstrated to be negligible [55]. In the Regge pole, also known as the complex angular momentum (CAM), method the important and revealing energy-dependent Regge trajectories are also calculated. Their effective use in low-energy electron scattering has been demonstrated in [55,56], for example.

As in [57], we consider the incident electron to interact with the complex heavy system without consideration of the complicated details of the electronic structure of the system itself. Thus, the robust Avdonina–Belov–Felfli potential [58], which embeds the vital core–polarization interaction is used:

$$U(r) = -\frac{Z}{r(1 + \alpha Z^{1/3}r)(1 + \beta Z^{2/3}r^2)} \tag{2}$$

In Equation (2) Z is the nuclear charge, α and β are variation parameters. For small r, the potential describes Coulomb attraction between an electron and a nucleus, $U(r) \sim -Z/r$, while at large distances it has the appropriate asymptotic behavior, *viz.* $\sim -1/(\alpha\beta r^4)$ and accounts properly for the polarization interaction at low energies. Notably, for an electron, the source of the bound states giving rise to Regge trajectories is the attractive Coulomb well it experiences near the nucleus. The addition of the centrifugal term to the well "squeezes" these states into the continuum [54,59]. For larger CAM, λ the effective potential develops a barrier. Consequently, a bound state crossing the threshold energy $E = 0$ in this region may become an excited state or a long-lived metastable state. As a result, the highest "bound state" formed during the collision is identified with the highest excited state, here labeled as EXT-1, see Table 1. As E increases from zero, the second excited state may form with the anionic BE labeled, EXT-2. For the metastable states, similar labeling is used as MS-1, MS-2, etc. However, it should be noted here that the metastable states are labeled relative to the anionic ground state. The CAM methods have the advantage that the calculations are based on a rigorous definition of resonances, viz. as singularities of the S-matrix [49,50]. It is noted here that $1/(\mathrm{Im}\ \lambda)$ also determines the angular life of a resonance [50,60].

The strength of this extensively studied potential, Equation (2) [61,62] lies in that it has five turning points and four poles connected by four cuts in the complex plane. The presence of the powers of Z as coefficients of r and r^2 in Equation (2) ensures that spherical and non-spherical atoms and fullerenes are correctly treated. Small and large systems are also appropriately treated. The effective potential $V(r) = U(r) + \lambda(\lambda + 1)/2r^2$ is considered here as a continuous function of the variables r and complex λ. The numerical calculations of the TCSs, limited to the near-threshold energy region, namely below any excitation thresholds to avoid their effects, are obtained by solving the Schrödinger equation as described in [54], see also [63]. The parameters "α" and "β" of the potential in Equation (2) are varied, and with the optimal value of $\alpha = 0.2$, the β-parameter is further varied carefully until the dramatically sharp resonance appears in the TCS. This is indicative of stable negative ion formation during the collision and the energy position matches with the measured EA of the atom/fullerene molecule; see for example the Au and C_{60} fullerene TCSs in [1]. This has been found to be the case in all the atoms and fullerenes investigated thus far.

Here, we consider the presence of a sufficiently narrow resonance, which allows the collision partners to form a long-lived intermediate complex that rotates as it decays at zero scattering angle to preserve the total angular momentum. If the complex has a large angular life, namely Im $\lambda << 1$, it will return to forward scattering many times. For the resonance to contribute to the TCS two resonance conditions must be satisfied: (1) Regge trajectory, namely Im λ versus Re λ stays close to the real axis and (2) the real part of the Regge pole is close to an integer.

3. Results

To better understand and appreciate the problem we are discussing here, we first consider the TCSs of the standard atomic Au and C_{60} fullerene molecule, given in Figure 1 of Ref. [1]. Clearly seen from the figure is that in the Au TCSs (left panel of Figure 1) there are three dramatically sharp resonances representing stable negative ion formation in the ground (2.26 eV), metastable (0.832 eV), and excited (0.326 eV) states of the formed negative ions during the collision. In the C_{60} fullerene TCSs (right panel of Figure 1) there are five dramatically sharp resonances, corresponding to the ground state BE (2.66 eV), two metastable BEs (1.86 eV and 1.23 eV), and two excited state BEs (0.378 eV and 0.203 eV). In both cases, the measured EAs of the Au atom and the C_{60} fullerene molecule correspond to the anionic BEs of the attached electron in the ground states of the formed anions during the collision, see also Table 1. Importantly, the delineation of the dramatically sharp resonances in the TCSs of both Au and C_{60} ensures the correct interpretation of what is being measured. It is noted here that both the Au and C_{60} TCSs also abound in SRs and R-T minima. Similarly, for Pt, At, and C_{70} the measured EAs correspond to the BEs of the electron when it is attached to the ground states of the formed negative ions (see Table 1 here). Indeed, this excellent agreement gives great credence to the Regge pole analysis to produce rigorous and unambiguous BEs without any assistance from either experiment or any other theory, as well as to our interpretation of the EAs of these complex systems, viz. as corresponding to the ground state BEs of the formed negative ions during the collisions.

Recall that the primary objective of this paper is to subject the measured and/or calculated EAs of the investigated atoms and fullerene molecules in this paper to the Regge pole-calculated ground, metastable, and excited state negative ion BEs of the formed anions during the collisions for unambiguous interpretation of the EAs. Here, the presented Figures 1 and 2 of the TCSs demonstrate the clear delineation of the dramatically sharp resonances leading to the unambiguous determination of the ground, metastable, and excited state BEs of the formed negative ions during the collisions. Table 1 summarizes the BEs of the various atoms and fullerenes, extracted from the TCSs of the atoms of interest here and compares them with the measured/calculated EAs.

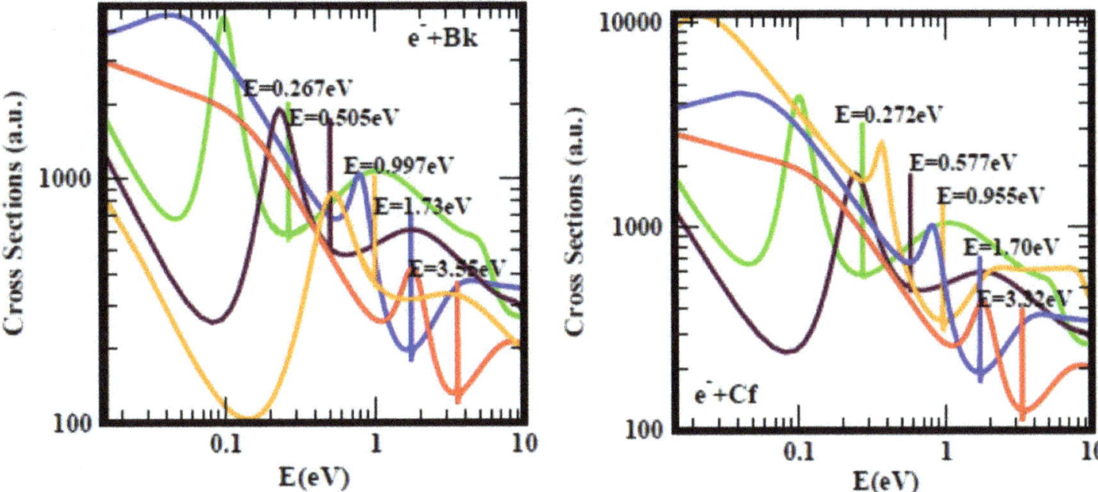

Figure 1. Total cross sections (a.u.) for electron elastic scattering from atomic Bk (**left panel**) and Cf (**right panel**) are contrasted. For both Bk and Cf the red, blue, and orange curves represent TCSs for the ground and the two metastable states, respectively, while the brown and the green curves correspond to excited state TCSs. The dramatically sharp resonances in the TCSs of both figures correspond to the Bk⁻ and Cf⁻ negative ions formed during the collisions. Importantly, the flip over of the near-threshold R-T minimum from the Bk TCSs to an SR very close to threshold in the Cf TCSs occurs here.

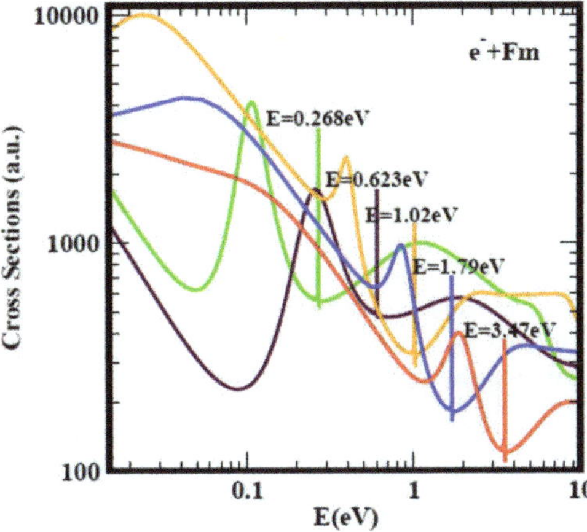

Figure 2. *Cont.*

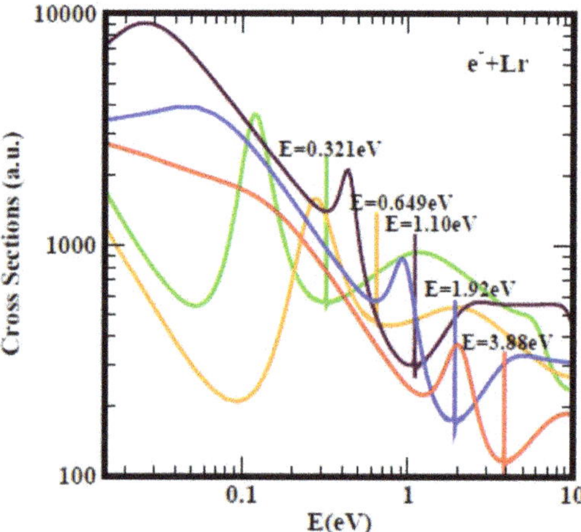

Figure 2. Total cross sections (a.u.) for the actinide atoms Fm (**top panel**) and Lr (**bottom panel**). In both panels the red curves represent the ground states. In Fm the blue and the orange curves and in Lr the blue and the brown curves are the metastable TCSs while the brown and the green curves in Fm and the orange and the green curves in Lr represent the excited state TCSs. The energy positions of the dramatically sharp lines correspond to electron BEs of the formed negative ions during collisions. The orange curve in the TCSs for Fm and the brown curve in those for Lr correspond to the polarization-induced TCSs with the deep R-T minima flipped over to SRs close to threshold (see also Figure 1 above).

The ambiguous and confusing EAs of the actinide atoms in particular, from Th through Lr have necessitated the careful evaluation/assessment of the measured and/or calculated EAs in order to determine their meaning. As seen from Table 1, it is particularly difficult to make sense of the existing theoretical EAs of the Bk, Cf, Fm, and Lr actinide atoms. Therefore, here, we demonstrate the importance and effective use of the Regge pole-calculated BEs by comparing them with the EAs of the listed atoms in Table 1, particularly the four actinide atoms above. To this end, we have grouped our discussions for convenience as follows: 3.1 Bk and Cf atoms: the interest in them is that a recent experiment probed their structure and dynamics using a nanogram matter; 3.2 Fm and Lr atoms: being at the end of the actinide series, have several calculated EAs available to compare with; 3.3 Relativistic effects in electron affinity calculations.

As seen from Table 1, for atomic Nb the measured and the calculated EAs agree very well, and with the Regge pole metastable BE of 0.902 eV, the ground and the excited state BEs are 2.48 eV and 0.356 eV, respectively. The latest measured EA (0.116 eV) [16] of Eu is in outstanding agreement with the Regge pole BE of 0.116 eV and with the MCDF-RCI EA of 0.117 eV [22]. Yet, the Regge pole value corresponds to an excited state BE of Eu. Notably, the previously measured EA of 1.053 eV [17] and the Regge pole-metastable BE of 1.08 eV for Eu agree excellently. Furthermore, the 1.029 eV measured EA of Tm [18] and the Regge pole BE of 1.02 eV also agree excellently. From Table 1, since the measured EA of Hf [15] is close to the RCI EA [44] and the Regge pole BE of an excited state [45], the measured EA of Hf is identified with the BE of an excited state of Hf contrary to the cases of the Au, Pt, and At atoms as well as of the fullerenes. Consequently, from the measured EAs of Nb, Eu, Tm, and Hf the simple question follows: Does the measured EA of these atoms correspond to the BE of the attached electron in an excited state of the formed anion during the collision?

For Th, the measured and the calculated EAs [19] 0.608 eV and 0.599 eV, respectively, are close to the Regge pole SR at 0.61 eV and the second excited state BE of 0.549 eV [46]. The measured EAs of U 0.315 eV [20] and 0.309 eV [21] with the calculated EA of 0.232 eV [21] are close to each other and to the Regge pole BE of the first excited state of the formed U⁻

anion, the MCDF-RCI EA value of 0.175 eV [47] and the 0.373 eV GW EA [24]. However, the Regge pole BE value of the second excited state, 0.507 eV agrees very well with the EA calculated by the GW method 0.53 eV [24]. Here, we are confronted with the inconsistency in the identification of what is actually being measured.

3.1. Bk and Cf Atoms

The selection of the Bk and Cf TCSs presented in Figure 1 is motivated by the rigorous probing of their electronic structure and dynamics through the Regge pole-calculated R-T minima and SRs, revealing their sensitivity [1]. Importantly, the flipping over of the deep R-T minimum in the Bk TCSs to an SR very close to the threshold occurs in the metastable TCS of the Cf atom: see the orange curves in Figure 1. The results of probing their electronic structure and dynamics by the experiment [31] as well as our validation of the observation [1] should help in the measurements of the EAs of Bk and Cf. In Table 1 we have presented our Regge pole anionic BEs for both Bk and Cf and compared them with the existing theoretical EAs: these are riddled with uncertainty and lack reliability. The same conclusion applies to the data of the remaining atoms.

3.2. Fm and Lr Atoms

There are no measured EAs for these atoms; therefore, reliable predictions of their EAs are essential. As the size of the actinide atoms approaches that of the Lr atom, their electronic structure becomes less complicated and the theoretical EAs are expected to become less uncertain. Figure 2 compares the TCSs of the Fm and Lr atoms, selected because of the availability of several theoretical EAs to compare with the Regge pole-calculated BEs. The TCSs of both atoms are characterized by well-delineated dramatically sharp resonances, representing in each atom a ground, two metastable, and two excited state BEs, see also Table 1. Worth remarking here is that in both the TCSs of Fm and Lr the sharp resonances of interest here, although well-delineated from each other, appear close to SRs; this can create problems in the identification of the BEs as seen in Figure 2.

For Fm, it is seen that the Regge pole BEs of the highest excited anionic state (0.268 eV) and the second excited anionic state (0.623 eV) agree rather well with the existing theoretical EA values of 0.354 eV [24] and 0.597 eV [24], respectively. Importantly, the reason why these theoretical EAs differ from each other is that they correspond to different anionic states: this demonstrates the need for rigorous values of the EAs for Fm and the other actinide atoms in general to guide measurements and/or calculations. We note here that the Fm TCSs still exhibit fullerene behavior [64], while in Lr the fullerene behavior has completely disappeared (see green curves in both figures of Figure 2). This almost atomic behavior of the Lr TCSs makes the Lr much easier to handle theoretically as evidenced below.

The importance of the Lr TCSs is two-fold: 1) The electronic structure of Lr is relatively simple and various sophisticated theoretical methods have calculated its EA, see Table 1. On comparing the TCSs of the Fm and Lr atoms, we see that the characteristic SRs appear very near threshold in both TCSs. Since the Lr atom is the last of the actinide series, sophisticated theoretical methods can be expected to obtain better values for the EA of Lr. Indeed, as seen in Figure 2, the TCSs are characterized by well-delineated ground (3.88 eV), two metastable (1.92 eV; 1.10 eV), and two excited (0.321 eV; 0.649 eV) state anionic BEs. Once the determination has been made regarding what anionic state BE was measured/calculated, then by comparison with the above values, an unambiguous EA determination can be obtained as was performed with the Au, Pt, At, and the fullerene molecules in Table 1. However, several calculated EAs are available; we will attempt to make sense of their meaning. Firstly, the 0.310 eV [25], 0.295 eV [23], and the Abs (−0.313 eV) [24] EAs can safely be identified with the Regge pole's highest excited state BE (0.321 eV) and secondly, the EA values of 0.465 eV [23] and 0.476 eV [26] could probably be identified with the Regge pole BE of the second excited state 0.649 eV; they could also correspond to the nearby shape resonance at 0.451 eV, see Figure 2.

Indeed, from Table 1 it is clear that for the actinide atoms, the various sophisticated theoretical methods calculate only the BEs of the formed negative ions in excited states and equate them with the EAs. There is nothing wrong with this viewpoint, except that a rigorous definition of the EA is required for consistency throughout the tabulated atoms and fullerenes in Table 1. This would then mean using the BEs in column BEs, EXT-1 for the corresponding EAs of all the atoms and fullerene molecules in Table 1. Unfortunately, this would contradict the established meaning of the EAs as found for the Au, Pt, and At atoms as well as for the C_{60} and C_{70} fullerenes. It is further noted here that for the carefully measured and calculated EA of At, various sophisticated theoretical EAs [32–35] agree excellently among themselves and with the measured EA as well as with the Regge pole ground state anionic BE and not with the metastable or the excited state anionic BEs. Unfortunately, it is a formidable task for most theoretical methods to calculate the BEs of metastable states, let alone the ground state BEs of the tabulated systems in Table 1.

These results are also important in guiding sophisticated theoretical methods on the importance of polarization interaction. This has been demonstrated unequivocally in the TCSs calculation of the Bk and Cf atoms, particularly in the At atom. When the β-parameter of Equation (2) was 0.042 we obtained the ground state anionic BE value of 2.51 eV [65], which was close to the then known theoretical EA value of 2.80 eV [66]. However, careful refinement of the β-parameter to 0.04195 yielded the ground state BE value of 2.42 eV, in excellent agreement with the measured EA of 2.416 eV [7] and the theoretical values [32–35]. Here, it is noted that the small Im (λ) decreased from 1.07×10^{-5} to 8.9×10^{-6}, indicative that the BE value of 2.42 eV corresponds to the longest-lived resonance as expected (see the importance of the use of Im (λ) in the Regge pole analysis in Ref. [55]). Indeed, a scientific will is needed to respond critically to the question: why are the EAs of Au, Pt, and At as well as those of the C_{60} and C_{70} fullerene molecules identified with the BEs of the formed negative ions in the ground states while for Th and U as well as for the other actinide atoms the EAs correspond to the Regge pole-calculated BEs of excited states? Notably, the Regge pole BEs are available to guide the EA measurements, regardless of their interpretation, namely whether they are viewed as corresponding to the BEs of electron attachment in the ground or excited states.

3.3. Relativistic Effects in Electron Affinity Calculations

The EA provides a stringent test of theoretical calculations when the calculated EAs are compared with those from reliable measurements. Unfortunately, low-energy electron interactions with heavy multi-electron atoms and fullerene molecules are characterized generally by the presence of many intricate and diverse electron configurations. These lead to computational complexity that, for a long time, made it virtually impossible for sophisticated theoretical methods to reliably predict the electron binding energies of the formed negative ions during collisions. Thus, the electron affinities calculated using many structure-based theoretical methods tend to be riddled with uncertainties making them difficult to interpret, particularly for the actinide atoms, see Table 1.

One of the most important and revealing investigations of the importance of Regge trajectories in low-energy electron collisions using the ABF potential, Equation (2), was carried out by Thylwe [56]. For the Xe atom, Regge trajectories calculated using the Dirac relativistic and non-relativistic methods were contrasted near the threshold and found to yield essentially the same Re $\lambda(E)$ when the Im $\lambda(E)$ was still very small, see Figure 2 of [56]. This implies the insignificant difference between the relativistic and non-relativistic calculations at low-energy electron scattering, which is the condition of our calculations here.

Most of the sophisticated theoretical calculations of the EAs include relativistic effects at various levels of approximations, see for example their comparisons in the calculation of the EA of At in Ref. [34]. Since many of these methods are tailored to reproduce the measurements very well, it is difficult to determine what essential physics is incorporated in the calculation of the EAs. For instance, Wesendrup et al. [36] carried out large-scale fully relativistic Dirac–Hartree–Fock and MP2 as well as nonrelativistic pseudo-potential calculations and obtained the EAs of Au as 2.19 eV and 1.17 eV, respectively. In addition, Cole and Perdew [27] employed relativistic and nonrelativistic calculations obtaining the EA of atomic Au as 2.5 eV and 1.5 eV, respectively. These EAs should be compared with the nonrelativistic Regge pole-calculated BE value of 2.263 eV [1,38] and the measured EAs presented in Table 1. Accordingly, it can be safely concluded that in the energy regime, $0.0 \leq E \leq 10.0$ eV the essential physics embedded in the Regge pole method such as electron–electron correlations and core–polarization interaction is adequate for the reliable prediction of the EAs of multi-electron atoms and fullerene molecules. Importantly, an impressive agreement with the measured EAs of Au [2–4] has been obtained by the relativistic coupled cluster calculations with variational quantum electrodynamics [37]; it also supports the Regge pole-calculated BE value.

The calculation of the EA of Nb ($Z = 41$) in [27] used a gradient-corrected exchange-correlation functional and a Nb scalar relativistic core, obtaining an EA value of 0.82 eV, which compares very well with the Regge pole metastable BE value of 0.902 eV and the measured EA values of 0.917 eV [13] and 0.894 eV [14]. These results for Nb further demonstrate the great need to ascertain precisely the state from whence the photodetachment process originates. The Eu atom, with a relatively high Z of 63, but a small, measured EA of 0.116 eV [16], provides a stringent test of the nonrelativistic Regge pole method when its BE value of 0.116 eV [43,55] is contrasted with the MCDF-RCI calculated EA value of 0.117 eV [22]. The theoretical results were calculated around 2009 while the experimental EA was measured in 2015. For the highly radioactive At atom the recently measured EA [7], which employed the coupled cluster method, agreed excellently with the Regge pole BE and the EAs from various sophisticated theoretical calculations, including the multiconfiguration Dirac–Hartree–Fock values [32–35], see Table 1 for comparisons. Furthermore, in [32,34] extensive comparisons among various sophisticated theoretical EAs have been carried out as well.

4. Summary and Conclusions

For the multi-electron atoms and the fullerene molecules considered in this paper, we presented rigorous and unambiguous ground, metastable, and excited state negative ion BEs extracted from the Regge pole-calculated TCSs. We then compared our BEs with the existing measured and/or calculated EAs as shown in Table 1. We found that for the Au, Pt, and At atoms as well as the C_{60} and C_{70} fullerene molecules, our ground state anionic BEs matched excellently with the measured EAs, implying that the measured EAs of these systems correspond to the BEs of the electron when it is attached in the ground states of the formed negative ions during the collision. However, for the lanthanide atom Eu, our excited state BE is in outstanding agreement with both the recently measured and the MCDF-RCI calculated EA values. For both the Eu and Tm atoms, good agreement between the Regge pole-metastable BEs and the previously measured EAs [17,18] has been realized. Overall, our excited state BEs are closer to the measured and/or calculated EAs of the Hf and the actinide atoms. This implies that for these atoms the measured and/or calculated EAs correspond to the BEs of the electron when it is attached in the excited states of the formed negative ions, contrary to the cases of the Au, Pt, and At atoms as well as the C_{60} and C_{70} fullerene molecules.

In conclusion, for the actinide atoms Bk, Cf, Fm, and Lr, we presented the Regge pole-calculated electron elastic TCSs demonstrating their richness in very sharp resonances representing negative ion formation in the ground, metastable, and excited states. From the positions of the dramatically sharp resonances in the TCSs, we extracted the BEs of the formed negative ions during the collisions. These BEs have been compared with the existing theoretical EAs to understand and make sense of these data since they are generally riddled with uncertainties, particularly those of the Bk and Cf atoms. There are no measured EAs for these atoms to compare with our BEs. It is hoped that the results of this paper will inspire and assist both measurements and theory in the determination of the long-overdue unambiguous and reliable EAs of the actinide atoms. It is noted that the TCSs of these actinide atoms also abound in SRs and R-T minima, making it challenging to extract the EAs. It is hoped that the actinide atoms will be subjected to similar investigations as in [32,34] for unambiguous and reliable EAs. The great strength of the Regge pole analysis is in its use of Im $\lambda(E)$ to differentiate among the ground, metastable, and excited negative ionic states, with the ground state having the smallest Im $\lambda(E)$, indicative of the longest-lived negative ionic state.

Author Contributions: Z.F. and A.Z.M. are responsible for the conceptualization, methodology, investigation, formal analysis, and writing of the original draft, as well as rewriting and editing. A.Z.M. is also responsible for securing the funding for the research. All authors have read and agreed to the published version of the manuscript.

Funding: Research was supported by the U.S. DOE, Division of Chemical Sciences, Geosciences and Biosciences, Office of Basic Energy Sciences, Office of Energy Research, Grant: DE-FG02-97ER14743. The computing facilities of the National Energy Research Scientific Computing Center, also funded by the U.S. DOE are greatly appreciated.

Institutional Review Board Statement: Not applicable.

Informed Consent Statement: Not applicable.

Data Availability Statement: Not applicable.

Conflicts of Interest: The authors declare no conflict of interest or state.

References

1. Msezane, A.Z.; Felfli, Z. Recent Progress in Low-Energy Electron Elastic-Collisions with Multi-Electron Atoms and Fullerene Molecules. *Atoms* **2022**, *10*, 79. [CrossRef]
2. Hotop, H.; Lineberger, W.C. Dye-laser photodetachment studies of Au$^-$, Pt$^-$, PtN$^-$, and Ag$^-$. *J. Chem. Phys.* **2003**, *58*, 2379–2387. [CrossRef]
3. Andersen, T.; Haugen, H.K.; Hotop, H. Binding Energies in Atomic Negative Ions: III. *J. Phys. Chem. Ref. Data* **1999**, *28*, 1511–1533. [CrossRef]
4. Zheng, W.; Li, X.; Eustis, S.; Grubisic, A.; Thomas, O.; De Clercq, H.; Bowen, K. Anion photoelectron spectroscopy of Au$^-$(H$_2$O)$_{1,2}$, Au$_2$$^-$(D$_2$O)$_{1\text{-}4}$, and AuOH$^-$. *Chem. Phys. Lett.* **2007**, *444*, 232–236. [CrossRef]
5. Gibson, D.; Davies, B.J.; Larson, D.J. The electron affinity of platinum. *J. Chem. Phys.* **1993**, *98*, 5104–5105. [CrossRef]
6. Bilodeau, R.C.; Scheer, M.; Haugen, H.K.; Brooks, R.L. Near-threshold laser spectroscopy of iridium and platinum negative ions: Electron affinities and the threshold law. *Phys. Rev. A* **1999**, *61*, 012505. [CrossRef]
7. Leimbach, D.; Karls, J.; Guo, Y.; Ahmed, R.; Ballof, J.; Bengtsson, L.; Pamies, F.B.; Borschevsky, A.; Chrysalidis, K.; Eliav, E.; et al. The electron affinity of astatine. *Nat. Commun.* **2020**, *11*, 3824. [CrossRef]
8. Huang, D.-L.; Dau, P.D.; Liu, H.T.; Wang, L.-S. High-resolution photoelectron imaging of cold C$_{60}$$^-$ anions and accurate determination of the electron affinity of C$_{60}$. *J. Chem. Phys.* **2014**, *140*, 224315. [CrossRef]
9. Brink, C.; Andersen, L.H.; Hvelplund, P.; Mathur, D.; Voldstad, J.D. Laser photodetachment of C$_{60}$$^-$ and C$_{70}$$^-$ ions cooled in a storage ring. *Chem. Phys. Lett.* **1995**, *233*, 52–56. [CrossRef]
10. Wang, X.-B.; Ding, C.F.; Wang, L.-S. High resolution photoelectron spectroscopy of C$_{60}$$^-$. *J. Chem. Phys.* **1999**, *110*, 8217–8220. [CrossRef]
11. Boltalina, O.V.; Sidorov, L.N.; Sukhanova, E.V.; Skokan, E.V. Electron affinities of higher fullerenes. *Rapid Commun. Mass Spectrom.* **1993**, *7*, 1009–1011. [CrossRef]
12. Palpant, B.; Otake, A.; Hayakawa, F.; Negishi, Y.; Lee, G.H.; Nakajima, A.; Kaya, K. Photoelectron spectroscopy of sodium-coated C$_{60}$ and C$_{70}$ cluster anions. *Phys. Rev. B.* **1999**, *60*, 4509. [CrossRef]
13. Luo, Z.; Chen, X.; Li, J.; Ning, C. Precision measurement of the electron affinity of niobium. *Phys. Rev. A* **2016**, *93*, 020501. [CrossRef]
14. Feigerle, C.S.; Corderman, R.R.; Bobashev, S.V.; Lineberger, W.C. Binding energies and structure of transition metal negative ions. *J. Chem. Phys.* **1981**, *74*, 1580–1598. [CrossRef]
15. Tang, R.; Chen, X.; Fu, X.; Wang, H.; Ning, C. Electron affinity of the hafnium atom. *Phys. Rev. A* **2018**, *98*, 020501. [CrossRef]
16. Cheng, S.-B.; Castleman, A.W. Direct experimental observation of weakly-bound character of the attached electron in europium anion. *Sci. Rep.* **2015**, *5*, 12414. [CrossRef]
17. Davis, V.T.; Thompson, J.S. An experimental investigation of the atomic europium anion. *J. Phys. B* **2004**, *37*, 1961. [CrossRef]
18. Davis, V.T.; Thompson, J.S. Measurement of the electron affinity of thulium. *Phys. Rev. A* **2001**, *65*, 010501. [CrossRef]
19. Tang, R.; Si, R.; Fei, Z.; Fu, X.; Lu, Y.; Brage, T.; Liu, H.; Chen, C.; Ning, C. Candidate for Laser Cooling of a Negative Ion: High-Resolution Photoelectron Imaging of Th$^-$. *Phys. Rev. Lett.* **2019**, *123*, 203002. [CrossRef]

20. Tang, R.; Lu, Y.; Liu, H.; Ning, C. Electron affinity of uranium and bound states of opposite parity in its anion. *Phys. Rev. A* **2021**, *103*, L050801. [CrossRef]

21. Ciborowski, S.M.; Liu, G.; Blankenhorn, M.; Harris, R.M.; Marshall, M.A.; Zhu, Z.; Bowen, K.H.; Peterson, K.A. The electron affinity of the uranium atom. *J. Chem. Phys.* **2021**, *154*, 224307. [CrossRef]

22. O'Malley, S.M.; Beck, D.R. Valence calculations of lanthanide anion binding energies: 6p attachments to 4fn6s2 thresholds. *Phys. Rev. A* **2008**, *78*, 012510. [CrossRef]

23. O'Malley, S.M.; Beck, D.R. Valence calculations of actinide anion binding energies: All bound 7p and 7s attachments. *Phys. Rev. A* **2009**, *80*, 032514. [CrossRef]

24. Guo, Y.; Whitehead, M.A. Electron affinities of alkaline-earth and actinide elements calculated with the local-spin-density-functional theory. *Phys. Rev. A* **1989**, *40*, 28. [CrossRef] [PubMed]

25. Eliav, E.; Kaldor, U.; Ishikawa, Y. Transition energies of ytterbium, lutetium, and lawrencium by the relativistic coupled-cluster method. *Phys. Rev. A* **1995**, *52*, 291. [CrossRef]

26. Borschevsky, A.; Eliav, E.; Vilkas, M.J.; Ishikawa, Y.; Kaldor, U. Transition energies of atomic lawrencium. *Eur. Phys. J. D* **2007**, *45*, 115–119. [CrossRef]

27. Cole, L.A.; Perdew, J.P. Calculated electron affinities of the elements. *Phys. Rev. A* **1982**, *25*, 1265. [CrossRef]

28. Calaminici, P.; Mejia-Olvera, R.J. Structures, Frequencies, and Energy Properties of Small Neutral, Cationic, and Anionic Niobium Clusters. *Phys. Chem. C* **2011**, *115*, 11891–11897. [CrossRef]

29. Felfli, Z.; Msezane, A.Z.J. Conundrum in Measured Electron Affinities of Complex Heavy Atoms. *J. At. Mol. Condens. Matter Nano Phys.* **2018**, *5*, 73–80. [CrossRef]

30. Felfli, Z.; Msezane, A.Z. Low-Energy Electron Elastic Total Cross Sections for Ho, Er, Tm, Yb, Lu, and Hf Atoms. *Atoms* **2020**, *8*, 17. [CrossRef]

31. Müller, A.; Deblonde, G.J.P.; Ercius, P.; Zeltmann, S.E.; Abergel, R.J.; Minor, A.M. Probing electronic structure in berkelium and californium via an electron microscopy nanosampling approach. *Nat. Commun.* **2021**, *12*, 948. [CrossRef]

32. Si, R.; Froese Fischer, C. Electron affinities of At and its homologous elements Cl, Br, I. *Phys. Rev. A* **2018**, *98*, 052504. [CrossRef]

33. Li, J.; Zhao, Z.; Andersson, M.; Zhang, X.; Chen, C. Theoretical study for the electron affinities of negative ions with the MCDHF method. *J. Phys. B* **2012**, *45*, 165004. [CrossRef]

34. Borschevsky, A.; Pašteka, L.F.; Pershina, V.; Eliav, E.; Kaldor, U. Ionization potentials and electron affinities of the superheavy elements 115–117 and their sixth-row homologues Bi, Po, and At. *Phys. Rev. A* **2015**, *91*, 020501. [CrossRef]

35. Sergentu, D.; David, G.; Montavon, G.; Maurice, R.; Galland, N. Scrutinizing "Invisible" astatine: A challenge for modern density functionals. *J. Comput. Chem.* **2016**, *37*, 1345–1354. [CrossRef]

36. Wesendrup, R.; Laerdahl, J.K.; Schwerdtfeger, P. Relativistic effects in gold chemistry. VI. Coupled cluster calculations for the isoelectronic series AuPt−, Au2, and AuHg+. *J. Chem. Phys.* **1999**, *110*, 9457–9462. [CrossRef]

37. Pašteka, L.F.; Eliav, E.; Borschevsky, A.; Kaldor, U.; Schwerdtfeger, P. Relativistic Coupled Cluster Calculations with Variational Quantum Electrodynamics Resolve the Discrepancy between Experiment and Theory Concerning the Electron Affinity and Ionization Potential of Gold. *Phys. Rev. Lett.* **2017**, *118*, 023002. [CrossRef] [PubMed]

38. Felfli, Z.; Msezane, A.Z.; Sokolovski, D. Near-threshold resonances in electron elastic scattering cross sections for Au and Pt atoms: Identification of electron affinities. *J. Phys. B* **2008**, *41*, 105201. [CrossRef]

39. Nagase, S.; Kabayashi, K. Theoretical study of the lanthanide fullerene CeC82. Comparison with ScC82, YC82 and LaC82. *Chem. Phys. Lett.* **1999**, *228*, 106–110. [CrossRef]

40. Zakrzewski, V.G.; Dolgounitcheva, O.; Ortiz, J.V. Electron propagator calculations on the ground and excited states of C60−. *J. Phys. Chem. A* **2014**, *118*, 7424–7429. [CrossRef]

41. Felfli, Z.; Msezane, A.Z. Simple method for determining fullerene negative ion formation. *Eur. Phys. J. D* **2018**, *72*, 78. [CrossRef]

42. Tiago, M.L.; Kent, P.R.C.; Hood, R.Q.; Reboredo, F. A Neutral and charged excitations in carbon fullerenes from first-principles many-body theories. *J. Chem. Phys.* **2008**, *129*, 084311. [CrossRef] [PubMed]

43. Felfli, Z.; Msezane, A.Z.; Sokolovski, D. Complex angular momentum analysis of low-energy electron elastic scattering from lanthanide atoms. *Phys. Rev. A* **2010**, *81*, 042707. [CrossRef]

44. Pan, L.; Beck, D.R. Calculations of Hf− has only one bound state, electron affinity and photodetachment partial cross sections. *J. Phys. B At. Mol. Opt. Phys.* **2010**, *43*, 025002. [CrossRef]

45. Felfli, Z.; Msezane, A.Z.; Sokolovski, D. Strong resonances in low-energy electron elastic total and differential cross sections for Hf and Lu atoms. *Phys. Rev. A* **2008**, *78*, 030703. [CrossRef]

46. Felfli, Z.; Msezane, A.Z. Negative Ion Formation in Low-Energy Electron Collisions with the Actinide Atoms Th, Pa, U, Np and Pu. *Appl. Phys. Res.* **2019**, *11*, 52. [CrossRef]

47. Dinov, K.D.; Beck, D.R. Electron affinities and hyperfine structure for U− and U I obtained from relativistic configuration-interaction calculations. *Phys. Rev. A* **1995**, *52*, 2632. [CrossRef]

48. Frautschi, S.C. *Regge Poles and S-Matrix Theory*; Benjamin: New York, NY, USA, 1963; Chapter X.

49. D'Alfaro, V.; Regge, T.E. *Potential Scattering*; North-Holland: Amsterdam, The Netherlands, 1965.

50. Omnes, R.; Froissart, M. *Mandelstam Theory and Regge Poles: An Introduction for Experimentalists*; Benjamin: New York, NY, USA, 1963; Chapter 2.

51. Hiscox, A.; Brown, B.M.; Marletta, M. On the low energy behavior of Regge poles. *J. Math. Phys.* **2010**, *51*, 102104. [CrossRef]

52. Mulholland, H.P. An asymptotic expansion for $\Sigma(2n+1)\exp(\grave{A}\sigma(n+1/2)2)$. *Proc. Camb. Philos. Soc.* **1928**, *24*, 280–289. [CrossRef]
53. Macek, J.H.; Krstic, P.S.; Ovchinnikov, S.Y. Regge Oscillations in Integral Cross Sections for Proton Impact on Atomic Hydrogen. *Phys. Rev. Lett.* **2004**, *93*, 183203. [CrossRef] [PubMed]
54. Sokolovski, D.; Felfli, Z.; Ovchinnikov, S.Y.; Macek, J.H.; Msezane, A.Z. Regge oscillations in electron-atom elastic cross sections. *Phys. Rev. A* **2007**, *76*, 012705. [CrossRef]
55. Felfli, Z.; Msezane, A.Z.; Sokolovski, D. Resonances in low-energy electron elastic cross sections for lanthanide atoms. *Phys. Rev. A* **2009**, *79*, 012714. [CrossRef]
56. Thylwe, K.W. On relativistic shifts of negative-ion resonances. *Eur. Phys. J. D* **2012**, *66*, 7. [CrossRef]
57. Dolmatov, V.K.; Amusia, M.Y.; Chernysheva, L.V. Electron elastic scattering off A@C60: The role of atomic polarization under confinement. *Phys. Rev. A* **2017**, *95*, 012709. [CrossRef]
58. Felfli, Z.; Belov, S.; Avdonina, N.B.; Marletta, M.; Msezane, A.Z.; Naboko, S.N. Regge poles trajectories for nonsingular potentials: The Thomas-Fermi Potentials. In Proceedings of the Third International Workshop on Contemporary Problems in Mathematical Physics, Cotonou, Republic of Benin, 1–7 November 2003; Govaerts, J., Hounkonnou, M.N., Msezane, A.Z., Eds.; World Scientific: Singapore, 2004; pp. 217–232.
59. Sokolovski, D.; Msezane, A.Z.; Felfli, Z.; Ovchinnikov, S.Y.; Macek, J.H. What can one do with Regge poles? *Nucl. Instrum. Methods Phys. Res. Sect. B* **2007**, *261*, 133–137. [CrossRef]
60. Connor, J.N.L. New theoretical methods for molecular collisions: The complex angular-momentum approach. *J. Chem. Soc. Faraday Trans.* **1990**, *86*, 1627–1640. [CrossRef]
61. Belov, S.; Thylwe, K.-E.; Marletta, M.; Msezane, A.Z.; Naboko, S.N. On Regge pole trajectories for a rational function approximation of Thomas–Fermi potentials. *J. Phys. A* **2010**, *43*, 365301. [CrossRef]
62. Belov, S.; Avdonina, N.B.; Marletta, M.; Msezane, A.Z.; Naboko, S.N. Semiclassical approach to Regge poles trajectories calculations for nonsingular potentials: Thomas–Fermi type. *J. Phys. A* **2004**, *37*, 6943. [CrossRef]
63. Burke, P.G.; Tate, C. A program for calculating regge trajectories in potential scattering. *Comput. Phys. Commun.* **1969**, *1*, 97–105. [CrossRef]
64. Msezane, A.Z.; Felfli, Z. New insights in low-energy electron-fullerene interactions. *Chem. Phys.* **2018**, *503*, 50–55. [CrossRef]
65. Felfli, Z.; Msezane, A.Z.; Sokolovski, D. Slow electron elastic scattering cross sections for In, Tl, Ga and At atoms. *J. Phys. B* **2012**, *45*, 045201. [CrossRef]
66. Zollweg, R.J. Electron Affinities of the Heavy Elements. *J. Chem. Phys.* **1969**, *50*, 4251–4261. [CrossRef]

Article

Manifestations of Rabi Dynamics in the Photoelectron Energy Spectra at Resonant Two-Photon Ionization of Atom by Intense Short Laser Pulses

Nenad S. Simonović *, Duška B. Popović and Andrej Bunjac

Institute of Physics, University of Belgrade, Pregrevica 118, 11080 Belgrade, Serbia
* Correspondence: simonovic@ipb.ac.rs

Abstract: We study the Rabi flopping of the population between the ground and excited 2p states of the hydrogen atom, induced by intense short laser pulses of different shapes and of carrier frequency $\omega = 0.375$ a.u. which resonantly couples the two states, and manifestations of this dynamics in the energy spectra of photoelectrons produced in the subsequent ionization of the atom from the excited state. It is found that, for Gaussian, half-Gaussian and rectangular pulses, characterized by the same pulse area, the final populations take the same values and the spectra consist of similar patterns having the same number of peaks and approximately the same separation between the prominent edge (Autler–Townes) peaks. The additional analysis in terms of dressed states showed that the mechanism of formation of multiple-peak structures during the photoionization process is the same regardless of the pulse shape. These facts disprove the hypothesis proposed in earlier studies with Gaussian pulse, that the multiple-peak pattern appears due to dynamic interference of the photoelectrons emitted with a time delay at the rising and falling sides of the pulse, since the hypothesis is not applicable to either a half-Gaussian pulse that has no rising part or a rectangular pulse whose intensity is constant.

Keywords: Rabi dynamics; laser pulse; photoionization; photoelectron energy spectrum; Autler–Townes splitting; multiple-peak pattern; dressed states; dynamic interference

Citation: Simonović, N.S.; Popović, D.B.; Bunjac, A. Manifestations of Rabi Dynamics in the Photoelectron Energy Spectra at Resonant Two-Photon Ionization of Atom by Intense Short Laser Pulses. *Atoms* **2023**, *11*, 20. https://doi.org/10.3390/atoms11020020

Academic Editors: Himadri S. Chakraborty and Hari R. Varma

Received: 12 December 2022
Revised: 16 January 2023
Accepted: 18 January 2023
Published: 23 January 2023

1. Introduction

If an atom, initially being in its ground state, interacts with an alternating field that resonantly couples this state to an excited state, the population will be periodically transferred from one state to another. This effect was first described theoretically by Rabi, who applied it for fermions in rotating magnetic fields [1]. In general, the flopping of the population can be explained by the fact that the eigenstates of the Hamiltonian describing the bare atom are no longer stationary states if the atom interacts with the field. Another consequence of this fact is the splitting of the coupled atomic states into doublets of "dressed states", whose quasi-energies are separated by the value corresponding to the frequency of Rabi flopping (see, e.g., Ref. [2]). This splitting can be observed in the photoabsorption and photoionization spectra of atoms and molecules. Before the availability of coherent light sources, it was first detected using radiation from the radio frequency domain. In the original observation by Autler and Townes [3], a radio frequency source tuned to the separation between two doublet microwave absorption lines of the OCS molecule was used.

Despite theoretical predictions to observe Rabi dynamics at short wavelengths [4,5] and the availability of intense XUV light sources for more than a decade, direct observation of Rabi dynamics at such short wavelengths has been reported only recently [6]. In the actual experiment, applying intense XUV laser pulses from a free-electron laser with high temporal and spatial coherence, one-photon Rabi oscillations are induced between the ground state and an excited state in helium atoms (pump). Then, a second (probe) photon from the same pulse ionizes the atom from the excited state (resonant two-photon ionization) or, at higher intensities, two photons can do it from the ground state (nonresonant two-photon ionization). In both cases, the emitted photoelectrons coherently probe the underlying dynamics and the measured signal reveals an Autler–Townes (AT) doublet.

In the above experiment, the AT doublets were built at the resonant two-photon ionization for 1.5 completed Rabi cycles. However, theoretical analysis of the resonant multiphoton ionization for more than two completed Rabi cycles during the pulse predicts the appearance of a multiple-peak pattern in the photoelectron energy spectrum (PES) [7–10]. The number of peaks appearing in the pattern is essentially determined by the pulse area [7]. The area theorem (see Ref. [11] and references therein) actually, relates this quantity to the number of Rabi cycles during the pulse, but numerical calculations have shown that this number, the number of peaks in the radial density of photoelectrons and the number of peaks in the pattern coincide [9]. The coincidence between the first two numbers is easily explained by the propagation of the emitted bunches of photoelectrons, which are separated in time and, thus, separated in space, too. On the other hand, the explanation for the multiple-peak pattern in the spectrum is still under consideration. There is a general agreement that this pattern is a result of the superposition of the contributions of photoelectrons ejected via two dressed states during the pulse action. The situation is simplest in the case of photoionization by a rectangular pulse, where the two contributions have the forms of cardinal sine (sinc) functions of energy, shifted by the value of the corresponding Rabi frequency, and the multiple-peak pattern is a result of their overlap [7] (see also Section 3.3). Conversely, in the case of smooth pulses such as the Gaussian, it is not clear exactly what is happening. The analysis performed within the stationary phase approximation suggested that dynamic interference of the photoelectrons emitted with the same energy, but with a time delay at the rising and falling sides of the pulse, essentially determines the multiple-peak structure (modulations) in the PES [8,12,13]. However, this assumption has been questioned by analyzing the conditions for dynamic interference [14,15], where it was found that they are not always fulfilled, particularly in the case of photoionization from the hydrogen ground state.

To shed more light on the above issue, in this paper we investigate manifestations of Rabi dynamics in the photoelectron energy spectra calculated for resonant two-photon ionization of the hydrogen atom by intense short laser pulses of three different forms—Gaussian, half-Gaussian and rectangular ones. By choosing the carrier frequency of 0.375 a.u. that resonantly couples the hydrogen ground (1s) and excited 2p states, the pulse induces one-photon Rabi oscillations between these states, and a second photon from the same pulse subsequently ionizes the atom from the 2p state. The problem was previously studied by other authors, who also used different forms of the laser pulse (see Refs. [4,5,7–9]), but conditions for the dynamic interference were not considered. The paper is organized in the following way. In the next section, we briefly describe the computational method for calculating the populations of atomic states and the photoelectron energy spectra, based on the three-level model, and present results for resonant two-photon ionization of hydrogen by intense short laser pulses. In Section 3 we analyze the Rabi dynamics and the AT patterns in the spectra in terms of dressed states. A summary and conclusions are given in Section 4.

2. Calculation of Populations of Atomic States and Photoelectron Energy Spectra

The populations of atomic states during the interaction of the atom with the laser pulse, including their final values when the pulse has expired, and the photoelectron energy spectra were obtained by solving the time-dependent Schrödinger equation (in atomic units)

$$i\frac{\mathrm{d}}{\mathrm{d}t}|\psi(t)\rangle = (H_0 + z\mathcal{E}(t))|\psi(t)\rangle. \tag{1}$$

Here $|\psi(t)\rangle$ is the non-stationary atomic state at time t, H_0 is the Hamiltonian of the field-free (bare) atom, $\mathcal{E}(t)$ is the electric field component of the laser pulse and z is the projection of the electron–nucleus distance in the field direction. The term $z\mathcal{E}(t)$ describes the atom-field interaction in the dipole approximation using the length gauge. We consider a linearly polarized laser pulse, whose electric field component reads

$$\mathcal{E}(t) = \mathcal{E}_0 g(t) \cos \omega t, \tag{2}$$

where $\mathcal{E}_0$ is the peak value of the field strength, ω is the laser carrier frequency, and the function $g(t)$ determines the shape of the pulse envelope.

Below, we solve Equation (1), assuming that the atom is initially in its ground state, i.e., $|\psi(t_0)\rangle = |1s\rangle$, where t_0 is a time before the beginning of the interaction. Since the atom interacting with the field (2) has axial symmetry, the z-projection of the electron angular momentum l_z is a constant of motion and the magnetic quantum number m is a good quantum number for any field strength. Thus, the state $|\psi(t)\rangle$ is at any time t characterized by the value $m = 0$, which characterizes the ground state of the bare atom. Unless otherwise stated, atomic units (a.u.) are used throughout the paper.

2.1. The Three-Level Model

In the case of photoionization which goes via resonant or near-resonant excitation of an intermediate state, which here is 2p, the other excited states are nonessential and at weak fields the process can be adequately described within the three-level model. A computational method for solving Equation (1) within this model is presented in our recent paper [16], and in more detail in Ref. [8]. Here, we give only the basic expressions and the final set of relevant equations.

The atomic state at time t within the three-level model reads

$$|\psi(t)\rangle = C_{1s}(t)|1s\rangle + C_{2p}(t)e^{-i\omega t}|2p\rangle + e^{-2i\omega t}\int [C_{\varepsilon s}(t)|\varepsilon s\rangle + C_{\varepsilon d}(t)|\varepsilon d\rangle]d\varepsilon, \tag{3}$$

where $C_{1s}(t)$, $C_{2p}(t)$ and $C_{\varepsilon l}(t)$ are the time-dependent amplitudes for the population of the ground state $|1s\rangle$, intermediate state $|2p\rangle$ and continuum states $|\varepsilon l\rangle$ ($l = 0, 2$), respectively. The variables ε and l label the kinetic energy and orbital momentum of produced photoelectrons. The states $|2p\rangle$ and $|\varepsilon l\rangle$ have been multiplied with the phase factors $e^{-i\omega t}$ and $e^{-2i\omega t}$ in order to simplify the set of equations for the amplitudes.

If we set the ground state energy E_1 to zero, by inserting Equation (3) in the Schrödinger Equation (1) and applying the rotating wave approximation [2] and the local approximation [8,17], we obtain the set of equations for the amplitudes

$$i\dot{C}_{1s} = \frac{1}{2}\Omega_0^* g(t) C_{2p}(t),$$

$$i\dot{C}_{2p} = \frac{1}{2}\Omega_0 g(t) C_{1s}(t) + \left[E_2 - \frac{i}{2}\Gamma g^2(t) - \omega\right] C_{2p}(t), \tag{4}$$

$$i\dot{\tilde{C}}_\varepsilon = \frac{1}{2}\mathcal{E}_0 g(t) C_{2p}(t) + (\varepsilon - \varepsilon_0)\tilde{C}_\varepsilon(t),$$

where $\Omega_0 = D\mathcal{E}_0$ is the frequency of Rabi flopping between the populations of states 1s and 2p at the peak value of laser intensity, $\Gamma = 2\pi|d_{\varepsilon_0}\mathcal{E}_0/2|^2$ is the ionization rate of the intermediate (near-)resonant state 2p and $\tilde{C}_\varepsilon(t) = C_{\varepsilon s}(t)/d_{\varepsilon s} \equiv C_{\varepsilon d}(t)/d_{\varepsilon d}$ is the scaled amplitude for the population of continuum states. Here, $D = \langle 2p|z|1s\rangle$ and $d_{\varepsilon l} = \langle \varepsilon l|z|2p\rangle$ are the dipole transition matrix elements for the excitation of the 2p state and for its subsequent ionization, respectively, and $|d_\varepsilon|^2 = |d_{\varepsilon s}|^2 + |d_{\varepsilon d}|^2$. For a given carrier frequency of the laser pulse ω, the expected energy of photoelectrons is $\varepsilon_0 = 2\omega - I_p$, where $I_p = 0.5\,\text{a.u.} = 13.606\,\text{eV}$ is the ionization potential of the hydrogen atom. Note that, by taking $E_1 = 0$, the energies of the 2p and final continuum states are $E_2 = 0.375\,\text{a.u.} =$

10.204 eV and $I_p + \varepsilon$, respectively. Finally, let us state that the formal solution of the third of Equation (4) is

$$\tilde{C}_\varepsilon(t) = -\frac{i}{2}\mathcal{E}_0 \int_{-\infty}^{t} e^{-i(\varepsilon-\varepsilon_0)(t-t')} g(t') C_{2p}(t')\,dt'.\tag{5}$$

The quantities $|C_{1s}(t)|^2$ and $|C_{2p}(t)|^2$ can be interpreted, respectively, as the populations of atomic states $|1s\rangle$ and $|2p\rangle$ after the interaction of the atom with the laser field until time t. Thus, the populations of these states, after time t_{ex} when we assume that the laser pulse has expired, are $|C_{1s}(t_{\mathrm{ex}})|^2$ and $|C_{2p}(t_{\mathrm{ex}})|^2$. Analogously, the quantities $|C_{\varepsilon l}(t)|^2$ and $|C_{\varepsilon l}(t_{\mathrm{ex}})|^2$ represent the probability densities of finding the atomic electron in the continuum state $|\varepsilon l\rangle$ (here $l = 0, 2$) after the interaction of the atom with the laser field until time t and after the pulse has expired, respectively. Since the photoelectron yield at a given energy ε is proportional to the total probability density of finding the electron in continuum states corresponding to this energy, the PES is adequately represented by the distribution

$$w(\varepsilon) = |C_{\varepsilon s}(t_{\mathrm{ex}})|^2 + |C_{\varepsilon d}(t_{\mathrm{ex}})|^2 = |d_\varepsilon|^2 |\tilde{C}_\varepsilon(t_{\mathrm{ex}})|^2.\tag{6}$$

The values of the dipole matrix elements for transitions from the 1s to the 2p state and from the 2p state to continuum states are determined applying expressions given in Appendix A in Ref. [16]. The matrix element for the transition 1s $\to$ 2p is $D = 0.7449$ a.u., while the values of $|d_\varepsilon|^2$ are shown in Figure 2 in the same reference. The resonant excitation of the 2p state and the subsequent ionization occurs if the laser carrier frequency is $\omega = 0.375$ a.u., which coincides with the transition frequency between the 1s and 2p states (in the weak field limit). The photon energy corresponding to this frequency is 10.204 eV, and the expected kinetic energy of the ejected electrons is $\varepsilon_0 = 0.25$ a.u. $= 6.803$ eV. In this case, one has $|d_{\varepsilon_0}|^2 = 0.1663$ a.u. [16]. We will see later that the approximate results obtained using the three-level model, in which the exact values for $|d_\varepsilon|^2$ are replaced by the value of $|d_{\varepsilon_0}|^2$, as used in previous studies [8], are sufficient for a qualitative analysis of spectra.

2.2. Results

The populations of atomic states and the photoelectron energy spectra of the hydrogen atom exposed to the laser pulse of carrier frequency $\omega = 0.375$ a.u. have been calculated using the described method for three pulse shapes: (a) the Gaussian shape

$$g(t) = e^{-t^2/\tau^2}\tag{7}$$

with $\tau = 30$ fs, (b) the half-Gaussian shape

$$g(t) = \begin{cases} 0 & \text{for } t < 0, \\ e^{-t^2/\tau'^2} & \text{for } t > 0, \end{cases}\tag{8}$$

with $\tau' = 2\tau = 60$ fs, and (c) the rectangular shape

$$g(t) = \begin{cases} 1 & \text{for } |t| < \tau'', \\ 0 & \text{for } |t| > \tau'', \end{cases}\tag{9}$$

with $\tau'' = \tau\sqrt{\pi}/2 = 26.5868$ fs. The parameters τ, τ' and τ'' are chosen so that for a given value of $\mathcal{E}_0$ all three pulses have the same value of the pulse area [7]

$$\theta = \Omega_0 \int_{-\infty}^{+\infty} g(t)\,dt,\tag{10}$$

which here is $\theta = \sqrt{\pi}\,\Omega_0\tau = \sqrt{\pi}\,\Omega_0\tau'/2 = 2\,\Omega_0\tau''$. For times when the pulses (7)–(9) expire, we take $t_{\mathrm{ex}} = 3\tau$, $t'_{\mathrm{ex}} = 3\tau'$ and $t''_{\mathrm{ex}} = \tau''$, respectively. Let us state at this point that, referring to the area theorem [7,11], the number of Rabi cycles completed during the pulse is $N = \theta/(2\pi)$.

Figure 1 shows the evolution of the populations of states 1s and 2p, calculated for pulses of the above three shapes and peak intensity $I_0 = 1\,\mathrm{TW/cm^2}$ ($I_0 = \mathcal{E}_0^2/(8\pi\alpha)$, $\alpha = 1/137$) for which the pulse area is $\theta = 8.741$ and $N \approx 1.4$. One can see that, although the evolution is different, in accordance with the area theorem [11] the final populations for all three pulses (after they have expired) take the same values.

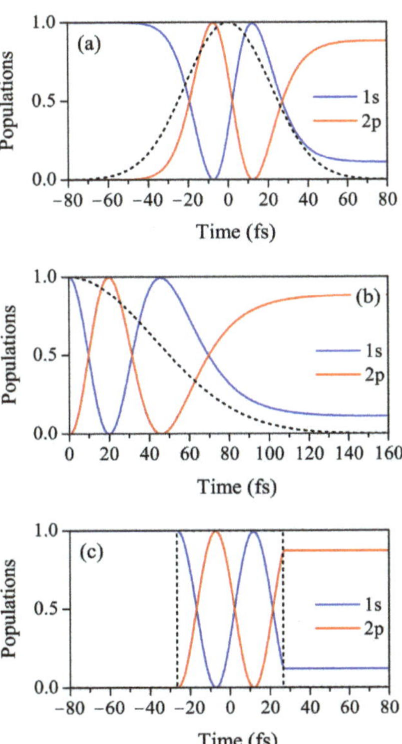

Figure 1. (Color online) The evolution of populations of the ground state (1s) and the excited 2p state during the process of resonant two-photon ionization of hydrogen by: (**a**) Gaussian laser pulse (7) with $\tau = 30\,\mathrm{fs}$, (**b**) half-Gaussian pulse (8) with $\tau' = 60\,\mathrm{fs}$ and (**c**) rectangular pulse (9) with $\tau'' = 26.5868\,\mathrm{fs}$, all of carrier frequency $\omega = 0.375\,\mathrm{a.u.} = 10.203\,\mathrm{eV}$, which is resonant for transition $1s \rightarrow 2p$ and peak intensity of $1\,\mathrm{TW/cm^2}$. The dashed lines represent the envelopes of the laser pulses. The parameters τ, τ' and τ'' are chosen so that all three pulses have the same value of the pulse area $\theta = 8.741$, for which the populations perform approximately 1.4 Rabi cycles.

Figure 2 shows the final populations of the states 1s and 2p as functions of I_0 in the domain of 10^9–$10^{13}\,\mathrm{W/cm^2}$. Again, in agreement with the area theorem, for each peak intensity the final populations of atomic states for the considered three pulses have the same values. Due to this fact, the blue and red lines in Figure 2 represent the populations of the ground and excited states, respectively, obtained for all three pulse shapes. The vertical dashed lines indicate the peak intensities at which an integer number of Rabi cycles during the pulse is completed: $I_0(N) = 0.517, 2.067, 4.650, 8.267, 12.917\,\mathrm{TW/cm^2}$ for $N = 1, 2, 3, 4, 5$, respectively.

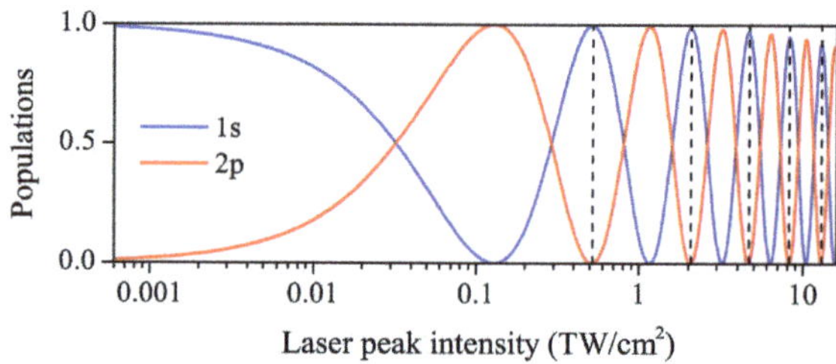

Figure 2. (Color online) Final populations of the ground state (1s) and the excited 2p state of hydrogen at the end of the process of its resonant two-photon ionization, as functions of the laser peak intensity. The results obtained using the Gaussian, half-Gaussian and rectangular pulses (Equations (7)–(9)) of carrier frequency $\omega = 0.375$ a.u. with $\tau = 30$ fs, $\tau' = 60$ fs and $\tau'' = 26.5868$ fs practically coincide and they are represented by common lines. The vertical dashed lines indicate the peak intensities at which an integer number of Rabi cycles during the pulse is completed.

Figure 3 shows the photoelectron energy spectra determined by solving the set of Equation (4) and applying Equation (6) with exact $|d_\varepsilon|^2$ values (solid red lines) and with $|d_\varepsilon|^2 \approx |d_{\varepsilon_0}|^2$ (dashed lines) for: (a) Gaussian pulse (7), (b) half-Gaussian pulse (8) and (c) rectangular pulse (9), with the peak intensities $I_0(N)$, $N = 1, \ldots, 5$. Note that the spectra obtained using the approximate value $|d_{\varepsilon_0}|^2$ are symmetric, but this is not the case when the exact values for $|d_\varepsilon|^2$ are used. The observed asymmetry, which is more pronounced at higher laser field intensities, has recently been studied in several publications [9,16,18]. For each value of I_0, the PES consist of a pattern exhibiting the AT splitting. The separation between the most prominent edge peaks (AT doublet) increases with the square root of I_0, i.e., linearly with the peak value of electric field strength. In addition, for the laser peak intensities when more than two Rabi cycles during the pulse are completed, our results confirm the previously reported appearance of a multiple-peak pattern in the calculated PES [7–10].

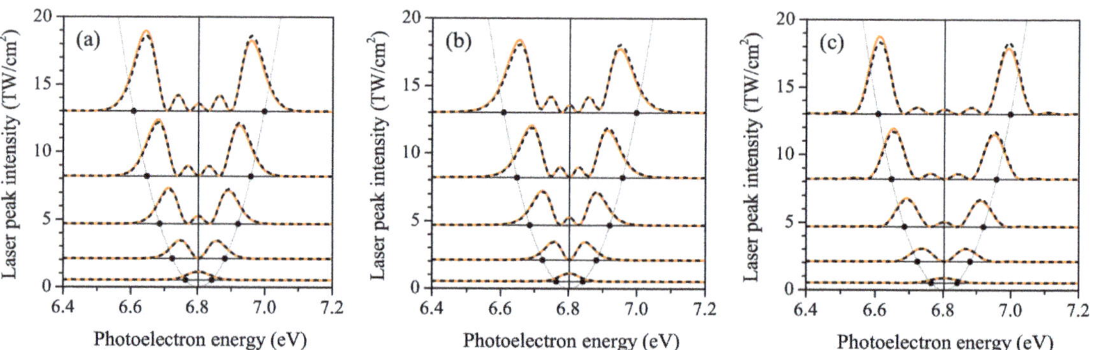

Figure 3. (Color online) Photoelectron energy spectra calculated using the three-level model (Equations (4)–(6)) with exact values of $|d_\varepsilon|^2$ (solid orange lines) and with the approximation $|d_\varepsilon|^2 \approx |d_{\varepsilon_0}|^2$ (dashed lines) for: (**a**) Gaussian pulse (7), (**b**) half-Gaussian pulse (8) and (**c**) rectangular pulse (9), all of carrier frequency $\omega = 0.375$ a.u. and the peak intensities marked in Figure 2 by vertical dashed lines. Black dots mark the real parts of $E_\pm(0) + \varepsilon_0$, whose separation ($\approx \Omega_0$) estimates the splitting of the resonant peak.

Demekhin and Cederbaum [8] analyzed the multiple-peak patterns in the PES obtained for the photoionization with the Gaussian pulse. They attributed the appearance of modulations inside the AT doublets to the dynamic interference of two photoelectron waves with the same kinetic energy emitted at two different times during the pulse—at the time when the pulse is growing and at the time when it decreases. Our calculations, however, show that similar modulations also exist in the case of photoionization with the half-Gaussian pulse, that has no growing part, as well as at the photoionization with the rectangular pulse, whose intensity is constant. Thus, we conclude that the dynamic interference cannot be the principal reason for the modulations in the calculated spectra. This conclusion is supported by the analysis of the conditions for dynamic interference [14,15], where it was found that they are not fulfilled in the case of resonant photoionization via the 2p state (see Figure 3c in Ref. [14]).

3. Analysis of the AT Patterns in Terms of Dressed States

3.1. Dynamics of the Ground and Intermediate States

As the amplitude $\tilde{C}_\varepsilon(t)$ does not appear in the first two of Equation (4), the dynamics of the 1s and 2p states within the three-level model is formally decoupled from the dynamics of continuum states. The equations for these two states can be written in the matrix form

$$i\frac{\mathrm{d}}{\mathrm{d}t}\begin{pmatrix} C_{1s}(t) \\ C_{2p}(t) \end{pmatrix} = \begin{pmatrix} 0 & \frac{1}{2}\Omega_0^* g(t) \\ \frac{1}{2}\Omega_0 g(t) & -\frac{i}{2}\Gamma g^2(t) \end{pmatrix}\begin{pmatrix} C_{1s}(t) \\ C_{2p}(t) \end{pmatrix}. \tag{11}$$

This matrix equation represents the time-dependent Schrödinger equation that describes the resonantly coupled dynamics of the 1s and 2p states in the basis of the same states in the interaction picture [19]. Using Dirac's formalism, this equation reads

$$i\frac{\mathrm{d}}{\mathrm{d}t}|\psi_b(t)\rangle = \mathcal{H}|\psi_b(t)\rangle, \tag{12}$$

where $|\psi_b(t)\rangle = e^{iH_0 t}\left[C_{1s}(t)|1s\rangle + C_{2p}(t)e^{-i\omega t}|2p\rangle\right] = C_{1s}(t)|1s\rangle + C_{2p}(t)|2p\rangle$ is the bound part of the state (3) in the interaction picture and $\mathcal{H}$ is the interaction Hamiltonian, whose representations in the actual basis are

$$|\psi_b(t)\rangle \rightarrow \begin{pmatrix} C_{1s}(t) \\ C_{2p}(t) \end{pmatrix}, \tag{13}$$

$$\mathcal{H} \rightarrow \begin{pmatrix} 0 & \frac{1}{2}\Omega_0^* g(t) \\ \frac{1}{2}\Omega_0 g(t) & -\frac{i}{2}\Gamma g^2(t) \end{pmatrix}. \tag{14}$$

Since the interaction picture hides the time dependence related to the unperturbed Hamiltonian H_0, the amplitudes $C_{1s}(t)$, $C_{2p}(t)$ and the Hamiltonian $\mathcal{H}$ are slowly varying quantities. By diagonalizing this Hamiltonian, one obtains two slowly varying complex eigenenergies (quasi-energies)

$$E_\pm(t) = \pm\frac{1}{2}\sqrt{\Omega_0^2 g^2(t) - \Gamma^2 g^4(t)/4} - \frac{i}{4}\Gamma g^2(t) \approx \pm\frac{1}{2}\Omega_0 g(t) - \frac{i}{4}\Gamma g^2(t), \tag{15}$$

which correspond to dressed states

$$|\pm\rangle \approx \frac{1}{\sqrt{2}}\left(|1s\rangle \pm |2p\rangle\right). \tag{16}$$

The approximate expressions are applicable if $\Omega_0 \gg \Gamma g(t)$, which is fulfilled if the pulses are not of excessive intensity. Using inverse relations $|1s\rangle = (|+\rangle + |-\rangle)/\sqrt{2}$, $|2p\rangle = (|+\rangle - |-\rangle)/\sqrt{2}$, the state $|\psi_b(t)\rangle$ can be written in the form

$$|\psi_b(t)\rangle = C_+(t)|+\rangle + C_-(t)|-\rangle, \tag{17}$$

where

$$C_\pm(t) = \frac{1}{\sqrt{2}}\left[C_{1s}(t) \pm C_{2p}(t)\right]. \tag{18}$$

Note that, due to the presence of an imaginary part in $E_\pm$, the dressed states $|\pm\rangle$ are decaying, i.e., they are two decoupled resonances. The real parts of the quasi-energies move adiabatically apart as the pulse arrives, and towards each other as the pulse expires. More precisely, according to Equation (15), their distance evolves as $E_+(t) - E_-(t) \approx \Omega_0 g(t)$.

3.2. Dynamics of Continuum States

By inserting Equation (17) into Equation (12) and applying the eigenvalue problem $\mathcal{H}|\pm\rangle = E_\pm|\pm\rangle$, one obtains equation $i\dot{C}_\pm = E_\pm(t)C_\pm(t)$, which can be solved analytically. Employing the initial conditions $C_\pm(-\infty) = 1/\sqrt{2}$, we find

$$C_\pm(t) = \frac{1}{\sqrt{2}} e^{-i\int_{-\infty}^{t} E_\pm(t')\,dt'} = \frac{1}{\sqrt{2}} e^{\mp i\Omega_0 \mathcal{J}_1(t)/2} e^{-\Gamma \mathcal{J}_2(t)/4}, \tag{19}$$

where $\mathcal{J}_n(t) = \int_{-\infty}^{t} g^n(t')\,dt'$.

From Equation (18), it follows that $C_{2p}(t) = [C_+(t) - C_-(t)]/\sqrt{2}$, which by substitution in Equation (5) gives

$$\tilde{C}_\varepsilon(t) = -\frac{i}{2\sqrt{2}}\mathcal{E}_0 \int_{-\infty}^{t} e^{-i(\varepsilon-\varepsilon_0)(t-t')} g(t')\,[C_+(t') - C_-(t')]\,dt'. \tag{20}$$

Finally, using Equation (6), we obtain

$$w(\varepsilon) = \frac{|d_\varepsilon|^2 \mathcal{E}_0^2}{8} \left| \int_{-\infty}^{+\infty} e^{i(\varepsilon-\varepsilon_0)t} g(t)\,[C_+(t) - C_-(t)]\,dt \right|^2$$

$$= \left| \frac{d_\varepsilon \mathcal{E}_0}{4} \int_{-\infty}^{+\infty} g(t)\,e^{-\Gamma \mathcal{J}_2(t)/4} \left[e^{i\phi_+(t)} - e^{i\phi_-(t)} \right] dt \right|^2, \tag{21}$$

where $\phi_\pm(t) = (\varepsilon - \varepsilon_0)t \mp \Omega_0 \mathcal{J}_1(t)/2$ are the phases of two oscillatory functions in the integrand. This formula gives exactly the same results as Equations (4)–(6), some of them shown in Figure 3, but it provides a deeper insight into the multiple peak structure of the PES.

The AT splitting of the resonant peak in the PES can be roughly estimated from the maximum distance between quasi-energies (15)

$$\Delta_{AT} \equiv \varepsilon_+ - \varepsilon_- \sim E_+(0) - E_-(0) \approx \Omega_0 g_0, \tag{22}$$

where $\varepsilon_\pm$ are the positions of the AT doublet peaks in the PES and $g_0 \equiv g(0)$ is the maximum value of the envelope $g(t)$ (usually $g_0 = 1$).

Figure 4 shows the time evolution of the photoelectron energy distribution, represented by $|\tilde{C}_\varepsilon(t)|^2$ using Equation (20), during the photoionization process of the hydrogen atom by: (a) Gaussian pulse (7), (b) half-Gaussian pulse (8) and (c) rectangular pulse (9), all of them having the carrier frequency of 0.375 a.u. and the peak intensity of 12.917 TW/cm^2, which leads to five Rabi cycles completed at the end of the pulse ($N = 5$). In all three cases, the number of Rabi cycles performed up to a given instant of time coincides with the number of peaks in the energy distribution at that instant. Thus, the mechanism of formation of a structure with multiple peaks is the same regardless of the shape of the pulse and, therefore, it cannot be dynamic interference of two photoelectron waves emitted during the rising and falling part of the pulse. The appearance of a multiple-peak pattern in the case of rectangular pulse, where an analytical solution is possible, is analyzed in the next subsection.

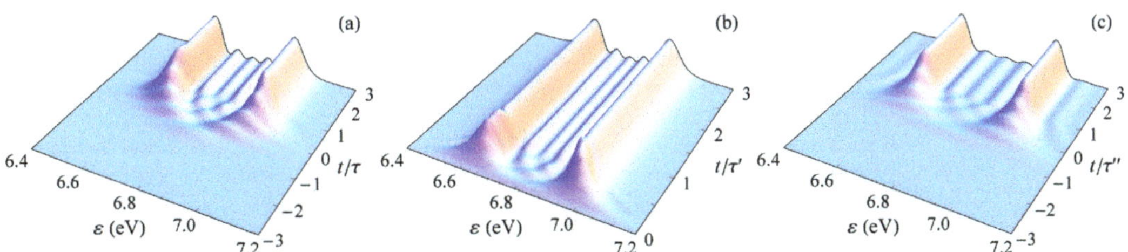

Figure 4. (Color online) Time evolution of the photoelectron energy distribution (in arbitrary units) during the photoionization process of the hydrogen atom by: (**a**) Gaussian pulse (7), (**b**) half-Gaussian pulse (8) and (**c**) rectangular pulse (9) of carrier frequency $\omega = 0.375$ a.u. and peak intensity of 12.917 TW/cm^2 at which the atom completes five Rabi cycles during the pulse.

3.3. Analytical Solution for Rectangular Pulse

The limits of the integral in Equation (21) for the rectangular pulse (9) are reduced to interval $[-\tau'', +\tau'']$, in which $\mathcal{J}_n(t) = \tau'' + t$ and $\phi_\pm(t) = (\varepsilon - \varepsilon_0 \mp \Omega_0/2)\,t \mp \Omega_0\tau''/2$, so that this integral can be solved analytically. Furthermore, since the ionization rate for the laser peak intensities considered here (up to 13 TW/cm^2) is small ($\Gamma < 10^{-4}$ a.u.), it can be neglected and the expression for energy distribution of photoelectrons to a good approximation becomes

$$w(\varepsilon) = \left| \frac{d_{\varepsilon_0}\mathcal{E}_0}{2} \left(e^{-i\Omega_0\tau''/2}\,\frac{\sin\delta_+\tau''}{\delta_+} - e^{i\Omega_0\tau''/2}\,\frac{\sin\delta_-\tau''}{\delta_-} \right) \right|^2, \qquad (23)$$

where $\delta_\pm = \varepsilon - \varepsilon_0 \mp \Omega_0/2$. The positions of the two main peaks of this distribution are very close to the positions of the main peaks of partial distributions

$$w_\pm(\varepsilon) = \left| \frac{d_{\varepsilon_0}\mathcal{E}_0}{2} \right|^2 \left(\frac{\sin\delta_\pm\tau''}{\delta_\pm} \right)^2, \qquad (24)$$

whose values are $\varepsilon_\pm = \varepsilon_0 \pm \Omega_0/2$. Since the zeros of functions $\sin(\delta_\pm\tau'')/\delta_\pm$ are $\delta_\pm = k\pi/\tau''$, where k are integers, and in agreement with the area theorem $\tau'' = N\pi/\Omega_0$, where N is the number of Rabi cycles during the pulse, the separation of two adjacent zeros is $\Delta\varepsilon = \pi/\tau'' = \Omega_0/N$. Thus, in the interval $(\varepsilon_-, \varepsilon_+)$, whose length here is $\Delta_{\mathrm{AT}} = \Omega_0$, there are exactly $N - 1$ zeros and N peaks (see Figure 5 for $N = 5$). The latter explains the coincidence between the number of Rabi cycles during the pulse and the number of peaks in the AT pattern in PES. Obviously, local peaks in distribution (23) also exist in partial distributions (24), i.e., they are not a product of dynamic interference.

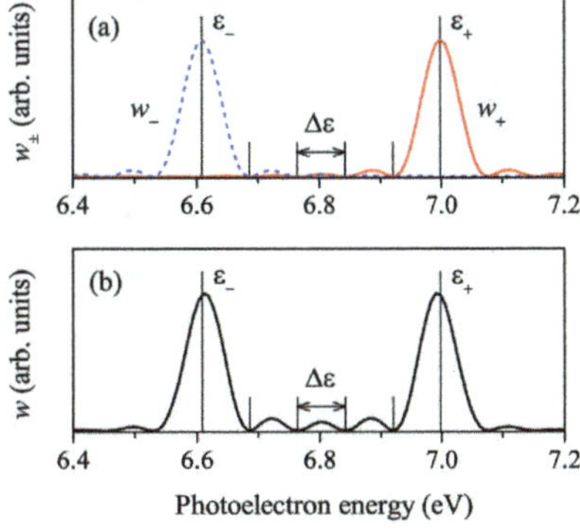

Figure 5. (Color online) (**a**) Partial distributions $w_\pm(\varepsilon)$ and (**b**) total distribution $w(\varepsilon)$ given by Equations (23) and (24), respectively, for the rectangular laser pulse with $N = \theta/(2\pi) = 5$.

4. Summary and Conclusions

In this paper, we studied the Rabi flopping of the population between the ground (1s) and excited 2p states of the hydrogen atom, induced by intense short laser pulses of different shapes and of carrier frequency $\omega = 0.375$ a.u., which resonantly couples these states, and manifestations of these dynamics in the energy spectra of photoelectrons produced in the subsequent ionization of the atom from its periodically populated/depopulated 2p state. Manifestations of the Rabi dynamics in the spectra are the AT splitting and multiple-peak structure of the AT pattern. The populations of states and spectra were calculated for three different pulse shapes—Gaussian, half-Gaussian and rectangular ones, whose pulse durations were tuned so that, for a given laser peak intensity, their pulse areas have the same value. It was found that, for these pulses, in accordance with the area theorem, the final populations (once the pulses have expired) are the same, and the spectra have similar forms in that they consist of AT patterns with the same number of peaks and with approximately the same separation between the prominent edge (AT) peaks. These facts essentially disprove the assumption that the multiple-peak pattern appears due to dynamic interference of the photoelectrons emitted with the same energy, but with a time delay at the rising and falling sides of the pulse [8,12,13], for the simple reason that a half-Gaussian pulse has no rising part, while the intensity of a rectangular pulse is constant. This conclusion is in agreement with the analysis of the conditions for dynamic interference [14,15], where it was found that they are not fulfilled in the case of resonant photoionization via the 2p state.

The additional analysis in terms of dressed states provided deeper insight into the structure of obtained spectra. This approach implies that the ionization occurs via dressed states, which directly explains the appearance of AT doublets in the PES. Here, the formula for the energy distribution of photoelectrons has the form of the time integral of the sum of two terms with different phase factors corresponding to two dressed states. In the case of rectangular pulse, this integral is analytically solvable and is reduced to the sum of two contributions that have the forms of sinc functions of energy, shifted by the value of the corresponding Rabi frequency. Then, the multiple-peak pattern is simply the result of their overlapping, which explains the matching of the number of completed Rabi oscillations with the number of peaks in the AT pattern. Analysis of the time evolution of the photoelectron energy distribution during the photoionization process showed that the mechanism of formation of multiple-peak structures is the same regardless of the pulse shape and is, therefore, not related to dynamic interference.

Author Contributions: Conceptualization, N.S.S.; methodology, N.S.S.; software, N.S.S. and A.B.; formal analysis, N.S.S., D.B.P. and A.B.; investigation, N.S.S. and D.B.P.; writing—original draft preparation, N.S.S.; writing—review and editing, N.S.S., D.B.P. and A.B. All authors have read and agreed to the published version of the manuscript.

Funding: This research received no external funding.

Data Availability Statement: All the data reported in this work are available from the correspondence author on reasonable request.

Acknowledgments: This work was supported by the COST Action No. CM18222 (AttoChem).

Conflicts of Interest: The authors declare no conflict of interest.

Abbreviations

The following abbreviations are used in this manuscript:

XUV	extreme ultraviolet
AT	Autler–Townes
PES	photoelectron energy spectrum

References

1. Rabi, I.I. Space Quantization in a Gyrating Magnetic Field. *Phys. Rev.* **1937**, *51*, 652. [CrossRef]
2. Steck, D.A. Quantum and Atom Optics. Revision 0.13.4. Available online: Http://steck.us/teaching (accessed on 24 September 2020).
3. Autler, S.H.; Townes, C.H. Stark Effect in Rapidly Varying Fields. *Phys. Rev.* **1955**, *100*, 703. [CrossRef]
4. LaGattuta, K.J. Above-threshold ionization of atomic hydrogen via resonant intermediate states. *Phys. Rev. A* **1993**, *47*, 1560–1563. [CrossRef] [PubMed]
5. Girju, M.G.; Hristov, K.; Kidun, O.; Bauer, D. Nonperturbative resonant strong field ionization of atomic hydrogen. *J. Phys. B* **2007**, *40*, 4165. [CrossRef]
6. Nandi, S.; Olofsson, E.; Bertolino, M.; Carlström, S.; Zapata, F.; Busto, D.; Callegari, C.; Di Fraia, M.; Eng-Johnsson, P.; Feifel, R.; et al. Observation of Rabi dynamics with a short-wavelength free-electron laser. *Nature* **2022**, *608*, 488. [CrossRef] [PubMed]
7. Rogus, D.; Lewenstein, M. Resonant ionisation by smooth laser pulses. *J. Phys. B* **1986**, *19*, 3051–3059. [CrossRef]
8. Demekhin, P.V.; Cederbaum, P.V. Coherent intense resonant laser pulses lead to interference in the time domain seen in the spectrum of the emitted particles. *Phys. Rev. A* **2012**, *86*, 063412. [CrossRef]
9. Müller, A.D.; Kutscher, E.; Artemyev, A.N.; Cederbaum, L.S.; Demekhin, P.V. Dynamic interference in the resonance-enhanced multiphoton ionization of hydrogen atoms by short and intense laser pulses. *Chem. Phys.* **2018**, *509*, 145. [CrossRef]
10. Tóth, A.; Csehi, A. Probing strong-field two-photon transitions through dynamic interference. *J. Phys. B At. Mol. Opt. Phys.* **2021**, *54*, 035005. [CrossRef]
11. Fischer, K.A.; Hanschke, L.; Kremser, M.; Finley, J.J.; Müller, K.; Vučković, J. Pulsed Rabi oscillations in quantum two-level systems: Beyond the area theorem. *Quantum Sci. Technol.* **2018**, *3*, 014006. [CrossRef]
12. Demekhin, P.V.; Cederbaum, L.S. Dynamic Interference of Photoelectrons Produced by High-Frequency Laser Pulses. *Phys. Rev. Lett.* **2012**, *108*, 253001. [CrossRef] [PubMed]
13. Demekhin, P.V.; Cederbaum, L.S. ac Stark effect in the electronic continuum and its impact on the photoionization of atoms by coherent intense short high-frequency laser pulses. *Phys. Rev. A* **2013**, *88*, 043414. [CrossRef]
14. Baghery, M.; Saalmann, U.; Rost, J.M. Essential Conditions for Dynamic Interference. *Phys. Rev. Lett.* **2017**, *118*, 143202. [CrossRef] [PubMed]
15. Jiang, W.C.; Burgdörfer, J. Dynamic interference as signature of atomic stabilization. *J. Opt. Express* **2018**, *26*, 19921. [CrossRef] [PubMed]
16. Bunjac, A.; Popović, D.B.; Simonović, N.S. Analysis of the asymmetry of Autler-Townes doublets in the energy spectra of photoelectrons produced at two-photon ionization of atoms by strong laser pulses. *Eur. Phys. J. D* **2022**, *76*, 249. [CrossRef]
17. Demekhin, P.V.; Cederbaum, L.S. Strong interference effects in the resonant Auger decay of atoms induced by intense X-ray fields. *Phys. Rev. A* **2011**, *83*, 023422. [CrossRef]
18. Zhang, X.; Zhou, Y.; Liao, Y.; Chen, Y.; Liang, J.; Ke, Q.; Li, M.; Csehi, A.; Lu, P. Effect of nonresonant states in near-resonant two-photon ionization of hydrogen. *Phys. Rev. A* **2022**, *106*, 063114. [CrossRef]
19. Meystre, P.; Sargent, M., III. *Elements of Quantum Optics*, 4th ed.; Springer: Berlin, Germany, 2007.

Opinion

The Spin-Orbit Interaction: A Small Force with Large Implications

Steven T. Manson

Department of Physics and Astronomy, Georgia State University, Atlanta, GA 30303, USA; smanson@gsu.edu

Abstract: The spin-orbit interaction is quite small compared to electrostatic forces in atoms. Nevertheless, this small interaction can have large consequences. Several examples of the importance of the spin-orbit force in atomic photoionization are presented and explained.

Keywords: spin-orbit interaction; atomic photoionization

Citation: Manson, S.T. The Spin-Orbit Interaction: A Small Force with Large Implications. *Atoms* **2023**, *11*, 90. https://doi.org/10.3390/atoms11060090

Academic Editor: Ulrich D. Jentschura

Received: 4 May 2023
Revised: 24 May 2023
Accepted: 25 May 2023
Published: 2 June 2023

1. Introduction

More than a half-century ago, Ugo Fano pointed out that the small spin-orbit interaction had significant implications for atomic physics [1]. Photoelectron spin polarization and the splitting of inner and outer atomic energy levels were considered in his comment. Over the intervening half-century, there have been a number of new aspects of the importance of the spin-orbit interaction that have been investigated, both experimentally and theoretically, that exemplify and amplify the earlier observations. In this short review, we shall discuss several more recent examples of the large influence of the spin-orbit interaction in various aspects of atomic photoionization.

2. Spin-Orbit Splitting of Cooper Minima

Cooper minima [2], zeros or near-zeros in dipole photoionization matrix elements are ubiquitous features in valence and near-valence shell photoionization cross sections of atoms over the entire periodic table [3]. These Cooper minima occur in ground state photoionization only in the $l \to l + 1$ dipole channels. Typically, the Cooper minima have significant influence on the energy dependence of the photoionization cross section and the spectral distribution of oscillator strength over a broad energy region around the location of the minimum [4]. From a simple single-particle point of view, a Cooper minimum occurs at an energy where the overlap between the initial and final state wave functions in the dipole matrix element is such that the positive and negative contributions just cancel each other out, resulting in a zero in the matrix element, as a function of energy. For photoionization from an nl atomic subshell with $l \neq 0$, the cross section never can go to zero because of the existence of the $l \to l - 1$ channel that does not have a Cooper minimum. However, for ns subshell photoionization, there is the possibility of a zero-cross section since no $l \to l - 1$ photoionization channel exists. However, owing to the spin-orbit interaction, a single $ns \to \varepsilon p$ transition splits into two.

This splitting of Cooper minima by the spin-orbit interaction was first found by Seaton [5], where the $ns \to \varepsilon p_{1/2}$ and $ns \to \varepsilon p_{3/2}$ dipole matrix elements in the alkali atoms exhibit their Cooper minima at *slightly* different energies so that the sum of the cross sections of the two channels never vanished, thereby leading to the non-zero photoionization cross section Cooper minimum observed experimentally in the alkali atoms.

For non-s states, the Cooper minima split into three owing to the spin-orbit splitting of the bound states in addition to the splitting of the continuum states [6–9]. These splittings of the Cooper minima are very much larger than the initial state spin-orbit splittings. For high-Z atoms, the splittings become quite significant indeed. For example, for the uranium atom, the calculated 6p spin-orbit splitting is 9.5 eV, but the Cooper minima are split by more than 200 eV [6]. From a physical standpoint, these effects result from the spin-orbit force being attractive for (both discrete and continuum) $j = l - 1/2$ states and repulsive for $j = l + 1/2$ states.

For superheavy elements, the splittings are magnified even further [10], as seen in Table 1 where the splittings of the $6s$ Cooper minima, calculated using the relativistic-random-phase approximation (RRPA), including coupling among all relevant channels, increase from 0.47 a.u. for Hg (Z = 80) to 167.50 a.u (more than 4 keV) for Og (Z = 118), the heaviest known atom. Now, for initial ns states, there is no spin-orbit splitting, so this effect is entirely the result of the spin-orbit interaction in the final continuum states. Thus, while the vast majority of studies of the spin-orbit interaction are for discrete (bound) states, it must be emphasized that there are important effects on continuum (unbound) state wave functions as well.

Table 1. Positions of Cooper minima in $6s$ subshells in photoelectron energy (a.u.).

Atom	$6s \rightarrow p_{3/2}$	$6s \rightarrow p_{1/2}$	Splitting
Hg (Z = 80)	4.14	3.67	0.47
Rn (Z = 86)	5.93	4.43	1.5
Ra (Z = 88)	6.38	3.88	2.5
No (Z = 102)	11.7	6.7	5
Cn (Z = 112)	24.82	4.82	20
Og (Z = 118)	171.02	3.52	167.5

In any case, all this is as a result of the spin-orbit interaction.

3. Photoelectron Angular Distributions from s-States

Within the framework of the dipole approximation, generally valid for low photon energy, the photoemission angular distribution of atomic subshell i for incident linearly polarized light is given by [11] the following:

$$\frac{d\sigma_i}{d\Omega} = \frac{\sigma_i}{4\pi}[1 + \beta_i P_2(\cos\theta)],$$

where σ_i is the subshell cross section, θ is the angle between the photon polarization and the photoelectron momentum. Non-relativistically, for ns subshells of closed-shell, $^1S_0$ atoms, $\beta_{ns} = 2$ and is energy-independent. This is because there is only one possible final state for the $ns \rightarrow \varepsilon p$ process, leading to a $^1P_1$ final state of the residual ion-plus-photoelectron system. Using a relativistic formulation, however, the possible transitions are $ns \rightarrow \varepsilon p_{1/2}$ and $ns \rightarrow \varepsilon p_{3/2}$, which can interfere, leading to an energy-dependent β_{ns} [11]. Looked at another way, the final states of the system are the possible J = 1 states $^1P_1$ and $^3P_1$, which are the eigenchannels of the final states. Clearly, the transition to the triplet final state involves a spin flip and can only be effected by the spin-orbit interaction.

An extremely useful way to look at photoelectron angular distributions involves the use of the angular momentum transfer analysis of Dill and Fano [12]. The angular momentum transfer is defined generally as $\mathbf{j_t} = \mathbf{J_c} + \mathbf{s} + \mathbf{J_0}$, where $\mathbf{J_c}$ and $\mathbf{J_0}$ are the angular momenta of the ion core and initial state, respectively, and $\mathbf{s}$ is the photoelectron spin. The utility of this analysis is that there is a β for each allowed value of $\mathbf{j_t}$ and these add up incoherently to calculate the observed β. Now, it turns out that the transition to the $^1P_1$ corresponds to $\mathbf{j_t} = 0$ and leads to $\beta_{ns} = 2$, but the transition to $^3P_1$ corresponds to $\mathbf{j_t} = 1$, which is what is known as a parity-unfavored transition, leading to $\beta_{ns} = -1$ [11,12]. Then, β_{ns} is a linear combination of these values, 2 and -1, weighted by their cross sections, i.e.,

$$\beta_{ns} = [2\sigma(^1P_1) - \sigma(^3P_1)]/[\sigma(^1P_1) + \sigma(^3P_1)].$$

These effects are particularly enhanced near Cooper minima, where the singlet cross section becomes quite small. As an example, in Figure 1, the situation for Xe 5s, calculated using the fully relativistic RRPA, is shown [13]. It is seen that the cross section exhibits a Cooper minimum and, in that energy region, β_{ns} is seen to deviate from the value of 2 and to be strongly energy-dependent. Parenthetically, note that this behavior has also been validated in the laboratory [13].

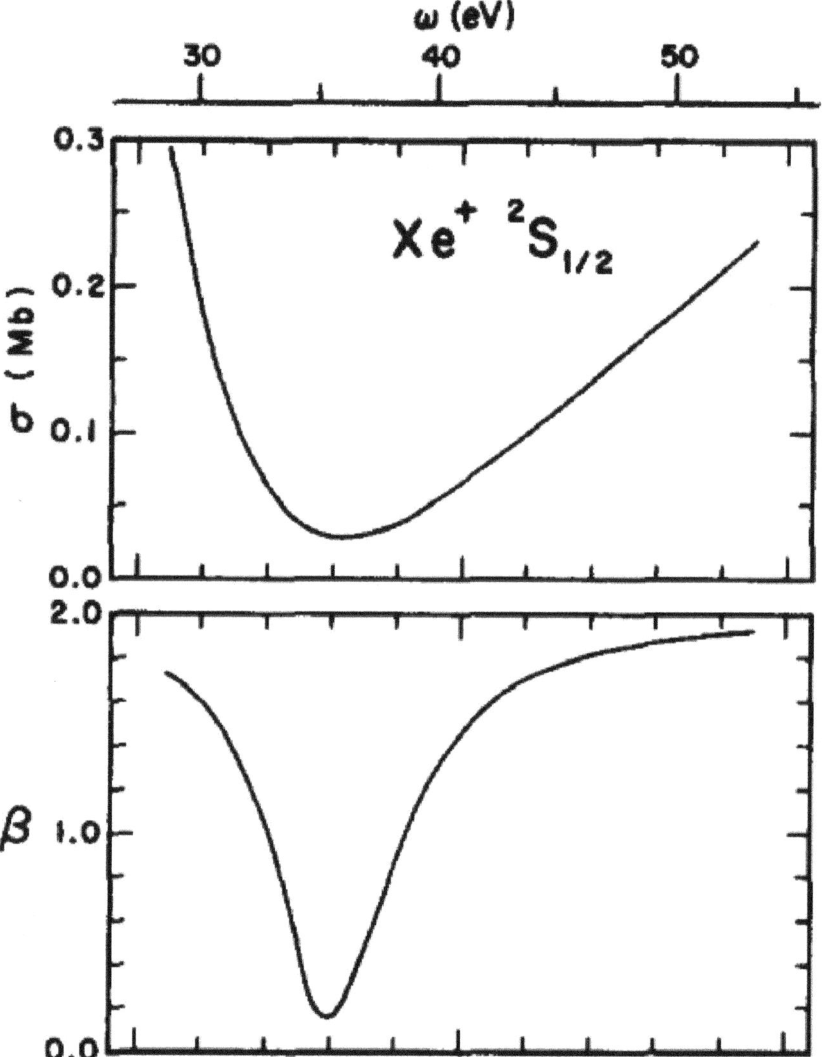

Figure 1. Xe 5s photoionization cross section (upper curve) and β parameter (lower curve) vs. photon energy, ω, calculated using relativistic-random-phase approximation (RRPA) [13].

It is thus evident that the small spin-orbit force changes the *ns* photoelectron angular distribution markedly.

4. Branching Ratios at High Energy

Atomic nl subshells with $l \neq 0$ are split into doublets with $j = l \pm 1/2$ owing to the spin-orbit interaction; these splittings are quite small compared to the binding energies of the nl subshells. Thus, in a photoionization process at a given photon energy, the photoelectrons from the split subshells have *slightly* different energies and this gives the branching ratios of the $j = l \pm 1/2$ cross sections an energy dependence, even if the dynamics of the two are the same; this is known as the *kinetic energy effect*. At high energies, where the cross sections vary slowly with energy, this small energy difference is no longer of any consequence, and it was expected earlier that the branching ratios for the $j = l + 1/2/j = l - 1/2$ cross should approach the non-relativistic value of $(l + 1)/l$, which simply reflects the occupation numbers of the spin-orbit-split nl subshell [14]. However, it was later shown that relativistic interactions, particularly spin-orbit, affect not only the energies, causing a splitting, but the initial state wave functions too, and this causes the ratio to drop below the statistical value at high energies [15,16]. In addition, this prediction has recently been verified experimentally [17]. In fact, the nl_j wave functions for $j = l \pm 1/2$ are essentially the same for intermediate and large r, but differ considerably for small r. The Dirac equation shows that the ratios of the $nl_{l-1/2} : nl_{l+1/2}$ probability densities diverge as Z^2/r^2 as $r \to 0$ [18].

Now, the relevant region for the dipole matrix element moves to smaller and smaller r with increasing energy, and this can be understood both mathematically and physically [19]. From a mathematical standpoint, with increasing photoelectron energy, the continuum wave function (the final state of the photoelectron after photoabsorption) becomes increasingly oscillatory, resulting in a net cancellation of the matrix element beyond the first node of the continuum wave function. This node moves towards the nucleus with increasing energy, thereby causing the matrix element to be generated in a region increasingly close to the nucleus as the energy increases. From a physical point of view, both energy and linear momentum must be conserved in the photoionization process. High-energy photoabsorption entails a lot of linear momentum which must be transferred to the residual atom, where most of the mass is at the nucleus. Thus, to take up this momentum, the absorption is most likely to take place near the nucleus, i.e., at small r. From these arguments, it is evident that the branching ratios do not reach a limit but continually decrease as a function of photon energy.

As an example, the theoretical results for Kr $2p$, $3p$, $4p$ and $3d$ branching ratios are shown in Figure 2 [19], where it is seen that all of the branching ratios decrease with energy. These calculations were performed using RRPA which was modified to be able to deal with high energies. This required modifying the integration mesh by increasing both the number and density of the mesh points to be able to deal accurately with the extremely oscillatory high-energy continuum wave functions. At the highest energies, the np ratios are between 1.7 and 1.8, well below the statistical value of 2.0 for np-states. The $3d$ branching ratio is also decreasing below the statistical value of 1.5 but much more slowly. This occurs because in nd-states, where the main transition is $nd \to \varepsilon f$; the f-wave centrifugal barrier keeps the continuum wave function away from the small-r region where the initial state wave functions differ.

Parenthetically, also seen in Figure 2 is the fact that in the vicinity of inner-shell thresholds, the branching ratios experience excursions from smooth behavior. This is due to interchannel coupling between the inner-shell photoionization channels and the channels involved in the branching ratio; this is evident by noting that this structure completely disappears when the interchannel coupling interactions are omitted, as shown in Figure 2. In any case, this is another example of the small spin-orbit force having a significant effect.

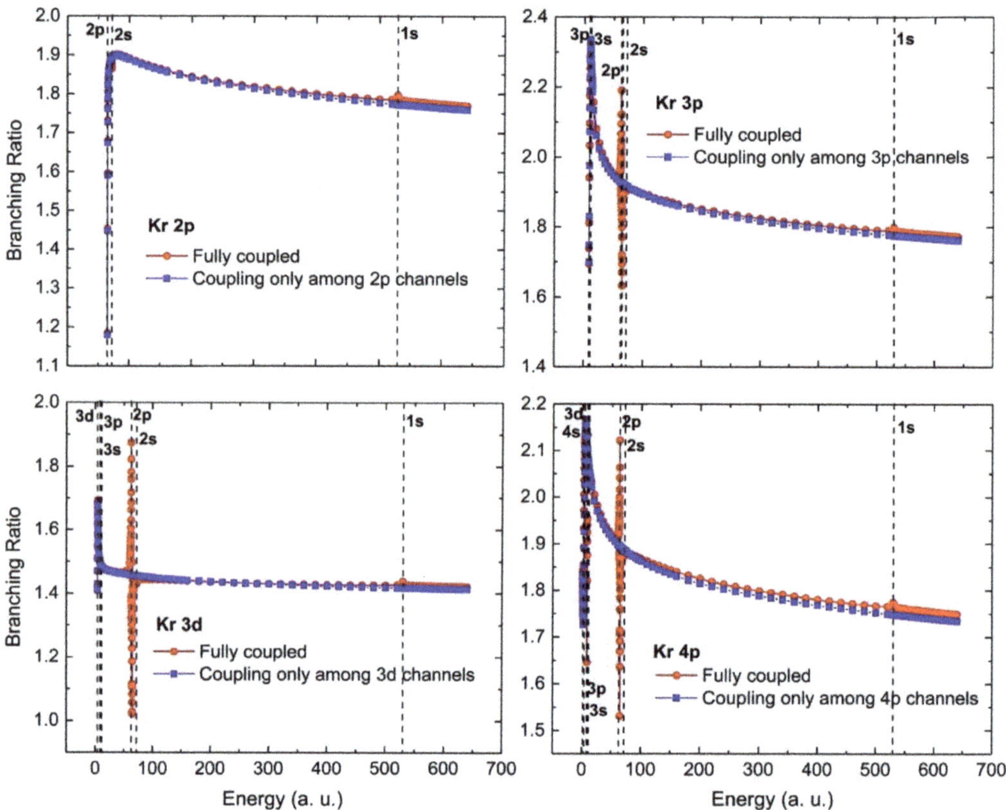

Figure 2. Photoionization cross section branching ratios for Kr $np_{3/2}/np_{1/2}$ and $3d_{5/2}/3d_{3/2}$ calculated using RRPA with full coupling (red dots) and with only intrashell coupling as indicated (blue squares). The vertical dashed lines indicate the thresholds.

5. Final Remarks

The above examples are in no way exhaustive; they are illustrative of a few of the consequences of the small spin-orbit force on the atomic photoionization process. These suggest that the spin-flip channels, engendered by the spin-orbit force, will also be important in attosecond photoemission time delay, which has been the focus of quite a number of investigations over the past decade or so (see, for example, [20–22] and references therein), particularly in the neighborhood of Cooper minima where, the non-spinflip channel amplitudes become quite small. In addition, as pointed out by Fano [1], there are also implications in other aspects of atomic physics. Furthermore, there is nothing special about atoms; the same implications are also true for atomic ions (both positive and negative), molecules and condensed matter, i.e., over a broad range of AMO physics and chemistry. This note is to remind us that as calculations and experiments become more detailed and dig deeper into AMO structure and processes, in many cases, it can be crucial to include the spin-orbit interaction into the mix to properly calculate and understand what might be going on.

Funding: This work was supported by the US Department of Energy, Office of Basic Sciences, Division of Chemical Science, Geosciences and Biosciences under Grant No. DE-FG02-03ER15428.

Conflicts of Interest: The author declares no conflict of interest.

References

1. Fano, U. Spin-Orbit Coupling: A Weak Force with Conspicuous Effects. *Comments At. Molec. Phys.* **1970**, *2*, 30–36.
2. Cooper, J.W. Photoionization from Outer Atomic Subshells. A Model Study. *Phys. Rev.* **1962**, *128*, 681–693. [CrossRef]
3. Manson, S.T. Systematics of zeros in dipole matrix elements for photoionizing transitions: Nonrelativistic calculations. *Phys. Rev. A* **1985**, *31*, 3698–3703. [CrossRef] [PubMed]
4. Starace, A.F. Theory of Atomic Photoionization. In *Handbuch der Physik*; Mehlhorn, W., Ed.; Springer: Berlin/Heidelberg, Germany, 1882; Volume 32, pp. 1–121.
5. Seaton, M.J. A comparison of theory and experiment for photo-ionization cross-sections II. Sodium and the alkali metals. *Proc. Roy. Soc. Lond. Sect. A* **1951**, *208*, 418–430.
6. Kim, Y.S.; Ron, A.; Pratt, R.H.; Tambe, B.R.; Manson, S.T. Relativistic Effects in the Photoionization of High Z Elements: Splitting and Shifts in Minima. *Phys. Rev. Lett.* **1981**, *46*, 1326–1329. [CrossRef]
7. Deshmukh, P.C.; Radojevic, V.; Manson, S.T. Relativistic Splitting of Cooper Minima in Radon: A Relativistic Random Phase Approximation Study. *Phys. Lett. A* **1986**, *117*, 293–296. [CrossRef]
8. Deshmukh, P.C.; Tambe, B.; Manson, S.T. Relativistic Effects in the Photoionisation of Heavy Atoms: Cooper Minima. *Aust. J. Phys.* **1986**, *39*, 679–686. [CrossRef]
9. Yin, R.Y.; Pratt, R.H. Survey of relativistic Cooper minima. *Phys. Rev. A* **1987**, *35*, 1149–1153. [CrossRef] [PubMed]
10. Baral, S.; Saha, S.; Dubey, K.A.; Jose, J.; Deshmukh, P.C.; Razavi, A.K.; Manson, S.T. Unusual behavior of Cooper minima of ns subshells in high-Z atoms. *Phys. Rev. A* **2022**, *105*, 062819. [CrossRef]
11. Manson, S.T.; Starace, A.F. Photoelectron Angular Distributions: Energy Dependence for s Subshells. *Rev. Mod. Phys.* **1982**, *54*, 389–406. [CrossRef]
12. Fano, U.; Dill, D. Angular Momentum Transfer in the Theory of Angular Distributions. *Phys. Rev. A* **1972**, *6*, 185–192. [CrossRef]
13. Huang, K.-N.; Johnson, W.; Cheng, K. Theoretical photoionization parameters for the noble gases argon, krypton, and xenon. *At. Data Nucl. Data Tables* **1981**, *26*, 33–45. [CrossRef]
14. James, A.R.; Samson, J.; Gardner, L.; Starace, A.F. 2P3/2: 2P1/2 partial photoionization cross-section ratios in the rare gases. *Phys. Rev. A* **1975**, *12*, 1459–1463.
15. Ron, A.; Kim, Y.S.; Pratt, R.H. Subshell branching ratios of partial photoionization cross sections. *Phys. Rev. A* **1981**, *24*, 1260–1263. [CrossRef]
16. Kim, Y.S.; Pratt, R.H.; Ron, A. Nonstatistical behavior of photoeffect subshell branching ratios at high energies. *Phys. Rev. A* **1981**, *24*, 1889–1893. [CrossRef]
17. Püttner, R.; Martins, J.B.; Marchenko, T.; Travnikova, O.; Guillemin, R.; Journel, L.; Ismail, I.; Goldsztejn, G.; Koulentianos, D.; Céolin, D.; et al. Nonstatistical Behavior of the Photoionization of Spin-Orbit Doublets. *J. Phys. B* **2021**, *54*, 085001. [CrossRef]
18. Bethe, H.A.; Salpeter, E.E. *Quantum Mechanics of One- and Two-Electron Atoms*; Springer: Berlin/Heidelberg, Germany, 1957; p. 63ff.
19. Munasinghe, C.R.; Deshmukh, P.C.; Manson, S.T. Photoionization branching ratios of spin-orbit doublets far above thresholds: Interchannel and relativistic effects in the noble gases. *Phys. Rev. A* **2022**, *106*, 013102. [CrossRef]
20. Pazourek, R.; Nagele, S.; Burgdörfer, J. Attosecond chronoscopy of photoemission. *Rev. Mod. Phys.* **2015**, *87*, 765–802. [CrossRef]
21. Deshmukh, P.C.; Banerjee, S.; Mandal, A.; Manson, S.T. Wigner-Eisenbud-Smith Time Delay in Atom-Laser Interactions. *Eur. Phys. J. Spec. Top.* **2021**, *230*, 4151–4164. [CrossRef]
22. Kheifets, A.S. Wigner time delay in atomic photoionization. *J. Phys. B* **2023**, *56*, 022001. [CrossRef]

Opinion

The Atom at the Heart of Physics

Jean-Patrick Connerade [1,2]

1   Quantum Optics and Spectroscopy Group, Physics Department, Imperial College London, London SW7 2BZ, UK; jean-patrick@connerade.com
2   European Academy of Sciences Arts and Letters (EASAL), 75006 Paris, France

Abstract: A number of reasons are advanced for which atoms stand at the heart of research in the physical sciences. There are issues in physics which are both fundamental and only partly resolved or, at least, imperfectly understood. Rather than chase them towards higher and higher energies, which mainly results in greater complexity, it makes sense to restrict oneself to the simplest systems known, held together by the best understood force in nature, viz. those governed by the inverse square law. Our line of argument complements the adage of Richard Feynman, who asked: should Armageddon occur, is there a simple, most important idea to preserve as a testament to human knowledge? The answer he suggested is: the atomic hypothesis.

Keywords: atom; physics; cluster; many-body; quantum mechanics; chaos; endohedral

1. Introduction

Scientists, although it may not always be apparent, do follow trends, and scientific fashions, like others, come and go. We see the birth of new journals, covering areas nobody had named before and there is always a temptation to consider that a subject must appear strikingly new to be of interest. However, novelty is more elusive than it seems. Some areas use new words, but the key point is whether they involve new principles. In reverse, when considering the importance of a field of research, it is just as relevant to enquire how long it has been pursued fruitfully rather than always insisting on evident novelty. Thus, Richard Feynman [1], in his celebrated series of lectures, once asked a fundamental question. In the event of Armageddon, were all human knowledge to be threatened with extinction, is there one single idea which should be preserved for future inhabitants of our planet? The answer he suggested is *The atomic hypothesis*, namely that all matter is made of atoms. When one considers how ancient this idea is, stretching back at least as far as Democritos, issues of fashion may well appear secondary. There is, however, another aspect to consider. Once a problem is fully resolved, a subject previously regarded as very relevant may suddenly cease to be attractive. So, a periodic re-appraisal of central themes is a necessity in scientific research to make sure they remain relevant.

2. Concerning Unsolved Problems

Unanswered questions and unsolved problems are the true stimulus of scientific investigation. The discussion presented here is an attempt to consider why atomic physics is useful from the standpoint: how does it help us deal with so far incompletely resolved issues in science?

This form of discussion is more difficult than presenting results. Conventionally, through the corpus of research papers, scientists must present solutions, i.e., answers which advance our understanding. Discussing unsolved problems means approaching research from the opposite point of view. The importance of an area of work then stems from the difficulty of discovering answers or, to put it another way, 'so far unsolved' problems are regarded as very significant. If an area of investigation does not call for new methods or

Citation: Connerade, J.-P. The Atom at the Heart of Physics. *Atoms* **2023**, *11*, 32. https://doi.org/10.3390/atoms11020032

Academic Editors: Himadri S. Chakraborty and Hari R. Varma

Received: 21 December 2022
Revised: 1 February 2023
Accepted: 2 February 2023
Published: 6 February 2023

principles, it then implements nothing fundamentally 'new'. As a simple example, biochemistry is an important subject, because, despite great strides in contemporary research, nobody can claim to understand how a particular molecular structure comes to life while another, little different and of equivalent complexity, remains inert. (e.g., Salam [2]).

Model problems which can be solved exactly in physics are all very interesting, but turn out to have limited scope. Inevitably, they involve some kind of compromise with reality. Thus, an allegedly simple situation, such as the two-body problem (hydrogen) in quantum mechanics is closely related to Newton's exact solution of the two-body problem in celestial mechanics, but neither of them really exists in nature. Taking hydrogen first [3], there are different degrees of approximation. The Schrödinger equation possesses many wonderful properties, but there are also difficulties associated with its use. Its leading term is the nonrelativistic kinetic energy, which obeys the Galilean transformation law. The next term is a scalar electric potential, i.e., an incomplete electromagnetic term. One can insert the vector potential of electromagnetism into the first term, but the result is no substitute for a proper covariant equation.

In addition, there is the difficulty that time, in quantum mechanics, possesses no associated operator and is not 'quantised'. It appears in the Schrödinger equation as a classical parameter (see, e.g., [4] for discussion) and is therefore different from space in this respect. The next improvement is to replace the Schrödinger equation by the Dirac equation [5] which, at least formally, satisfies the Lorentz transformation, but this does not dispel the qualitative difference between space and time just noted. Furthermore, this is still not enough, because neither of these equations allow for the quantisation of radiation. Further progress takes us into quantum electrodynamics and quantum field theory, for which we must accept that no exact solution is known. One usually resorts to the so called Furry picture [6], which introduces radiation via a perturbative scheme, with a complexity increasing order by order as the calculations are improved and further extended. At present, this is seemingly the best one can do to compute the two-body problem in quantum mechanics.

3. Is There a 'Pure' Two-Body Problem?

One might at first suspect that the problem just described occurs only due to the radiation field. However, even in celestial mechanics, it turns out, first, that there is no such thing as an isolated two-body system in nature and, second, that the gravitational field is not the only force between two particles. The first of these problems was addressed by Poincaré in 1891 [7]. He considered the three-body problem and proved that the orbits do not close. They give rise to chaos (non-integrable solutions of the equations of motion), even allowing only for a pure gravitational field. The best one can do is to obtain very local solutions, such as the one discovered by Lagrange [8] in connection with the system known as 'the Greeks and the Trojans' in astronomy.

The second issue (i.e., the existence of further fields of force) emerges in high energy physics. Accelerators of ever greater sophistication allow all of the forces between a pair of particles to be explored by increasing their relative energies to extremely high values. Entirely novel systems of particles then appear. This has opened up a magnificent and inspiring intellectual adventure in modern physics, culminating in the unification of all the fundamental forces (with the sole exception of gravitation [9]. Its crowning achievement is the Weinberg–Salam theory. The next unification point (to include gravitation) would require an energy of 10^{17} GeV, way beyond what can be reached experimentally, except perhaps through astronomical observation of early (remote) stages of the universe. These are all wonderfully impressive developments, but they do not bring us closer to resolving the issue raised at the outset of the present comment.

Chasing the two-body problem towards higher and higher energies does not preserve the simplicity of the original two-body system. In fact, one of the main consequences is the production of many new particles, which of course can be classified through a novel form of spectroscopy. However, their presence complexifies the system. I will argue that, in quantum mechanics, even at low energies, the pure two-body problem is something of an illusion, despite the impression that 'exact solutions' exist. In reality, the closest we can get to it, and to extending it to a few bodies, occurs in atomic physics.

4. The Challenge of the 'Few-Body' Problem

So far, the problem uncovered by Poincaré in classical mechanics remains unsolved, even at low energies, and this transposes into quantum mechanics when the latter is set up following the path of Landau and Lifshitz [10,11]. One can well argue that, in addition to discovering new particles or unifying the fundamental forces, one should continue to study an energy range within which interactions remain limited to the best-documented interactions in physics (i.e., those governed by the inverse square law) and to investigate the effect of slowly increasing the number of interacting particles (essentially electrons, protons and neutrons) within a system. This brings us straight back to atomic physics and to the periodic table of elements as the set of basic situations to study. Atoms thus appear as the ideal testing ground for the many-body problem.

Although one refers to 'many-body' effects in atomic and molecular physics, this terminology is not really the most accurate. The word 'many' creates the impression that the difficulties involved necessarily increase with number. In fact this is not the case. For example, an infinite 'sea' of occupied states (as considered originally by Dirac to handle the negative energy states in his equation) minus only one particle behaves as just a single 'antiparticle', viz: the positron. Likewise, the fundamental antisymmetry of many-electron states lead to the discovery of 'closed electronic shells' and a single 'hole' in a closed shell is, again, similar in behaviour to a single 'antiparticle'. Thus, in some situations, quantum mechanics allows a more promising description of the many-body problem than classical mechanics. In a sense, this is a surprising consequence of a more sophisticated theory. 'Few-body problem' might be a better description.

Perhaps one should even restrict the definition of the most suitable energy range by excluding excitation energies high enough to produce electron–positron pairs. This would avoid not only single excitations of very high energy, but also multiphoton excitation by very intense laser fields and would stay more closely within the first order of the Furry picture [6].

5. The Awkward Connection between Classical and Quantum Mechanics

Even if we do restrict ourselves carefully, as just described, there are deeper issues to consider. As already noted, the three-body problem in classical mechanics cannot be solved exactly. This may seem semantic, since pretty accurate perturbative methods, well known to astronomers, can handle most practical problems when computing orbits. However, the issue looms again in the formulation of elementary quantum mechanics.

For this purpose, we need to specify the correct Hermitian operator to associate with each and every observable while avoiding an extensive and clumsy table as a separate postulate. A first approach, suggested by Bohr and Sommerfeld and refined by Landau and Lifshitz [10,11], was to study the so-called 'semiclassical limit' of quantum theory via the 'correspondence principle' through which the quantum and classical theories are supposed to merge. To be useful, the process would need to be applied 'backwards' i.e., from classical to quantum physics, since the classical problem is the one regarded as 'well-understood'. The procedure, however, involves integration around closed orbits of the underlying classical systems.

The difficulty, as Einstein famously objected, is: what should one do if the orbit never closes? Unfortunately, this is precisely the case for the few-body problem, as Poincaré [7] had discovered. So, the Bohr–Sommerfeld 'principle' actually fails in most situations except for a few ideal, integrable problems, such as the harmonic oscillator, Newton's two-body problem, etc. There are some complicated orbits (the Landau orbits) which resemble Lissajoux figures in phase space because they 'eventually' close, but these are not sufficient in number to account for all possible orbits of a non-integrable system.

Thus, 'difficult' situations, such as three-body problem, a pendulum with a magnet, etc., cannot strictly be handled in this way. In classical physics, a pen attached to a pendulum with a magnet underneath will write all over a piece of paper within the constraint imposed by its total energy and will never follow the same path twice, i.e., the orbit will never close. This leaves us with the complication that, despite the intuitively rather obvious correspondence principle, classical mechanics contains information which simply cannot be transferred to quantum mechanics. Why this happens remains a matter of opinion.

We end up with two physical theories connected by an imperfect correspondence. On the one hand, classical mechanics contains systems with both integrable and non-integrable solutions, while, on the other, quantum mechanics seemingly allows only one kind of solution. A legitimate question becomes: does the semi-classical limit of quantum mechanics recover *all* or can it recover *only a part* of classical mechanics?

To this one can add, starting out from the Dirac equation, that no semi-classical limit is known for this case, since the solutions involve spinors, and spin does not exist in classical physics. Hence, in the relativistic theory, the whole concept of the correspondence principle as the basis of a systematic method to set up quantum mechanics breaks down, which is why, in response to the question 'why is there no book by Landau and Lifshitz on relativistic quantum mechanics?', Lev Landau is said to have replied; "Because there is no such theory!"

6. The Structure of Empty Space

Another way of looking at the question is to ask what one would mean by orbits which do or do not 'close' in quantum mechanics. Would asking this very question imply a violation of the uncertainty principle? In quantum mechanics, should one consider phase-space itself as exhibiting a granular structure, with dimensions of individual grains determined by the magnitude of Planck's constant? Would it then suffice for the electron to return to within one such grain for an orbit to be regarded as 'closed'? This of course suggests a different definition of dynamical 'chaos' for classical and for quantum systems. There has been much discussion of the issue since the earliest experiments, by Garton and Tomkins [12], revealed the problem.

The refinement would be all well and good were it not that the theory of relativity requires space-time to be continuous and freely differentiable in the sense of classical mechanics. Hence, no doubt, Einstein's insistence that the Bohr–Sommerfeld quantisation was unsatisfactory. The nature of space (continuous or granular) becomes an awkward issue. It is even more so when we consider the difference between space (a true observable) and time (a classical parameter) in quantum mechanics, already noted above. There is perhaps no other situation in which the incompatibility of the two conceptions of empty space is so apparent as in atomic physics.

The problem of infinite divisibility, first raised by Pascal [13] in connection with the structure of atoms and of matter itself, re-emerges when one attempts to extend the equipartition theorem to microscopic systems. As commented by Dirac [5], it would ultimately imply infinite specific heats. Granularity is therefore also an essential ingredient in thermodynamics and this remark provided one of the earliest 'proofs' of the necessity of quantum mechanics.

7. Compressed Atoms

In the kinetic theory of gases, atoms and molecules also play an essential role without which the concept of pressure would remain undefined. It is assumed that they behave as point-like masses, bombarding one side of the walls of the container. The average force they exert is given as the origin of pressure, which becomes a macroscopic thermo-dynamic variable. Unfortunately, this picture, as so often happens in physics, becomes less straightforward as corrections to the ideal gas law are introduced. The first, due to van der Waals, involves attributing intrinsic volume to the atoms or molecules, but this volume is considered as 'fixed'. The reason for imposing this restriction is to preserve the consistency of the formalism, but it is obvious that a physical volume cannot remain fixed as the pressure increases. The real situation must of necessity be more complex.

Within the Thomas–Fermi model of the atom [14,15], the effect of externally applied pressure is readily understood as the compression of an electron 'gas'. One can simulate it by changing the external boundary condition and studying how the total energy and occupied volume are related. Within the Schrödinger picture also, the electron cloud is a kind of 'fluid' with the Schrödinger equation as an equation of state. Changes in the external boundary conditions then act like the external piston. The model just described was developed by Hellman [16], Feynman [17] and Feynman et al. [18]. Within it, there are difficult issues relating to the different definitions of probability in thermodynamics and in quantum physics which have to be reconciled.

This takes us to the area of microscopic thermodynamics and confined atomic systems ([19] and refs. therein). The natural starting point is still the Thomas–Fermi model of the atom because it provides the atom with a well-defined volume to start with. However, there is nothing to prevent extending the idea to the Hartree–Fock and Dirac–Fock equations (see [20] and earlier refs. therein). One readily establishes that the periodic table for atoms under compression is not the same as for free atoms because the order of filling is modified and actually approaches the ideal and complete *aufbau* principle (see [21]) more and more closely as the pressure is increased. A whole new chemistry opens up for study, for underlying reasons which stem from atomic physics, but it remains necessary to perform extensive ab initio calculations to account for them in detail (see e.g., [22]). One can think of many experimental applications (bubbles in solids, clusters, polaronic insertion of ions, atoms under extreme pressure, etc.).

8. Endohedral Confinement

Closely related to the atom under pressure is another novel area of research, namely the atom endohedrally confined within a hollow molecule, the most typical example being the metallofullerene. There are basically two conceptual approaches for such systems. The first is to attempt full molecular calculations, from which geometrical structures and symmetries can in principle be deduced (e.g., [23]). The second is to approximate the confining molecule semi-empirically as a hollow, spherical potential shell whose properties can be deduced experimentally from electron scattering experiments [24]. Apart from greater simplicity, the latter approach allows one to include some important quantum effects, such as the occurrence of confinement resonances [25].

9. Many-Body Theories of the Atom

Returning to the (unsolved) many-body problem, three general remarks can be made. The first, as noted above, is that many-electron states in quantum mechanics obey the Pauli principle. There is no such principle in classical physics, so we have good reason to hope for a better understanding of the many-body problem. The periodic table informs us about the properties of closed shells. They imply that atoms return regularly to nearly spherical shapes at each period as the number of electrons is increased, which is the crucial simplifying feature.

A first step towards the many-body theory of the atom is of course to solve the coupled system of 'independent electron' Schrödinger equations by the Hartree–Fock method [26], refined by the introduction of mixing between configurations. This method has been successfully extended to the Dirac equation [27] despite a complication originally pointed out by Brown and Ravenhall [28] for the multiconfigurational Dirac–Fock method: namely that configuration mixing, in this case, might involve negative energy states which cannot all be 'filled' simultaneously, in which case the variational principle upon which the Hartree–Fock method rests for convergence would collapse. This is somewhat controversial because properly converged multiconfigurational Dirac–Fock solutions have been obtained [27] and correspond very well with experimental data. It would be desirable for practitioners of the method to dispel any residual uncertainty surrounding this alleged 'dissolution into the negative energy continuum' if it is indeed a real effect.

Plasmon excitations (i.e., oscillations of closed shells) in free atoms can be computed either by the many-body perturbation theory (MBPT Kelly [29]) or by the random phase approximation with exchange (RPAE Amus'ya et al. [30]). These two theories are not equivalent, even when the perturbative expansions are performed on the same independent electron atomic basis. In the MBPT, all of the terms identified by their Feynman graphs are summed up to a given order, but the summation cannot be extended to the high order, as the computations become progressively more and more extensive. In the RPAE, only two classes of diagrams are treated (the forward bubble diagrams and their exchange equivalents) but they are summed to the infinite order. Obviously, the two approaches cannot be equivalent. These are the two theories we have at our disposal, neither of which is 'complete'. Both are useful, generally in different situations, the MBPT being more appropriate for open-shell systems and the RPAE for closed shell or half-closed shell atoms.

This ambivalence, again, can be taken to express the fact that we have no general solution of the many-body problem in quantum mechanics. The study of plasmon effects (giant resonances) in free atoms and in atoms trapped in different environments is one of the more promising areas for developing and improving theoretical tools to handle many-body systems.

The formation of negative ions by addition of an electron to a neutral atom also goes beyond the convergence capabilities of the Hartree–Fock basis. The polarisation of atomic shells by the extra electron is the mechanism involved. In this situation, a new model has been developed based on the many-body Dyson equation [31] which holds great promise for systematic computations of different negative ion species.

10. Wigner Scattering Theory and the Wigner Time delay

Atoms, of course, involve no new forces as compared to other physical systems and, in line with the economy of principles which should ultimately underpin a general understanding of nature, would be best described within the same, single, conceptual framework as all the other physical systems. The theory which best accomplishes this is the Wigner scattering theory [32,33], because it is extremely general in its formulation. It applies to all branches of physics where quantum scattering occurs and does not even require an explicit solution of the Schrödinger equation, but only postulates the existence of a differential equation of the Schrödinger type, together with the boundary conditions usual in quantum mechanics.

Even in this general context, however, atoms still have a very special role to play, by virtue of the asymptotic inverse square law of force, which allows the external K-matrix to be inverted analytically [21,34,35]. This situation is unique, on a par with Kepler's laws of planetary motion in a central inverse square field of force.

Scattering is, of course, not an instantaneous process because it involves the propagation of a scattered wave. This implies a time delay which, as shown originally by Wigner, is given by the derivative of the phase shift of the scattered wave with respect to energy. In the context of condensed matter or of large molecules or clusters, Wigner time delays are readily measurable by short pulse laser techniques. For individual atoms, this can also be true. A pioneering example is the work of Bourgain et al. [36] on the resonance line of a single trapped Rb atom, for which a time resolution of 256 ps proves adequate. The really interesting situation, however, is for interacting autoionizing resonances [35], where the time scales generally become much shorter (in the attosecond range) so that experimentation has only become feasible recently by ultrashort pulse technology.

11. Atomic Clusters

Traditionally, the transition from the free atom to the solid state has always been imagined by 'piling up' atoms or attempting to model infinite sequences similar to crystals. More recently, it has been shown [37] that this description gives an incomplete picture of the transition from the free atom to the solid. In reality, when atoms are piled together one by one, they first form clusters and several different transition points occur, depending on which physical variable is under study. Thus, the emergence of solid state properties occurs in different ranges as the size of a cluster is increased (e.g., [37–41]) etc.

Again, via the physics of clusters, the properties of atoms are central to a good understanding of condensed matter, achieved by adding them together one by one. Experimentally, the evolution of clusters as a function of size is now accessible by the study of mass-selected clusters. This is particularly interesting in the context of the present article for metallic clusters, because they possess delocalised electrons (precursors of the conduction bands in solids) and form closed electronic shells similar in principle to those of noble gas atoms.

12. Mie and Shape Resonances—Wigner Time Delays

The dynamics of such shells turn into the Mie resonances of classical electrodynamics [42]. In quantum systems with closed shells, they become shape or giant resonances. As such, they involve the collective pulsation of several electrons, i.e., an intrinsic 'many-body' effect which, however, is very short-lived as it is strongly damped. It can be calculated by the many-body perturbation theory (RPAE or MBPT) in non-relativistic or in relativistic versions and can also be modelled by using an effective atomic potential, which must include the influence of the centrifugal barrier, i.e., the angular momentum term in the radial Schrödinger equation. They are due to the shape of this effective potential, hence the name.

The physics of atoms with d and f subshells and the study of metallic clusters with delocalised electrons forming closed shells have revealed for both atoms and metallic clusters, the importance of the collective many-body phenomena or plasmon excitations [43]. By analogy with nuclear physics and in connection with the sum rule for a given atomic shell, they are termed 'giant resonances' when they exhaust most of the available oscillator strength available within a single feature. A peculiar property of these excitations for atoms is that they occur deep inside the system and are able to survive in different phases, from the free atom to clusters, molecules and solids [44], in contrast with other atomic states which are destroyed. This opens up new possibilities for extending and adapting many-body theories within different environments.

Giant resonances, or rather their Fourier transforms, yielding observable Wigner time delays, are also relevant in attosecond spectroscopy. As noted above, this new range of time intervals has recently become accessible to ultrafast laser experiments. A fine example is by Biswas et al. [45]. The observation of photoionization and the corresponding time-resolved atomic spectra provide complementary information which may eventually help to discriminate between the predictions of different models, such as the RPAE, RRPAE and MBPT theories and pseudopotential models.

13. Cooling, etc.

Last but by no means least, an area not covered in the present Comment, because it is a huge subject in its own right and would require a good deal more space, is the theme of atomic cooling and trapping, the Bose–Einstein condensation and all of the effects described by the Gross–Pitaevsky theory of ground-state bosonic fluids [46,47]. A Bose–Einstein condensate is a gas of bosons which are all in the same quantum state, described by a single wavefunction. The Gross–Pitaevsky equation is essentially a transposition of the Hartree–Fock theory to the ground state of a quantum system of identical bosons using a pseudopotential interaction. Both in terms of the general principles involved and the fluids to which they are applied, atomic physics is central also to this extensive area of study.

14. Conclusions

In summary, atomic physics remains a privileged testing ground for the fundamental problems of physics which are so far incompletely resolved, extending from the many-body theory to relativistic mechanics, the nature of 'empty' space and the principles of quantum field theory, as well as the full connection between quantum mechanics and classical physics, also including thermodynamics. Thus, the atom, as a system, remains very much at the heart of contemporary research in physics and chemistry.

Funding: This research did not require any funding contract.

Data Availability Statement: All the data referred to is available in the open literature.

Conflicts of Interest: No conflict of interest are involved.

References

1. Feynman, R.P. *The Feynman Lectures on Physics*; Feynman, R., Robert, B., Eds.; Leighton and Matthew Sands; Addison Wesley: New York, NY, USA, 1964; Volume 3.
2. Salam, A. The role of Chirality in the Origin of Life. *J. Mod. Evol.* **1991**, *33*, 105. [CrossRef]
3. Series, G.W. *The Spectrum of Atomic Hydrogen*; World Scientific Publishing Co.: Singapore, 1988.
4. Briggs, J.S.; Rost, J. Time dependence in quantum mechanics. *Eur. Phys. J. D* **2000**, *10*, 311. [CrossRef]
5. Dirac, P.A.M. *The Principles of Quantum Mechanics*; Oxford University Press: Oxford, UK, 1930.
6. Furry, W.H. Bound States and Scattering in Positron Theory. *Phys. Rev.* **1951**, *81*, 115. [CrossRef]
7. Poincaré. *Sur le Problème des Trois Corps*; Bulletin astronomique, Observatoire de Paris Année; Paris Observatory Bulletin: Paris, France, 1891; pp. 12–24.
8. Lagrange, J.-L. Oeuvres de Lagrange sur GDZ (Göttingen). 1772. Available online: http://gdz.sub.uni-goettingen.de/dms/load/toc/?PID=PPN308899466 (accessed on 30 January 2023).
9. Salam, A. Gauge Unification of Fundamental Forces. *Rev. Mod. Phys.* **1980**, *52*, 525. [CrossRef]
10. Landau and Lifshitz. *Physics Textbooks Series: Classical Mechanics*, 3rd ed.; Vol. 3—Landau, Lifshitz—Identifier landau-and-lifshitz-physics-textbooks-series; Identifier-ark ark:/13960/t7sn92122; Pergammon Press: Oxford, UK, 1976.
11. Landau and Lifshitz. *Physics Textbooks Series: Quantum Mechanics—Non-Relativistic Theory*, 3rd ed.; Vol 4—Landau, Lifshitz Identifier landau-and-lifshitz-physics-textbooks-series; Identifier-ark ark:/13960/t7sn92122; Pergammon Press: Oxford, UK, 1991.
12. Garton, W.R.S.; Tomkins, F.S. Diamagnetic Zeeman effect and magnetic configuration mixing in long spectral series of Ba I. *Astrophys. J.* **1969**, *158*, 839. [CrossRef]
13. Pascal, B. *Les Pensées de M. Pascal sur la Religion et Sur Quelques Autres Sujets, Qui ont été Trouvées Après sa Mort parmy Ses Papiers*; Port-Royal des Champs: Paris, France, 1670.
14. Thomas, L.H. The Calculation of Atomic Fields. *Math. Proc. Camb. Philos. Soc.* **1927**, *23*, 542–548. [CrossRef]
15. Fermi, E. Eine statistische Methode zur Bestimmung einiger Eigenschaften des Atoms und ihre Anwendung auf die Theorie des periodischen Systems der Elemente. *Z. Phys.* **1928**, *48*, 73–79. [CrossRef]
16. Hellmann, H. *Einführung in die Quantenchemie*; Franz Deuticke: Leipzig, Germany, 1937; p. 285.
17. Feynman, R.P. Forces in Molecules. *Phys. Rev.* **1939**, *56*, 340–343. [CrossRef]
18. Feynman, R.P.; Metropolis, N.; Teller, E. Equations of State of Elements Based on the Generalized Fermi-Thomas. *Theory Phys. Rev.* **1949**, *75*, 1561.
19. Jaskólski, W. Confined many-electron systems. *Phys. Rep.* **1996**, *27*, 1–56. [CrossRef]
20. Connerade, J.-P. Confining and compressing the atom. *Eur. Phys. J. D* **2020**, *74*, 211. [CrossRef]
21. Connerade, J.-P. *Highly Excited Atoms*; Cambridge University Press: Cambridge, UK, 1998; ISBN 9780511524.
22. Rahm, M.; Cammi, R.; Ashcroft, N.W.; Hoffmann, R. Squeezing All Elements in the Periodic Table: Electron Configuration and Electronegativity of the Atoms under Compression. *J. Am. Chem. Soc.* **2019**, *141*, 10253. [CrossRef] [PubMed]
23. Madjet, M.E.A.; Ali, E.; Carignano, M.; Vendrell, O.; Chakraborty, H.S. Ultrafast Transfer and Transient Entrapment of Photoexcited Mg Electron in Mg@C_{60}. *Phys. Rev. Lett.* **2021**, *126*, 183002. [CrossRef]
24. Dolmatov, V.K.; Baltenkov, A.S.; Connerade, J.-P.; Manson, S.T. Structure and photoionization of confined atoms. *Radiat. Phys. Chem.* **2004**, *70*, 417–433. [CrossRef]
25. Connerade, J.-P.; Dolmatov, V.K.; Manson, S.T. On the nature and origin of confinement resonances. *J. Phys. B At. Mol. Opt. Phys.* **2000**, *33*, 2279. [CrossRef]
26. Froese–Fischer, C. *The Hartree–Fock Method for Atoms, a Numerical Approach*; Wiley: New York, NY, USA, 1977.
27. Grant, I.P.; Quiney, H.M. Application of relativistic theories and quantum electrodynamics to chemical problems. *Int. J. Quantum. Chem.* **2000**, *80*, 283–297. [CrossRef]
28. Brown, G.E.; Ravenhall, D.G. On the Interaction of Two Electrons. *Proc. R. Soc. London* **1951**, *A208*, 552.
29. Kelly, H.P. Many-Body Perturbation Theory Applied to Atoms. *Phys. Rev.* **1964**, *136*, B896. [CrossRef]
30. Amus'ya, M.Y.; Amus'ya, N.A.; Cherepkov, L.V. Chernysheva. *Zh. Eksp. Teor. Phys.* **1961**, *60*, 160.
31. Chernysheva, L.V.; Gribakin, G.F.; Ivanov, V.K.; Kuchiev, M.I. Many-body calculation of negative ions using the Dyson equation. *J. Phys. B At. Mol. Opt. Phys.* **1988**, *21*, L419–L425. [CrossRef]
32. Wigner, E.P.; Eisenbud, L. Higher Angular Momenta and Long Range Interaction in Resonance Reactions. *Phys. Rev.* **1947**, *72*, 29. [CrossRef]

33. Wigner, E.P. Lower Limit for the Energy Derivative of the Scattering Phase Shift. *Phys. Rev.* **1955**, *98*, 145. [CrossRef]
34. Lane, A.M. The application of Wigner's R-matrix theory to atomic physics. *J. Phys. B At. Mol. Phys.* **1986**, *19*, 253. [CrossRef]
35. Connerade, J.-P. Wigner scattering theory for systems held together by Coulombic forces. *Eur. Phys. J. D* **2020**, *74*, 107. [CrossRef]
36. Bourgain, R.; Pellegrino, J.; Jennewein, S.; Sortais, Y.R.P.; Browaeys, A. Direct measurement of the Wigner time-delay for the scattering of light by a single atom. *Opt. Lett.* **2013**, *38*, 1963–1965. [CrossRef]
37. Cohen Marvin, L.; Knight Walter, D. The Physics of Metal Clusters. *Phys. Today* **1990**, *43*, 42. [CrossRef]
38. Bréchignac, C.; Broyer, M.; Cahuzac, P.; Delacretaz, G.; Labastie, P.; Wolf, J.P.; Wöste, L. Probing the transition from van der Waals to metallic mercury clusters. *Phys. Rev. Lett.* **1988**, *60*, 275. [CrossRef]
39. Lyalin, A.; Solov'yov, I.A.; Solov'yov, A.V.; Greiner, W. Strontium clusters: Electronic and geometry shell effects. In *Latest Advances in Atomic Cluster Collisions: Structure and Dynamics from the Nuclear to the Biological Scale*; Solov'yov, A.V., Ed.; Imperial College Press: London, UK, 2008; pp. 105–127.
40. March, N.H.; Angiella, G.G.N. *Exactly Solvable Models in Many-Body Theory*; World Scientific Press: London, UK; Singapore, 2016.
41. Forte, G.; Grassi, A.; Lombardo, G.M.; Pucci, R.; Angilella, G.G.N. From Molecules and Clusters of Atoms to Solid State Properties. In *Many-Body Approaches at Different Scales*; Angilella, G., Amovilli, C., Eds.; Springer: Cham, Switzerland, 2018. [CrossRef]
42. Mie, G. Beiträge zur Optik trüber Medien speziell kolloidaler Goldlösungen (contributions to the optics of diffuse media, especially colloid metal solutions). *Ann. Phys.* **1908**, *25*, 377. [CrossRef]
43. de Heer, W.A.; Selby, K.; Kresin, V.; Masui, J.; Vollmer, M.; Chatelain, A.; Knight, W.D. Collective dipole oscillations in small sodium clusters. *Phys. Rev. Lett.* **1987**, *59*, 1805. [CrossRef]
44. Connerade, J.-P.; Esteva, J.-M.; Karnatak, R.C. (Eds.) *Giant Resonances in Atoms, Molecules, and Solids*; NATO Advanced Study Institute Plenum Press: New York, NY, USA, 1987.
45. Biswas, S.; Trabattoni, A.; Rupp, P.; Magrakvelidze, M.; Madjet ME, A.; De Giovannini, U.; Castrovilli, M.C.; Galli, M.; Liu, Q.; Månsson, E.P.; et al. Attosecond correlated electron dynamics at C_{60} giant plasmon resonance. *At. Phys.* **2022**, arXiv:2111.14464.
46. Gross, E.P. Structure of a quantized vortex in boson systems. *Il Nuovo Cimento* **1961**, *20*, 454–457. [CrossRef]
47. Pitaevskii, L.P. Vortex lines in an imperfect Bose gas. *Sov. Phys. JETP* **1961**, *13*, 451–454.

MDPI
St. Alban-Anlage 66
4052 Basel
Switzerland
www.mdpi.com

Atoms Editorial Office
E-mail: atoms@mdpi.com
www.mdpi.com/journal/atoms

9 783725 802371